METHODS IN PHARMACOLOGY AND TOXICOLOGY

Series Editor
Y. James Kang
Department of Medicine
University of Louisville School of Medicine
Prospect, Kentucky, USA

For further volumes: http://www.springer.com/series/7653

Molecular Histopathology and Tissue Biomarkers in Drug and Diagnostic Development

Edited by

Steven J. Potts

Flagship Biosciences, LLC, Westminster, CO, USA

David A. Eberhard

University of North Carolina, Chapel Hill, NC, USA

Keith A. Wharton, Jr.

Novartis Institutes for BioMedical Research, Cambridge, MA, USA

Humana Press

Editors
Steven J. Potts
Flagship Biosciences, LLC
Westminster, CO, USA

David A. Eberhard
University of North Carolina
Chapel Hill, NC, USA

Keith A. Wharton, Jr.
Novartis Institutes for BioMedical Research
Cambridge, MA, USA

ISSN 1557-2153 ISSN 1940-6053 (electronic)
Methods in Pharmacology and Toxicology
ISBN 978-1-4939-2680-0 ISBN 978-1-4939-2681-7 (eBook)
DOI 10.1007/978-1-4939-2681-7

Library of Congress Control Number: 2015939268

Springer New York Heidelberg Dordrecht London

Printed on acid-free paper

Humana Press is a brand of Springer
Springer Science+Business Media LLC New York is part of Springer Science+Business Media (www.springer.com)

Dedication

Chris Callahan, M.D., Ph.D.

This book is dedicated to the memory of our dear friend and esteemed colleague, Chris Callahan, M.D., Ph.D., who suffered an untimely death from a progressive brain tumor in 2011. Already recognized as a leader early in his career, Chris held the position of Scientist and Investigative Pathologist at Genentech/Roche at the time of his passing at the young age of 46. Chris's professional contributions pervade this book's themes of molecular histopathology and tissue-based biomarkers, and several of its contributors trained with or worked alongside Chris at various stages of his career. More importantly, Chris represented a generation of visionary physician-scientists who began their training in the 1980s, the last "pre-genome" decade of human history, with the belief that the discovery of molecules and pathways governing normal development would provide key insights into human disease. While now considered dogma, at that time only scattered yet tantalizing hints existed to indicate that alterations of these same handful of pathways, deeply conserved during our evolution, caused (and were druggable targets of) diverse human diseases.

Chris completed his B.S. at Brown University in 1987, and earned his M.D. and Ph.D. degrees as a graduate student in John Thomas's lab at UC San Diego and Salk Institute performing groundbreaking work in Drosophila neurobiology. His accomplishments include the molecular cloning of *derailed*, a receptor tyrosine kinase crucial for axon guidance [1]. Recognizing the central role of pathology in identifying molecular and cellular mechanisms of disease, Chris moved to Stanford University to pursue residency training in anatomic pathology, ultimately serving as Attending Physician and Acting Assistant Professor of Pathology while engaged in postdoctoral research in Dermatology with Tony Oro (also a former fellow M.D./Ph.D. student with Chris at UC San Diego). Extending Tony's postdoctoral work with Matthew Scott that implicated aberrant Hedgehog signaling in the most common human tumor, basal cell carcinoma [4], Chris went on to discover a critical role for a Hedgehog pathway target gene Mtss1 (Missing in Metastasis) in the regulation of signaling and cancer progression [3].

Tragically, Chris's cancer diagnosis came at a most inopportune time—just as he sought to start his own lab. Following aggressive surgery and chemotherapy, he returned to the bench within days to continue his passion. By this time, Hedgehog pathway inhibitors were being developed for oncology indications, and Chris felt the best opportunity to apply his skills and talents to directly benefit cancer patients was to continue his career as a pathologist in biopharma. Chris joined Genentech, supporting drug development projects while directing several research projects aimed at understanding the mechanisms of Hedgehog pathway activation and therapy resistance in human cancer. Chris was a key contributor to the R&D team efforts that led to the FDA approval of vismodegib, the first-in-class Hedgehog pathway inhibitor for clinical use in advanced basal cell carcinoma. The several hats Chris wore in these efforts included basic research and medical scientist, translational biomarker and companion diagnostics R&D investigator, and pathology advisor to the development team. Chris investigated cancer mechanisms to the end of his life, with his final senior author paper on the role of Hedgehog signaling in tumor-stroma interactions published in the *Proceedings of the National Academy of Sciences* just a few months before his death [4].

Chris's approach to life inspired those around him, as these quotes from two of his Genentech colleagues attest:

> *(I have) never met anyone with more selfless dedication, engagement, focus, and commitment to his work and family than Chris. Chris loved his role as a pathologist at Genentech and immersed himself in it 100 %. He knew that his work was helping to transform the lives of patients, and felt fortunate that he could do so by engaging in professional activities he loved most—basic hypothesis-driven research and scientific collaboration.*

> *Chris knew, more clearly than most of us do, that his time with his family, his friends, and his work was limited. He relished that time, and he shared it generously with others. Chris had a selfless, collegial enthusiasm for his work. He was absolutely committed to his colleagues and to the projects he supported; he would do everything he could to maximize the chances for their success. He had a great appreciation of, and loved sharing subtle details of biology—not to advertise his brilliance, but because he trusted you would find them as satisfying and wonderful as he did. When he dropped off his sons at school in the morning he would tell them, 'Have fun, learn a lot, and be kind.' He fully modeled that advice.*

Beyond describing Chris to a tee, these reminiscences illustrate three qualities of a successful anatomic pathologist in biopharma that emerge as themes throughout this book: a focus on the biology of disease, a passionate curiosity, and a collaborative mindset. Although new insights about disease emerge daily, and opportunities for new discoveries have never been greater than the present time, some things haven't changed: A century and half ago, Rudolf Virchow, founder of cellular pathology, said "*If we would serve science, we must extend her limits, not only as far as our own knowledge is concerned, but in the estimation of others.*" [5]. Chris exemplified this Virchowian ideal.

Comprehending the biology of disease requires integration of knowledge from diverse disciplines, of which we concede the histopathology "stock in trade" of fixed and stained tissue is only one part. As a role model, Chris excelled at the challenges faced by anatomic pathologists embedded in drug and diagnostic industries, chief among them to bridge the power (and limitations) of histopathology methods and knowledge with those from a growing number of technology-driven disciplines, including genomics, protein biochemistry, quantitative image analysis, and in vivo imaging of cells to whole animals. Chris saw firsthand that delivering a diagnostic test or new therapy to patients requires diverse skills, far beyond what any one individual could possibly master. Whether performed in industry or research institutes, drug development requires coordinated efforts by multidisciplinary teams, and so the pathologist on the team must persuasively communicate with members from diverse backgrounds and viewpoints in order to foster collaboration, and ultimately, progress. Rare talents like Chris, with curious and creative minds, able to integrate emerging data and knowledge across disciplines, are poised to see old problems in new ways and develop novel, important hypotheses that demand investigation. As Virchow said, the pathologist must push the science—"*. . .extend her limits. . .*"—for all to see. Chris fearlessly pursued multidisciplinary investigations in fruit flies, mice, and human systems in order to understand core biologies and their alterations in human disease. We don't yet know if flies or mice will benefit from the fruits of Chris's research, but humanity has already benefited, and for that we are most grateful.

Chris and his wife Andrea are the proud parents of two boys, Nathan and Ryan.

Acknowledgements

Special thanks to Tony Oro, Cary Austin, and UCSD, Stanford, and Genentech colleagues for their contributions to this dedication.

Cambridge, MA, USA — ***Keith A. Wharton, Jr.***
Chapel Hill, NC, USA — ***David A. Eberhard***

References

1. Callahan CA, Bonkovsky JL, Scully AL, Thomas JB (1996) derailed is required for muscle attachment site selection in Drosophila. Development 122(9):2761–2767
2. Oro AE, Higgins KM, Hu Z, Bonifas JM, Epstein EH, Jr., Scott MP (1997) Basal cell carcinomas in mice overexpressing sonic hedgehog. Science 276(5313):817–821
3. Callahan CA, Ofstad T, Horng L, Wang JK, Zhen HH, Coulombe PA, Oro AE (2004) MIM/BEG4, a Sonic hedgehog-responsive gene that potentiates Gli-dependent transcription. Genes Dev 18 (22):2724-2729. doi:10.1101/gad.1221804
4. Chen W, Tang T, Eastham-Anderson J, Dunlap D, Alicke B, Nannini M, Gould S, Yauch R, Modrusan Z, DuPree KJ, Darbonne WC, Plowman G, de Sauvage FJ, Callahan CA (2011) Canonical hedgehog signaling augments tumor angiogenesis by induction of VEGF-A in stromal perivascular cells. Proc Natl Acad Sci U S A 108 (23):9589-9594. doi:10.1073/pnas.1017945108
5. Virchow R (1858) Cellular pathology (trans: Chance F). Edwards Brothers, Inc., Ann Arbor, MI

Preface

I've just sucked one year of your life away... What did this do to you? Tell me. And remember, this is for posterity so be honest. How do you feel?

–*Count Rugen, antagonist in the 1987 movie The Princess Bride*

In the movie *The Princess Bride*, the hero, Westley, has just been subjected to The Machine, a torture device that sucks years of life out of the victim. Like Westley, who cries and moans in pain in response to Count Rugen's query, anyone embarking on, or reflecting upon, a multi-year project knows the pathologic feeling of time spent on a lengthy and complex project, whether it is a book, a drug, or a film.

Feature films aspiring for blockbuster status can consume $100 million or more in production costs and 3 years just to get to production stage—all to entertain people for a mere 2 hours. Yet this amount of money pales in comparison to the economic realities of producing a new therapeutic that might address an unmet medical need for thousands or even millions of people. By most measures, developing a new drug in 2015 costs at least ten times more than a blockbuster movie. Three years in production is feature film fiction compared to the industry average of ~14 years for drug development.

The feature film and the pharmaceutical industries face similar challenges: years between the initial idea and a revenue-generating product, huge multifaceted teams, millions of dollars invested in multiple projects, only a few of which succeed, and the hope of the occasional blockbuster that must finance the failures of the rest.

At each phase in drug development, from early discovery through IND (Initial New Drug Application) to NDA (New Drug Application), the promise of efficacy is balanced against the penalty of toxicity. While there are many ways that efficacy and toxicity can be evaluated in animals and in people, the highest concentration of information relevant to many diseases remains the lesional tissue sample, microscopic examination of which provides a foundation to understand disease and the effect of therapy. Increasingly, whole microscopic slide imaging is used, providing at least an order of magnitude higher resolution of cellular context than current noninvasive in vivo radiological imaging techniques. However, microscopic data requires many players to extract its maximum value: histology (preparing the tissue sample) and pathology (interpreting the tissue sample), in addition to experts in disease-specific biology. Tissue-based studies help to understand how candidate therapies act in animals and humans, and this work is often performed by small biotech companies, large pharmaceutical companies, academic medical centers, commercial reference laboratories, and government entities. Each actor has a critical role to play in the process and in the development of the final product.

We anticipate those who will most benefit from reading this book will be embedded in government-sponsored academic research, diagnostics, or biopharmaceuticals, but we have strived to make the chapters accessible and interesting to a wide audience. Due to shrinking government budgets for basic research, more academic researchers are responding to grant announcements and pharmaceutical partnerships that drive them deeper into drug development. With the growth of companion diagnostics, experts in disease diagnostics will find useful information in this volume about co-development of diagnostic and therapeutic products, though the nature and timelines of the diagnostic industry are very different from those of drug development, creating some unanticipated but, on deeper consideration, not so surprising challenges. Our analogy to the film industry provides caution to those entering pharmaceutical drug development: While the biology underlying drug development may be familiar—and thus appear simple—to those outside the biopharma industry, one cannot overstate the complexities of drug development. One should approach the study of the biopharma industry, and one important part of it—tissue histopathology—with the same caution one might approach completely unfamiliar territory: with curiosity and respect for lessons learned by experience. Scientists from the diagnostics industry are forewarned: while pharma and diagnostics have shared biology, they are as dissimilar as are the pharma and moviemaking industries.

We often celebrate 2 hours of entertainment more than we give pause to acknowledge medicines that positively impact human lives. Recent progress in hepatitis C, cystic fibrosis, and tumor immunology has been nothing short of astounding. Our industry can be the best at times and the worst at times, but for many of us there is no more satisfying endeavor than the opportunity to design therapeutics that have the potential to save and improve lives.

It is our heartfelt belief that the biopharmaceutical industry makes positive and lasting contributions to humanity. With the human genome completed just over a decade ago, comparative genomics studies have revealed an array of druggable targets whose manipulation is at the root of most therapy development programs today. In the not-so-distant future, we are poised to witness dramatic improvements in the treatment of a myriad of severe and debilitating diseases including infectious diseases, intractable and largely incurable cancers, as well as autoimmune and genetic diseases. Histopathology is central to this effort, yet it is often relegated to a checkbox activity that is not given proper scrutiny or thought. Our authors and editorial team, consisting of experts in histopathology, have written from the trenches of diagnostic practice and pharmaceutical drug development, aiming to educate pharmaceutical and academic scientists how to best use tissue in drug development.

Most pathologists and histologists are, by their very nature, humble and not oriented to marketing their wares. This book aims to help make their contributions to drug development better understood as well as to identify best practices and new applications for their trade. The book's dedication highlights the contribution made by an exemplary individual—a pathologist no longer in our midst—whose example continues to motivate us.

Audience

This book is intended for three audiences. First and foremost, it is written for all scientists and managers in the Biopharma industry who must interact directly or indirectly with tissue samples but whose primary training did not include pathology or other skills of tissue interpretation; second, for pathology professionals and tissue scientists who will find some of the examples of applications of their trade in drug development by their peers and colleagues helpful; and third, for the many academic groups funded by government entities to become more engaged in all stages of drug development.

Information Content of a Tissue Biopsy

Both clinicians and the general public expect the pathologist's interpretation to be the "gold standard" of disease diagnosis—the absolute truth. "What were the path results?" "Do I have cancer or not? What kind?" In most cases, pathological interpretation of a relevant tissue sample is the final arbiter of truth. In both efficacy and toxicology studies, there is much information to be gleaned from local tissue environment and context, with the spatial and temporal characteristics of the cells in their organ preserved. While we strive to link cell and tissue-level resolution with complex datasets that derive from -omics analyses such as next-generation DNA and RNA sequencing, many forget that immunohistochemistry has provided single cell analysis of protein distribution—albeit a single protein at a same time—for decades. Technology is always creating new approaches to interrogate tissue samples, both within biopharma and in academia, but for a technology to be implemented in clinical practice, it requires confirmation of utility and definition of its limits and context that often comes years after the technology has lost its newness. But the expectation that magically circulating in the blood is information content that will replace the efficacy and toxicology information content of a tissue biopsy is a fantasy of Hollywood proportions. Many people, even within the biopharma industry, are not aware that every organ of every animal used in a GLP (Good Laboratory Practice) toxicology study must be examined under the microscope by one or more veterinary pathologists. The FDA and other regulatory agencies remain aware of the importance of the tissue microenvironment to drug development.

How Important Is Histopathology in Drug Development?

How do we assign an economic value to tissue analysis in drug development? One approach is to estimate the number of slides generated or read per year, which then can be converted to dollars. This can be estimated in several different ways, working from the number of pathologists who engage in testing or the number of drugs that require testing of tissue exposed to them. There are approximately 600 veterinary pathologists who work primarily in pharmaceutical research, and probably another ~40 M.D. pathologists exclusively working in pharmaceutical research. The vast majority of slides utilized in the pharmaceutical industry are in support of GLP toxicological pathology studies for IND submissions. Previous studies have estimated that on average one pathologist supports $1B of drug product development [1]. We can assume each pathologist reviews ~10,000 slides a year (~50 slides a day × ~200 working days a year) and yields an estimate of nearly ~7 million glass slides evaluated each year worldwide in drug development. However, this number does not account for non-boarded pathologists and scientists who review slides, so it may likely be closer to ten million glass slides per year. Even with a rough estimate of $150 per glass slide for creation (histology) and reading (pathology), this equates to $1.5 billion spent annually on histopathology within the pharmaceutical industry.

Organization and Goals

While much of the focus of pathology in the biopharmaceutical industry is on oncology programs, there is growing recognition of the value of pathology outside of oncologic disease. Consequently, the chapters have been deliberately selected to include other disease areas, including chapters addressing nonalcoholic steatohepatitis, arthritis, celiac disease, myeloproliferative disorders, neurology, and wound healing. As part of a methods series, this volume is designed to provide practical wisdom and examples that others can follow and apply as part of drug development. Some chapters are case studies in specific techniques or disease areas where tissue biopsies can be utilized effectively, while others are literature reviews, and still others are a summary of the authors' decades of collective experience with tissue in an area of drug development.

While a great deal of pathology-related efforts in drug development are by necessity geared toward formal toxicologic evaluation under GLP requirements, the editors have deliberately not focused on toxicologic pathology as a discipline. This is certainly not reflective of the relative importance of safety testing, but was rather for two other reasons: First, toxicologic pathology in drug development has been comprehensively covered for over a decade in a regularly updated excellent handbook for practitioners [2] as well as in specialty journals. Second, the pace of change in toxicologic pathology is slow in comparison to other more technology-driven aspects of pathology commonly used in drug development. The IND approval process allowing first-in-human studies is a highly regulated endeavor, stipulated by GLP standards, that can take on a life (and a career) of its own. Of necessity, innovation occurs slowly—and often reactively—in this field.

This book, which focuses on pursuits within industry that are variably termed "experimental" or "investigative" pathology, or "translational medicine," thus deals more with efficacy studies that bridge from early stage discovery to formal clinical trials. An introduction to the field of anatomic pathology and its application to the biopharma industry is first provided, based on the author's experience both in industry and through teaching medical students. This is followed by a personal narrative by a leading biopharma pathologist on the nuances of communicating pathology results to non-pathologist colleagues, and then by a chapter on planning and outsourcing histopathology-based investigations in clinical trials. Two leading experts in inflammatory disease then provide a specific example of how histopathology can be leveraged to better understand rheumatoid arthritis.

The second section focuses on applications of tissue image analysis–whole microscopic slide imaging and computer algorithms to quantitatively measure what the pathologist qualitatively observes. The chapters cover a variety of disease areas and concepts including angiogenesis, hepatic fibrosis, and celiac disease, providing a glimpse of future applications of digital pathology.

The third section discusses molecular histopathology, divided into in situ hybridization (mRNA and DNA), sequencing, and genomics. The reader will find state-of-the-art reviews with methodology on in situ hybridization as well as next-generation sequencing of tissue samples.

The fourth section covers companion diagnostics. The first two chapters describe preanalytic variables and then the adaptation of HER2 IHC scoring systems commonly used in breast cancer to gastrointestinal tumors. Three chapters then discuss the development of companion diagnostics from an industry standpoint, the relationship between the reference lab, diagnostic partner and the pharma client, and regulatory aspects of medical device submissions. A success story of a multianalyte IHC companion diagnostic is then presented. Finally, two biostatisticians discuss statistical approaches to cut-point analysis in digital pathology and applications in IHC companion diagnostics.

We hope that this volume will serve to inform and enlighten both tissue-focused and non-tissue-focused drug development scientists about better use and interpretation of the multidimensional data contained in a tissue biopsy. The hunt for new therapies remains one of the most exciting and meaningful pursuits in the twenty-first century, and so the evaluation of a tissue biopsy remains a central and challenging part of that pursuit.

Westminster, CO, USA *Steven J. Potts*
Cambridge, MA, USA *Keith A. Wharton, Jr.*
Chapel Hill, NC, USA *David A. Eberhard*

References

1. Potts SJ, Young GD, Voelker FA (2010) The role and impact of quantitative discovery pathology. Drug Discov Today **15**(21–22): 943-50
2. Haschek WM, Rousseaux CG, Wallig MA (2002) Handbook of toxicologic pathology, Vol 1. Academic, San Francisco

Contents

Contributors

JON ASKAA • *Medical Prognosis Institute, Hørsholm, Denmark*
ROBERT G. AUDET • *Development Engineering Sciences LLC, Flagstaff, AZ, USA*
JOSHUA C. BLACK • *Flagship Biosciences, Westminster, CO, USA*
PIERRE BROUSSET • *Department of Pathology, CHU Toulouse-Purpan, Toulouse, France*
PATRICK CAPLAZI • *Department of Research Pathology, Genentech, Inc., South San Francisco, CA, USA*
MIU CHAU • *Genentech, Inc., South San Francisco, CA, USA*
MAUREEN CRONIN • *Strategic Information Management, Celgene Corporation, San Francisco, CA, USA*
LAURI DIEHL • *Department of Research Pathology, Genentech, Inc., South San Francisco, CA, USA*
ROBERT B. DILLER • *Development Engineering Sciences LLC, Flagstaff, AZ, USA; Department of Biological Sciences, Center for Bioengineering Innovation, Northern Arizona University, Flagstaff, AZ, USA*
JOHANNA K. DISTEFANO • *TGEN, Phoenix, AZ, USA*
DAVID A. EBERHARD • *Department of Pathology and Laboratory Medicine, Department of Pharmacology and Lineberger Comprehensive Cancer Center, University of North Carolina, Chapel Hill, NC, USA; Laboratory Corporation of America (LabCorp), Research Triangle Park, NC, USA*
A. ETMAN • *University of Utah & ARUP Laboratories, Salt Lake City, UT, USA*
TOMMY FU • *Celgene Corporation, Summit, NJ, USA*
HUMPHREY GARDNER • *Translational Medicine, Early Clinical Development, AstraZeneca R&D, Waltham, MA, USA*
RANDY D. GASCOYNE • *Department of Pathology and the Center for Lymphoid Cancer, Organization British Columbia Cancer Agency, Vancouver, BC, Canada*
ARKADY GUSEV • *Biomarker Development, Translational Medicine, Novartis Institutes for BioMedical Research, Cambridge, MA, USA*
ERIK HAGENDORN • *Flagship Biosciences, Westminster, CO, USA*
CARLA HEISE • *Celgene Corporation, Summit, NJ, USA*
DAVID G. HICKS • *Surgical Pathology Unit, Department of Pathology and Laboratory Medicine, University of Rochester Medical Center, Rochester, NY, USA*
RAINER HILLENBRAND • *Biomarker Development, Translational Medicine, Novartis Institutes for BioMedical Research, Basel, Switzerland*
AARON HUBER • *School of Medicine and Dentistry, University of Rochester, Rochester, NY, USA*
XIAOYU JIANG • *Biomarker Development, Translational Medicine, Novartis Institutes for BioMedical Research, Cambridge, MA, USA*
ROBERT S. KELLAR • *Development Engineering Sciences LLC, Flagstaff, AZ, USA; Department of Biological Sciences, Center for Bioengineering Innovation, Northern Arizona University, Flagstaff, AZ, USA; Department of Mechanical Engineering, Center for Bioengineering Innovation, Northern Arizona University, Flagstaff, AZ, USA*
MARK KOCKX • *HistoGeneX, Antwerp, Belgium*
JOSEPH S. KRUEGER • *Flagship Biosciences, Westminster, CO, USA*

JEFF L. KUTOK • *Infinity Pharmaceuticals, Inc., Cambridge, MA, USA*
NICHOLAS D. LANDIS • *Flagship Biosciences, Westminster, CO, USA*
HOLGER LANGE • *Flagship Biosciences, Westminster, CO, USA*
CAMILLE LAURENT • *Department of Pathology, CHU Toulouse-Purpan, Toulouse, France*
BENJAMIN H. LEE • *Oncology Translational Medicine/Oncology Business Unit, Novartis Institutes for BioMedical Research, Cambridge, MA, USA*
MICHAEL D. LUNT • *School of Medicine and Dentistry, University of Rochester, Rochester, NY, USA*
MATTHEW J. MCGINNISS • *Medical Laboratory, Genoptix, a Novartis Company, Carlsbad, CA, USA*
DALE MONGEON • *Biomarker Development, Translational Medicine, Novartis Institutes for BioMedical Research, Cambridge, MA, USA*
PIERRE MOULIN • *Discovery and Investigative Pathology, Preclinical Safety, Novartis Institutes for BioMedical Research, Basel, Switzerland*
MARCUS OTTE • *Oridis Biomarkers, Graz, Austria*
MIRZA PELJTO • *Flagship Biosciences, Westminster, CO, USA*
SHERRIE L. PERKINS • *University of Utah & ARUP Laboratories, Salt Lake City, UT, USA*
STEVEN J. POTTS • *Flagship Biosciences, Westminster, CO, USA*
JOSEF T. PRCHAL • *University of Utah & ARUP Laboratories, Salt Lake City, UT, USA*
MOHAMED E. SALAMA • *Department of Pathology, ARUP Reference Lab Research Institute, University of Utah, Salt Lake City, UT, USA*
STEFANIE DE SCHEPPER • *Immunohistochemistry, HistoGeneX, Antwerp, Belgium*
GRAHAM W. SLACK • *Department of Pathology and the Center for Lymphoid Cancer, Organization British Columbia Cancer Agency, Vancouver, BC, Canada*
MAHIPAL V. SURANENI • *Flagship Biosciences, Boulder, CO, USA; Flagship Biosciences, Westminster, CO, USA*
CHRISTOPHER UNG • *HistoGeneX, Antwerp, Belgium*
KEITH A. WHARTON JR. • *Discovery and Investigative Pathology, Preclinical Safety, Novartis Institutes for BioMedical Research, Cambridge, MA, USA*
CHRISTA L. WHITNEY-MILLER • *School of Medicine and Dentistry, University of Rochester, Rochester, NY, USA*
HYNEK WICHTERLE • *Department of Pathology, Columbia University, New York, NY, USA; Department of Neurology, Columbia University, New York, NY, USA; Department of Neuroscience, Columbia University, New York, NY, USA*
BIN YE • *Biomarker Development, Translational Medicine, Novartis Institutes for BioMedical Research, Cambridge, MA, USA; Beijing Shenogen Biomedical Co., Beijing, China*
G. DAVID YOUNG • *Flagship Biosciences, Westminster, CO, USA*

Methods in Pharmacology and Toxicology (2015): 1–26
DOI 10.1007/7653_2014_33

Published online: 22 January 2015

Histopathology: A Canvas and Landscape of Disease in Drug and Diagnostic Development

Keith A. Wharton Jr.

Abstract

The aims of diagnostic and therapeutic development are to accurately diagnose and cure disease, respectively. For the past century and a half, histopathology—the microscopic examination of cells and tissues—has been considered a "gold standard" for the diagnosis of many diseases. As an introduction to this volume on molecular histopathology and tissue biomarkers in drug and diagnostic development, I explore the relationship between histopathology and the nature of disease itself. A lack of agreement on the meaning of "disease" has led to widespread and indiscriminate use of the term. Here, I propose that the term "disease" be reserved for conditions where there exists some knowledge of alterations in cells or their products that participate in cause-effect relationships in lesional (diseased) tissue. This is a definition that simultaneously lends legitimacy to the term's use while enabling revision and testing of hypotheses based on rapidly emerging scientific knowledge. With this perspective, histopathology, as a preferred means to visualize and depict the cellular events that constitute disease as it impacts tissue structure and function, will remain essential to develop new diagnostic tests and targeted therapies for the foreseeable future.

Key words Disease, Histopathology, Biomarker, Lesion, Diagnosis, Illness, Condition, Disorder, Pathogenesis, Feedback, Crosstalk, Clinical trial, Diagnostic test, Drug development

1 Introduction

If thought corrupts language, language can also corrupt thought.

—George Orwell

I rob banks because that's where the money is.

—Willie Sutton

The concept of "disease" is central to understanding health in all forms of life as well as for optimal delivery of health care. All human cultures have words or symbols that represent disease or related ideas. From our earliest experiences, we develop an understanding of its general nature, but the term is increasingly adapted to suit circumstances at hand. Diagnostic criteria, a set of features that define each disease, range from simple to complex. For some well-understood diseases, diagnosis is clear-cut, relying on defined and measurable criteria. For poorly understood or controversial diseases, observation and professional judgment play a major role, creating opportunities for uncertainty, bias, and misdiagnosis. Health care providers make treatment decisions and bill for reimbursements based on an incomplete and evolving understanding of many diseases. Perusal of http://www.clinicaltrials.gov/, the USA FDA's Web site that discloses the essentials of all human clinical trials, reveals that enrollment criteria and response variables (i.e., trial endpoints) for many diseases consist of features learned from the medical history, or of clinical observations and measurements with uncertain relevance to the disease processes under investigation. To the afflicted individual, having a particular disease can alter self-identity and behavior, and can impart social stigma and/or advantage. Disease, frankly, is a big deal. Given that shared goals of biomedical research, the biopharmaceutical industry, and the health care industry as a whole are accurate diagnosis and treatment of disease, as an introduction to this timely book devoted to molecular histopathology and tissue biomarkers in drug and diagnostic development, it seems reasonable to first consider the nature of disease itself.

Now, early in the twenty-first century, humanity finds itself squarely in the middle of the molecular era of disease, full of optimism that, from our recently sequenced genomes, cures for our myriad diseases are imminent. Investigations from diverse fields of biomedicine have revealed over the past 50 or so years the molecular underpinnings, or at least a framework, for thousands of hitherto mysterious and sometimes misclassified diseases—spanning from the rare and precisely defined to the common and heterogeneous. For example, there are at least 456 distinct genetic diseases that affect the skeletal system [1], a majority for which causative gene mutations have been identified. We have learned that the culprits in many human (and animal) diseases are corrupted versions of vital processes deeply conserved in our evolutionary history—alterations in specific proteins, protein complexes, macromolecular structures, and organelles that comprise molecular communication pathways with developmental, homeostatic, and adaptive functions. In some cases, we know which cells misbehave and how, and we can replicate key features of the disease in wild-type or genetically altered animals. Since organisms are composed of massive networks of interconnected, compartmentalized molecular pathways, the task that remains for each disease, in each patient, is to understand the dynamic, physical nature of each pathway alteration and then devise strategies to oppose the abnormal network of behaviors that promote disease without compromising functions of the remaining pathways required for life.

2 What Is Histopathology?

Histopathology is the microscopic visualization of tissue to describe how cellular and tissue morphology is altered by manifestations of disease (http://en.wikipedia.org/wiki/Histopathology). It is distinct from techniques that use (and destroy) tissue as a raw material for nucleic acid sequencing or bioassays to measure specific molecules or their concentrations, each of which separate the data generated from its endogenous tissue context. Histopathology thus provides a visual, descriptive picture of cells and tissues, fixed or frozen in time, in their natural context—a landscape of sorts. When viewed using the microscopic techniques of histopathology, normal tissue might resemble a well-ordered city when viewed from the sky, whereas tissue from a destructive autoimmune disease might look like a war zone, and a cancer might kindle images of prisoners escaping Alcatraz or Godzilla tipping skyscrapers. Histopathology is central to patient care, whether to discern normal tissue from benign or malignant disease, to ascertain surgical tissue margins, or to investigate cause of death. Among all diagnostic assays, histopathology remains unparalleled in its ability to discriminate among a seemingly infinite variety of disease states, especially in the presence of a much greater excess of normal tissue. For example, in an appropriate biopsy, a trained pathologist can diagnose prostate cancer by recognizing less than ten malignant (cancerous) cells among thousands of nonmalignant cells. Histopathology is also a crucial tool to assess efficacy and possible safety hazards of candidate human therapies in preclinical (i.e., animal) studies. Despite the regulatory requirement for histopathology-based interpretation of tissues in animal toxicology studies, collecting tissues as part of human clinical trials remains a rare practice, except in oncology trials. (A later chapter in this book explores issues to consider when incorporating tissue and histopathology-based measurements in human clinical trials.)

Anatomic pathologists are the sages of histopathology. They are mostly physicians (M.D., D.O., or equivalent) or veterinarians (D.V.M. or equivalent) who have completed one or more anatomic pathology fellowships and often have research experience in investigative pathology or related scientific disciplines. They train over years and decades to render interpretations and diagnoses based on observable and, to an outsider, sometimes arcane criteria. Despite the fact that *Pathology*, translated from its Greek origins, literally means the study of *suffering*, pathology emerged as a medical specialty, in part, to provide an *objective* diagnosis of disease, unbiased by patient presentation or a physician's subjective (but often crucial) knowledge about a patient's history. While a computer, IBM's Watson, can win the TV game show Jeopardy, no computer is yet capable of replacing a diagnostic anatomic pathologist.

In this book, the editors and contributors adopt a far broader view of histopathology, to include any discipline or technique that enhances our ability to visualize and quantify the three-dimensional, physical nature of normal and diseased tissue, including advanced techniques in microscopy and cell biology, image segmentation and quantitation, omics-scale profiling of various molecular species, and "big data" analysis. Thus histopathology is not a static or archaic art based only on routine stains of tissue sections, but rather highly technical and continually evolving. These ancillary technologies thus add more, and often critical, detail to the landscape that is each disease. Modern histopathology is a firmly molecular pursuit, encompassing the detection of specific proteins by immunohistochemistry (IHC) and nucleic acids by in situ hybridization (ISH). The editors espouse the view that data generated from techniques that destroy tissue must be interpreted in the context of methods such as histopathology that preserve it, in order to best understand the manifestations of each disease in tissue and in the patient. Modern histopathology thus requires pathologists to collaborate with a wide variety of scientists and engineers who generate complementary, often deeper insights from diseased tissue, in order to create a coherent working model of each disease process.

3 What Is Disease?

Historical concepts of disease arose independently in diverse civilizations, including causation by external factors (e.g., gods, demons, spells, miasmas) and imbalances in internal qualities (e.g., the four humors). Largely incorrect in their primitive forms, each view has elements of truth: disease can be initiated by external factors (e.g., infectious agents, toxin exposure) and can originate or manifest as imbalances in internal qualities (e.g., persistent hormone production or accumulation of toxic metabolites, mutations, or rogue cells). Historical review of disease conceptions across cultures is outside the scope of this chapter, but several sources exist [2–4].

Clinical acumen and technology jointly drive the discovery and diagnosis of disease. Many diseases are—or were, at some point in history—named for the person, usually a physician, who first noticed a nonrandom aggregation of clinical or histopathological features in their patients; Alois Alzheimer and James Parkinson are two whose namesakes persist, though many names of diseases, proper and otherwise, have since been discarded in favor of more descriptive designations. The light microscope revealed fine details of healthy and diseased tissues in the mid-nineteenth century, revolutionizing disease diagnosis, classification, and nomenclature. Thousands of discrete disease entities are recognized by the International Classification of Diseases, ninth revision (ICD-9), a list subject to revision every few years based on emerging knowledge and expert consensus. Recognize, however, that such lists are made largely for the purposes of counting and billing, and their composition usually lags behind the most current thinking about disease. Today, a panoply of omics techniques (genomics, proteomics, etc.) and in vivo imaging modalities give unprecedented insight into the complexity of disease as it affects our bodies; these technologies challenge prevailing notions of disease, force continual reassessment of naming conventions, and can allow earlier, more accurate diagnosis and monitoring of disease. For example, diagnosis of Alzheimer's dementia, in the past requiring an autopsy, can now be made premortem with some certainty by combining cognitive testing with imaging and cerebrospinal fluid biomarker analysis [5]. Massively parallel (next generation) sequencing, the subject of another chapter in this book, stands to revolutionize disease diagnosis, classification, and monitoring in the next decade. Principles of systems biology are being applied to datasets derived from diseased tissue or fluids in an attempt to better classify, understand, and treat disease (e.g., ref. 6). Such studies create compelling and often testable hypotheses. However, the enormous sizes of datasets in comparison to the typically small numbers of diseased samples used to generate them have raised concern that resulting predictive models will be overfitted to the diseased sample set and thus not applicable to the population as a whole.

Many consider "disease" a more specific concept than simply the absence of health, and defining it, in general and for each specific case, is not a trivial exercise. One classic definition of disease, articulated by Sir John Russell Reynolds in 1866, is: "*any condition of the organism which limits life in either its power, enjoyment, or duration*" [7]. Merriam Webster (www.m-w.com) defines disease as, "*a condition of the living body or of one of its parts that impairs normal functioning and is typically manifested by distinguishing signs and symptoms.*" (Recall that *symptoms* are experienced by the patient, such as insomnia or pain, whereas *signs* are observed by the physician during physical examination; e.g., Courvoisier's sign is a palpable gallbladder in a patient with obstructive jaundice). Wikipedia (c. 2014) offers a wider, more inclusive definition: "*…any condition that causes pain, dysfunction, distress, social problems, or death to the person afflicted, or similar problems for those in contact with the person. In this broader sense, it sometimes includes injuries, disabilities, disorders, syndromes, infections, isolated symptoms, deviant behaviors, and atypical variations of structure and function.*" Unfortunately, here inclusion breeds imprecision, although given the mutable nature of Wikipedia the definition might be changed by the time you read this chapter!

Such definitions of disease, both classic and modern, are inadequate tools for thinking about and communicating modern concepts of disease. Too often language, entrenched by historical accidents of discovery, shapes the boundaries of our thought, when thought, shaped by scientific discovery, should enhance the accuracy and precision of language that describes the products of our thought. These observations suggest a new working definition of disease is needed—one that not only reflects scientific principles upon which experts can agree, but that also facilitates thinking about new ways to treat or cure disease. This (or any) attempt to create a single umbrella definition for disease might be regarded as futile, but one hopes the effort helps to focus scrutiny on disease-relevant events.

4 Three Concepts Related to Disease

Disease can be considered distinct from *illness*, which m-w.com defines as "*unhealthy in body or mind*," but commonly refers a patient's subjective experience of disease. Given that one can feel ill without having a known disease or have a severe disease with no ill feelings (e.g., early stage cancer), it is reasonable to assume that a defined disease need not be a prerequisite for illness, or at least temporarily ill feelings. Treating symptoms of disease, illness, and suffering are laudable goals, but the view of disease proposed here, absent knowledge about the disease process itself, might not inform the best way to manage illness. (Note that alternate definitions of "illness" are nearly synonymous with disease).

Disease is distinct from a medical *condition*, which implies deviation from normalcy but is not necessarily an abnormality. For example, pregnancy is a temporary condition, but not an abnormal one. Disease is also distinct from a *disorder*, a term related to disease but with the implication that much less is known about the nature of what's wrong. *Disorder* is often used to describe intrinsic abnormalities of brain function. For example, the Diagnostic and Statistical Manual of Mental Disorders (DSM), currently in its fifth revision, is an ongoing but imperfect attempt to classify disorders of thought, feeling, or behavior. A few entities described in the current DSM qualify as a "disease" by the standards of this chapter (e.g., those conditions highly associated with mutations in genes that act in known pathways impacting specific regions of the brain), but their mysterious nature precludes classification by criteria more precise than those based on constellations of behavior or symptoms. *Disorder* is also used as a term that refers to a family of diseases or their manifestations, e.g., *disorders* of metabolism.

Designation as "disease" remains controversial and is not without consequence. Ill feelings, a condition, or even normal physiological variation can be "medicalized" or "pathologized" into disease by motivated parties, including scientists seeking grant funding or biopharmaceutical companies seeking a new market niche. For example, there is ongoing debate whether certain life-cycle-associated conditions are bona fide diseases or are simply consequences of aging that many in Western society refuse to accept. Companies that make products such as nutritional supplements not subject to FDA approval craft language to imply that their product is beneficial ("*supports the health of...*" is a commonly used phrase) for this or that organ (and by implication, will oppose disease of that organ) while required by USA federal law to state that their "*product is not intended to diagnose, treat, cure, or prevent any disease.*" Carrying a diagnosis of a so-named disease has socioeconomic consequences. The American Medical Association's recent declaration that obesity is a disease drew attention to its increasing prevalence and impact on society's health [8], but at the same time affixed a label of questionable utility beyond possession of a "risk factor" for other diseases such as diabetes, hypertension, and myocardial infarction to millions of overweight people. Often, implicit in affixing the label of "disease" is absolution of blame for the affected individual's state of health or actions resulting from it. Addiction, personality disorders, or willful criminal acts: are they diseases or not? Any debate about the nature of disease can quickly transcend science, reaching the realm of philosophy and raising the question: Do we control the disease or does the disease control us? Highlighting our pervasive inability to agree on what constitutes disease, an insightful New Yorker magazine article [9] quotes E.M. Jellinek's mid-twentieth century work that advocated alcoholism as a disease: "*A disease is what the medical profession recognizes as such.*"

Given the stakes, there is need for a working definition of disease that simultaneously lends credibility to use of the term, while enabling revision of disease concepts based on emerging scientific knowledge. I propose that bona fide "disease" be considered *an abnormality of the organism, caused by abnormal cells or their products, leading to cause-effect relationships in lesional tissue that compromise function in a societal context or attenuate normal lifespan*. Through visualization of diseased cells and tissues, histopathology becomes both the canvas and the landscape upon which disease unfolds.

5 Organic Disease vs. Functional Disease

In line with the proposed definition of disease, the medical profession has long distinguished between *organic* and *functional* disease. *Organic* disease can cause a measurable physiological change, and thus changes in molecules, cells, tissues, or organs that give rise to the disease (or for which such changes are a consequence of the disease) can be identified and often measured with a diagnostic test. *Functional* diseases, sometimes called functional symptoms or disorders, can manifest as signs, symptoms, or somehow altered function (e.g., pain, fatigue, or other form of suffering) without apparent alterations in known tests. As science progresses, many functional disorders will join the ranks of organic diseases. For example, "inflammatory bowel disease" encompasses several organic diseases of the bowel in which alterations in tissue structure documented by histopathology generally correlate with the severity of clinical symptoms, whereas "functional bowel disease" (a.k.a. "irritable bowel syndrome"), despite sometimes equally debilitating symptoms, is characterized by an absence of specific abnormalities by histopathology, but has been associated with clinical depression. The organic/functional distinction can become blurred in certain instances: although we do not know exactly which cells are altered, and how, in the brains of patients with schizophrenia or autism, few doubt their organic nature. Some disorders currently classified as functional are associated with psychological or psychiatric conditions, raising the question of whether they are somatic (body) manifestations of an underlying psychiatric condition or disorder. Pathologists claim dominion over organic disease, leaving the more difficult questions of functional disease and its management to other specialists. A *diagnosis of exclusion* is one that is made after other causes are excluded: some diseases, like the multisystem granulomatous disease sarcoidosis, are clearly organic, whereas others not currently linked to definitive test results are classified as functional but may be no less a burden than organic disease to those afflicted.

6 Three More Concepts Related to Disease

Three more concepts deserve mention before I further examine the nature of disease.

A *syndrome* is a nonrandom association of several apparently unrelated features (signs, symptoms, test results) that suggest a common underlying cause or pathogenesis. Down syndrome, a constellation of developmental defects due to trisomy 21, and Acquired Immune Deficiency Syndrome (AIDS), caused by HIV infection-induced immune deficiency, are classic examples.

Pathognomonic describes the almost certain presence of a disease (or disorder) when a particular sign or symptom, or a combination thereof, is present, or with a particular diagnostic test result. In statistical terms, this means that the "positive predictive value" (the percentage of patients with a feature such as a positive test result that have the disease) of a particular sign, symptom, or test result is 100 %.

Two skin diseases illustrate these two points. Cutaneous xanthomas (small skin papules or nodules composed predominantly of macrophages stuffed with cholesterol) are pathognomonic for hyperlipoproteinemia; i.e., all individuals with xanthomas have some sort of alteration in lipoprotein metabolism.

Dozens to hundreds of cutaneous basal cell carcinomas in sun-exposed areas on a young adult are pathognomonic for Gorlin's basal cell nevus syndrome, caused by a heterozygous, germ line mutation in the *PTCH1* gene [10]. But, how specific the relationship is between the pathognomonic feature and the disease varies with the definition and cause of each disease: not all individuals with hyperlipoproteinemias have xanthomas, and there are many different causes of hyperlipoproteinemia. In contrast, the relationship between *PTCH1* mutation and Gorlin's syndrome is obligate, in that there appears to be no other genes that when singly mutant give rise to such a specific, striking array of clinical manifestations.

The third concept is *forme fruste*, which describes a very mild or atypical form of a disease. One example is discoid lupus, a variant of the autoimmune disease systemic lupus erythematosus (SLE) whose manifestations are restricted to skin. Sometimes, a forme fruste variant manifests in a distinct tissue or organ compared to the more typical or severe form of the disease. For example, cystic fibrosis, due to mutation in the CFTR channel protein, has severe, life-threatening manifestations in the lung and pancreas, but occasional patients with partial loss of function mutations in the same protein present with male infertility due to the requirement of CFTR for vas deferens function [11]. Hence, male infertility can be a forme fruste of cystic fibrosis. Since each of us harbors on the order of 400 defective genes in our genomes and inherits ~60 new mutations not present in our parents' DNA [12, 13], widespread use of massively parallel sequencing technologies in future diagnostic tests (the subject of another chapter in this book) is likely to reveal combinations of sequence polymorphisms or otherwise cryptic gene mutations as factors contributing to "forme frustes" of many diseases.

7 Etiology/Chain of Causation

The foundation of differential diagnosis—the process of determining which among a list of potential diseases a patient might have based on their unique presentation—is classification of disease by apparent cause or *etiology*. Broad, descriptive categories have traditionally been used to classify disease: these include neoplastic, infectious, genetic, vascular, and idiopathic—the latter signifying our ignorance of an underlying cause. Distinctions are made between *congenital* and *acquired* disease, and *genetic*, *familial*, and *sporadic* disease; these categories will not be further explored. Diseases, however (as defined), can be considered *primary* if they originate within the named organ, or *secondary* or even *tertiary* if causative factors originate from other organs. For example, *primary* hyperparathyroidism is caused by autonomous hypersecretion of parathyroid hormone (PTH) from the parathyroid gland (e.g., by a parathyroid cell-derived tumor or from nonmalignant parathyroid cells harboring a mutation that activates a signaling pathway, leading to PTH hypersecretion). Since the function of PTH is to increase serum calcium by acting on a variety of target tissues (bone, kidney, intestine), *secondary* hyperparathyroidism is characterized by increased parathyroid secretion from the parathyroid gland, usually due to the body's attempt to correct for low serum calcium. The manifestations of secondary hyperparathyroidism on target organs outside of the parathyroid gland are similar to those of primary hyperparathyroidism, yet the causes are very different.

Sometimes, a *presumptive* or *working* diagnosis of disease is made if time is limiting and the consequences of missing a diagnosis, and the appropriate intervention, are dire. For example, pneumonia is commonly treated with medical therapy based on clinical history and symptoms, often confirmed with an X-ray, but without identification of a causative infectious agent. Even a definitive diagnosis of a disease in a given patient should inspire further inquiry, sometimes urgently. For example, while iron deficiency anemia can be considered a disease—characterized by abnormal test results, signs, and symptoms—its presence beckons the search for a cause. Indeed, distinct, qualitative categories of disease, coupled with ordered (e.g., primary and secondary) events during disease progression, implies that a sequence of discrete events, a so-called *chain of causation*, is a more accurate way to describe disease. *Pathogenesis* is a term used to describe the mechanism and sequence of key events typical of a disease or in a given patient with a disease. A related term, *pathophysiology*, emphasizes alterations in normal physiology caused by a disease.

By analogy to a river, "upstream" and "downstream" refer to events that occur earlier or later in the chain of causation. Since correlation does not imply causation, we must be careful when describing a putative chain of disease causation that we do not assume one event causes another when both events are temporally distinct but are due to shared (or distinct), yet unknown, "upstream" causes. Nevertheless, in reference to a particular event in the disease (e.g., appearance of a sign or symptom, positive test result, evidence of disease progression, or even death), I borrow two terms from law, *proximate cause* and *ultimate cause*.

A proximate cause is an occurrence or entity most closely responsible for the event of interest. For example, a proximate cause of most myocardial infarctions (heart attacks) is rupture of the fatty clot-forming material contained within an atherosclerotic plaque into the lumen of a coronary artery, resulting in thrombus formation, arterial lumen occlusion, and necrosis of myocardium in the artery's territory. A proximate cause of the signs and symptoms in many autoimmune diseases appears to be rare but crafty populations of autoreactive lymphocytes that escape elimination by the immune system, secreting proteins (e.g., TNF-alpha, IL17) that drive progression of disease, wreaking havoc on various tissues and organs.

An *ultimate* cause (also known as *root* cause) is a further upstream event or entity "without which" the proximate cause or downstream events would not occur: Without atherosclerosis of coronary arteries, most myocardial infarcts would not occur, and without a mutation in the *CFTR* gene, cystic fibrosis (or its forme frustes) would not occur. Common ultimate causes are genetic mutations, environmental exposures, infectious agents, and injuries; without the inciting event, e.g., a mutation or exposure, the disease would likely not occur. For many diseases, ultimate causes remain elusive.

While ordering events in disease pathogenesis can be a satisfying exercise, it is often said that "*life itself is a terminal disease*," implying that for every event, further upstream causes exist. For example, origins of atherosclerosis remain under intense investigation, with evidence that genetic predisposition, environmental exposure (e.g., diet), and perhaps the interaction between inopportune infections and the immune system, combine to create the lesions of atherosclerosis [14]. There is epidemiologic evidence in human, and experimental evidence in lower organisms, that parental and even grandparental environment can affect health in adulthood, possibly through epigenetic changes in DNA that are passed through the germ line [15].

Indeed, the term *chain* of causation implies a *linear* order of events in disease pathogenesis that for many diseases is inaccurate; there are frequently nodes of *divergence* and *convergence* of events to suggest a *web* or *network* of causation is a more apt term. For example, "congestive heart failure" has secondary consequences in other organ systems due to sluggish, compromised blood flow, reduced oxygen delivery to tissues, and compensatory (and eventually decompensatory) hemodynamic changes—an example of divergence. Many causes of chronic lung injury ultimately manifest as a common histopathological picture of interstitial pneumonitis—an example of convergence. Diseases with a common ultimate cause, e.g., an inherited genetic mutation, can manifest differently even in identical twins, highlighting the crucial interaction between genes, environment, and other unknowns (epigenetics, serendipity) that contribute to the heterogeneous nature of nearly all diseases.

Within and between categories of disease, two additional concepts deserve mention, *lumping* and *splitting*. Lumping occurs when distinct diseases are shown to share a common or related pathogenesis or therapeutic vulnerability. There are several recent notable examples of lumping: growth of two distinct types of cancer, chronic myeloid leukemia (CML) and gastrointestinal stromal tumor (GIST), is driven by overactivity of a shared intracellular growth promoting protein kinase cascade, due to mutations in related receptors that are susceptible to common kinase inhibitors [16]. Splitting occurs when a disease that appears homogeneous by traditional (often histopathological) criteria but manifests heterogeneously in a diverse population is subdivided into discrete diagnostic categories, typically by insights gained from new technologies. One example of splitting is a type of cancer termed diffuse large B-cell lymphoma (DLBCL). Although a given patient's prognosis with DLBCL bears some relationship to traditional, clinically measurable attributes such as tumor size, extent of spread throughout the body, and symptoms, most DLBCLs are indistinguishable from one another by histopathology: they consist of abnormal appearing lymphoid tumor cells mixed with host inflammatory cells. Expression profiling of mRNAs from a large set of DLBCL tumor tissues revealed at least two subtypes with mRNA expression profiles that bear some resemblance to benign B lymphocytes at different developmental stages—the germinal B-cell (GBC) type and activated B-cell (ABC) type [17]. Splitting is important, because the GBC and ABC tumor types probably have distinct ultimate causes (i.e., "driver" events such as gene mutations that give rise to, or propagate, the disease), prognoses, and, presumably, therapeutic vulnerabilities that are under investigation. The right degree of splitting for a given disease is the essence of "personalized medicine."

Events that occur on a particular space or time scale in disease chain of causation can influence events that occur at vastly different space or time scales. Consequently, understanding chain of causation can be a multidisciplinary pursuit, spanning epidemiology and public health to molecular biophysics and electrophysiology. The "channelopathies," a heterogeneous group of diseases due to dysfunction of a variety of ion channels, are excellent examples of this principle [18]. Genetic mutation or acquired alteration in the hERG potassium channel subunit alters cardiac conduction that can manifest as chronically inefficient pumping of blood, or, in rare cases, sudden death [19]. Another class of diseases, characterized by alterations in extracellular matrix proteins such as fibrillin, causes give rise to mechanical fragility of blood vessels at sites of high shear stress with loss of vascular tone, followed by high blood pressure and the often lethal consequence of aortic dissection [20].

Feedback can be important to understand chain of causation in disease. Feedback normally acts at critical upstream or downstream "nodes" (control points) to amplify or dampen the activity of biological systems. Feedback maintains homeostasis and reversibly adapts a system to acute, repetitive, and chronic perturbations. Proteins, molecular machines and pathways, cells, organ systems, and our bodies as a whole have evolved a variety of homeostatic mechanisms to optimize physiology as we interact with our environment through our life cycle. Enzyme production and activity are modified in response to metabolic demand; inflammation escalates upon pathogen contact and then regresses upon clearance of the invader; and blood pressure rises with arousal then falls with perceived safety—all examples of feedback. Some types of feedback occur within seconds, others over years. In disease, altered feedback can be considered an ultimate and/or proximate cause of disease, and can lead to specific signs and symptoms. Understanding how feedback mechanisms are altered in disease can be crucial to developing effective targeting strategies, anticipating mechanisms of resistance to therapies, and managing side effects.

Another concept widely invoked in normal physiology and disease is *crosstalk*: how one defined molecular pathway influences the activity of other pathways. While crosstalk in electronics is generally undesirable, crosstalk between biological pathways enables system integration. Altered inputs and outputs of different molecular pathways are features of many diseases, leading to alterations in cell function, tissue composition, and organ physiology, and so opposition of pathway alterations is a common strategy to target disease. Metabolic pathways that provide fundamental nutrients such as glucose and fatty and amino acids to cells and tissues of the body are among the most "crosstalked" pathways, in part because so many factors must be considered when allocating cellular resources to energy utilization vs. storage. Diabetes is an example of a disease with altered feedback and crosstalk. In its common forms, it is characterized by failure to regulate serum glucose due to a deficit of, or altered responses to, insulin. Multiple inputs regulate insulin production and release from beta cells in the pancreatic islets, and the effect of circulating insulin in different cell types depends on inputs from other pathways. Crosstalk can be mediated by a direct physical interaction between two components of simultaneously active yet distinct signaling pathways, and the consequences of the interaction can be inhibitory or synergistic. Understanding at what "level" crosstalk between two pathways occurs in health or diseases, including the cell types and subcellular compartments where key interactions occur, can inform diagnostic and therapeutic strategies.

8 The Significance of the *Lesion*

If we must identify abnormalities in cells or their products to declare "disease," then it becomes necessary to pinpoint, visualize, describe, and eventually measure the abnormality, known as the *lesion*. In pathology circles, the *lesion* is a focus of thought about disease. Intuitively, a lesion is mass or lump visible to the naked eye, but it can also be microscopic (i.e., visible only with a microscope) or molecular (e.g., a gene mutation is a genetic "lesion"). In toxicology, the lesion is the histopathological change in tissue structure and cell composition induced by exogenous agents, and whether it is reversible or not—i.e., whether it resolves a suitable time after the drug is withdrawn—often predicts whether a drug can be safely administered to humans. Until we have "tricorders" (a la Star Trek) capable of disease diagnosis at a distance, in most cases it is only possible to study the nature of lesions in live patients by studying diseased tissue, a fluid sample containing analytes (molecules that can be identified and measured) that originated from the diseased tissue, or another tissue that is specifically altered as a consequence of the diseased cells or tissues. Beyond routine histopathologic examination, attempts to understand lesions in diseased tissue must pay strict attention to tissue procurement, handling, and processing in order to preserve any molecules of interest (i.e., the analytes to be measured). Lesions, or, properly stated, *visual representations* of lesions, can be monitored in live patients by a variety of imaging modalities including CT, MRI, PET, SPECT, etc. Identifying and characterizing culprit lesions in disease are key to understand chain of causation/pathogenesis as well as to devise therapeutic strategies.

Although a lesion can be a variety of sizes, Virchow—the "Father of Microscopic Pathology" who advocated for microscopy in disease diagnosis—espouses the central role of the cell in disease pathogenesis in his nineteenth century Pathology text:

> ...the chief point in this application of histology to pathology is to obtain recognition of the fact, that the cell is really the ultimate morphological element in which there is any manifestation of life, and we must not transfer the seat of action to any point beyond the cell. [21]

Virchow's quote is simultaneously prescient and timeless: he implies efforts to understand each disease should focus on the structure and function of cells. Thus, for each disease, the key questions become: *Which cells are diseased? What abnormalities exist within the diseased cells? How do diseased cells affect nearby cells to create the manifestations of disease? How can we detect the disease with a diagnostic test? How might we design therapies that reverse or mitigate disease-associated abnormalities in diseased tissues (efficacy) with acceptable consequences for nondiseased tissues (safety/toxicity)?*

9 Heterogeneity of Disease

Medical textbooks impart the illusion to students (and the lay public with unlimited internet access) that diseases, as they commence in patients, tidily fit into categories. Experienced clinicians know better, attesting to the unpredictable and heterogeneous nature of disease with the adage, "*diseases don't read the textbooks.*" Recall that despite the fact that all humans share ~99.9 % identical DNA, the ~0.1 % difference means there are ~*3 million* differences in the DNA sequence between two random individuals—more than

15 differences, on average, *per gene*. Heterogeneity of our DNA is merely the foundation of human disease heterogeneity; additional factors include environment, nutrition, and physical activity as well as psychological variables such as social support structure and emotional outlook. Serendipity, or statistical noise—simply bad luck—cannot be overlooked.

Two types of disease heterogeneity can be considered: (1) across categories of disease; and (2) within defined disease entities. *Heritability* of disease—the extent to which disease incidence, prevalence, consequences, and course are due to genetic factors—varies across disease categories and for each disease, and is often contrasted to *environmental* factors, whose exposure influences or is required for disease. At the extremes of the environment-genetics spectrum are infectious and Mendelian genetic diseases, respectively. Infectious diseases, by definition, occur only when the infectious agent—a virus, bacteria, fungus, etc.,—is present, although interventions (e.g., antibiotics) and host genetic variation, typically in genes that regulate the various aspects of the immune response, influence disease course and outcome. The thousands of described Mendelian genetic diseases are caused by single dominant or recessive mutations in specific genes, and when inherited, the culprit mutation can be traced through a family tree. Other diseases have a strong heritable component, consisting of variations in dozens to hundreds of genes. Disease categories in the middle of the spectrum, with contributions from both genetics and environment, include neurodegenerative, autoimmune, and neoplastic diseases. Twin concordance is one method to infer the extent of heritability for a disease, but massively parallel sequencing of large patient cohorts is providing new insights into complex, heterogeneous diseases. For example, recent reports have revealed that schizophrenia might be eight distinct diseases caused by variation in over 100 genes [22, 23]. Following the identification of causative genes for each disease, the next challenge is to place them into functional networks of gene products (typically proteins, but also including noncoding RNAs) acting within cells of lesional tissue.

10 Emerging Concepts of Disease—"New-Opathies"

Historically, diseases have been grouped by organ or organ system, with defined groups designated by appending the suffix "–opathies," meaning "abnormality of." Thus, a cardiomyopathy leads to failure of cardiac muscle, a nephropathy to renal failure, and a retinopathy to blindness. As a classification tool, the concept of -opathies has recently been extended to functional entities smaller than an organ or tissue, such as a genetically defined pathway, a cell type, an organelle, a macromolecular structure or complex, or even single proteins [24]. This idea is not new, as "hemoglobinopathies," a term first appearing in the biomedical literature following World War II, are a family of genetic diseases that manifest as anemias and are characterized by abnormal structure and/or production of the abundant oxygen-carrying protein hemoglobin. Several such "new-opathies" have emerged in recent years. Channelopathies were previously discussed; the laminopathies include a severe form of premature aging as well as a subtype of muscular dystrophy, both caused by distinct mutations in the nuclear lamina protein lamin A [25]; and telomeropathies manifest as diseases as diverse as pulmonary fibrosis and acute bone marrow failure. Given the known association between telomeres and the replicative potential of chromosomes, the findings implicate premature telomere shortening in undefined—possibly stem—cells as an underlying cause of these diverse conditions [26]. Linking of specific proteins and subcellular structures to pathogenesis of various diseases will likely reveal many more families of opathies and suggest completely new disease targeting strategies.

11 Omics Profiling and Subclassification of Disease

By itself, histopathology provides essential, but ultimately limited, information about disease. The significance of some histological features is uncertain, and diseases that appear homogeneous by microscopic examination can have highly variable courses or responses to treatment in patients. In the mid-1990s, microarray-based nucleic acid hybridization technologies were developed to measure the relative abundance of all mRNAs in a biological sample—a powerful disease-"splitting" tool. A common observation from these investigations, which continue to this day, has been that a variety of diseases, both neoplastic and non-neoplastic, can be subdivided into just a handful of categories based on a specific biologic feature, such as shared gene signatures (groups of RNAs, often with coordinate functions in a pathway, that are simultaneously up- or downregulated) from benign tissue and malignant counterparts, or up- or down-modulation of known biological pathways or processes. DLBCL was described above; breast cancer has since been divided into basal and luminal types whose mRNA expression profiles resemble normal basal or luminal cells in the breast duct, as well as into classes of tumors resembling other resident cell types [27]; and the childhood tumor medulloblastoma has been subdivided into at least four subtypes, each with prominent activation of Wnt, Hedgehog, or other pathways [28]. A variety of other omics profiling methods have been used to probe diseased tissues, including those that measure different RNAs; chromatin proteins or their posttranslational modifications; DNA content or sequence; various proteins; or other chemical species such as metabolites, lipids, or carbohydrates. One intuitive appeal of these profiling technologies is that the complexity of data they generate approaches that of diseased tissue itself, raising the hopes that culprit molecular species and possible drug targets can be identified. Datasets from these technologies can be analyzed in hypothesis-generating and hypothesis-testing modes. For them to be broadly useful for disease diagnosis requires interrogation of larger populations of patients and tissues, and ultimately, interpretation of the data with reference to events occurring in lesional cells and tissues. These technologies hold great promise to reveal ultimate causes of many diseases, but with few exceptions to date they have not been incorporated into routine clinical practice. Some of these reasons are explored in the chapter on next-generation sequencing in anatomic pathology practice, but they generally relate to an uncertain relationship between much of the data generated and its utility in medical decision-making.

12 Implications of Disease Diagnosis for Treatment Strategies

Developing new disease therapies within the current health care and regulatory environment is time consuming, expensive, and risky, with ~12 years required from target discovery to approval, greater than ~$1 billion in research and development (R&D) costs per approved drug, and approximately only 5–10 % of molecules in the earliest phase of clinical testing reaching the marketplace [29–31]. Regulatory approval to market a new therapy is not the end of the process: added to the usual challenges of demonstrating safety, efficacy, and manufacturability, successful therapeutic development now and in the future will require demonstration of—as well as ability and willingness to pay for—added clinical value in the form of slowed disease progression, enhanced function and quality of life, or prolonged lifespan.

The diversity of expertise, resources, and results that must align to bring a new medicine to market is staggering. Some say that a bit of luck is required. A recent review of AstraZeneca's terminated small molecule drug development programs from 2005 to 2010 suggests a "6R" framework for success: the therapy must hit the right target, the right patient, the right tissue, have the right safety profile and the right commercial potential, and must be developed within the right workplace culture [32]. For a new medicine approved by the FDA or other regulatory agency, depending on its value proposition (i.e., the quantified dollar value benefit of the therapy to the patient), profit margins can be initially high, benefitting investors and employees, but several factors are putting extreme pressure on the biopharmaceutical industry to "de-risk" each stage of the clinical development plan. These factors include competition with existing or emerging medicines; a shrinking reimbursement pie; the substantial R&D outlay required to validate new pathways and targets; and a necessity to demonstrate added value of a new medicine to regulators, physicians, patients, and payers. The visible failure of several late-stage clinical trials in recent years (*see* ref. 31) motivates all parties concerned to precisely define, and make as transparent as possible, the risk/reward equation for each candidate therapy as it progresses from discovery to marketplace. With any risk-taking, some failures will persist, but collective efforts could be wasted if we do not heed lessons learned from each trial. As disease definition is central to clinical trial design, I contend that accurate definition of disease, coupled with a hypothesis-neutral approach to trial design that includes, where feasible, interrogation of diseased tissue (a topic addressed in another chapter in this book), will be crucial for increasing success of drug development efforts in the future. For context, I briefly summarize the steps in drug development.

Biopharmaceutical development is often depicted as a pipeline, because a series of milestones typically has to occur in a particular order to justify resource allocation while engendering confidence that subsequent (and more expensive) milestones are likely to be achieved. From the onset, these activities include target identification and validation, molecule synthesis and lead optimization, as well as other crucially important but often underappreciated factors such as securing intellectual property and documenting feasibility of manufacture. In the discovery phase, the focus can be a disease or therapeutic area, a molecule, a pathway, or a platform technology (e.g., a drug screening or delivery system), but these activities are heterogeneous across industry, and they often define the personality of each company, large or small. When a promising therapy emerges from a discovery project, which usually includes testing for efficacy and safety in animals, a company or other entity (e.g., a hospital, university, or research institute) commits to "develop" a medicine in preparation to test it in man. The culmination of the process, at least in the USA, is approval of an "IND" (Investigational New Drug) application by the FDA, which allows testing in humans under defined conditions.

Following IND approval, a clinical trial usually commences. A clinical trial is, in essence, an experiment that determines whether a specific intervention influences an outcome. By necessity it should be designed to definitively answer a precisely worded question or set of questions, based on a well-reasoned hypothesis. While clinical trial science can be intimidatingly complex to an outsider, it is aimed at answering three questions: (1) Is the intervention effective? (2) Is it tolerable or safe? (3) What difference does it make compared to existing treatments? Once a green light for human dosing is granted by a regulatory agency, clinical testing typically occurs in three phases: Phase 1 assesses whether the medicine is safe, typically in healthy volunteers but sometimes in a diseased patient population. Phase 2, often split into two parts, first identifies or confirms a dosing regimen that should have an intended effect and then attempts to demonstrate that effect in a small number of diseased patients. Sometimes phase 2 trials yield statistically significant efficacy, and sometimes not, but their results are often used to design, and appropriately power, definitive phase 3 trials which test whether the therapy is likely to be effective in a diseased population. (For completeness, "Phase 0" trials study the administration of microdoses [subtherapeutic doses] in order to better understand how the body handles the drug, and "Phase 4" trials occur after drug approval in order to expand indications to other diseases or test additional hypotheses.) In the USA, an NDA (new drug application) includes all of the information that a regulatory body needs to assess whether a new therapy can be marketed. For certain diseases, particularly those that are devastating without impactful therapies, several accelerated regulatory pathways exist, some of which reduce the size and cost of a typical phase 3 trial. Adaptive trial designs, in which a patient might be assigned to a specific treatment arm based on the results of a diagnostic test, are increasingly employed to streamline the process and increase the likelihood of success in those patients most likely to benefit, or declare an early failure. Like biopharmaceutical development in general, trial design requires frank appraisal of the diseased population (and hence agreement on the definition of the disease itself), the proposed effect of the intervention, and operational feasibility. A candidate therapy might provide consistent and significant, but incremental, benefit to much or all of a heterogeneous population, or provide dramatic benefit to a known (or unknown) subset of diseased patients; each scenario requires distinct strategies to design a trial that will yield a definitive result.

The goal of many drug development programs is creation of a so-called "disease modifying therapy" (DMT), as opposed to a "symptomatic therapy." DMTs are typically aimed at opposing or compensating for either ultimate or proximate causes of disease, thereby slowing or practically halting disease progression. Prominent examples of DMTs that have dramatically improved disease outcomes in the past half century include appendectomy for acute appendicitis, antibiotics for certain infectious diseases, chemotherapy for cancer, and immunomodulatory therapies for autoimmune diseases. As concepts, DMTs and symptomatic therapies are not mutually exclusive, as an ideal DMT alleviates symptoms as well. (The converse can also be true, as nonsteroidal anti-inflammatory drugs (NSAIDs), among the most widely used symptomatic therapies, have been shown to be a DMT for a variety of diseases in which inflammation plays a crucial role: e.g., the inflammatory disease ankylosing spondylitis [33].)

With this framework, it becomes clear that defining the population of diseased patients, as well as devising interventions that break the disease's chain of causation, should be central to the design and ultimate success of any clinical trial. Indeed, given the overall rate of trial failure, many organizations have moved toward a "proof of concept" (POC) model of drug development in which a therapy is first demonstrated to be effective (or that the molecule or pathway the drug targets is impacted by the therapy) in a small, often narrowly defined and thus relatively homogeneous patient population. In such trials, not only are treatment responses typically more robust, but important data surrounding the safety of administering a new drug to humans is generated. These data can be used as a basis to design definitive trials and to expand indications to other diseases or categories of patients that might also benefit from the new therapy [34].

One point about the relationship between disease definitions and clinical trial design is worthy of mention. Among properly administered trials, the least desirable outcome is an inconclusive or marginal result that does not advance knowledge to inform the next trial or help the patient population. A more desirable, but still not preferred, outcome is a conclusively negative trial, i.e., one that supports the null hypothesis that the intervention has no effect on the outcome, but that does not advance knowledge about the disease, or the effects of treatment, to inform the next steps. Sometimes the answers to fundamental questions are not known when the trial is designed; these questions include whether the right target is being affected by the therapy or whether manipulation of the target will improve disease in patients. If the trial intervention is ultimately not of benefit and does not advance our understanding of the disease or the effect of the therapy in question, it is at best a waste of resources and at worst unethical to patients. Wise trial design advances knowledge of disease or therapeutic strategies even if the primary hypothesis is proved incorrect; the likelihood of such an outcome can be increased if thoughtful collection and interrogation of lesional tissue by the appropriate technology is performed as part of the trial, if not as a primary objective (i.e., the main trial endpoint) then as a secondary or exploratory objective. Because collecting lesional tissue from human patients is not always possible in a trial (e.g., many brain diseases), there is a great interest to discover and implement *biomarkers* of disease.

13 Biomarkers and Disease

Biomarkers are a hot topic among students and stakeholders of disease. A biomarker is an *objectively measured* parameter that serves as a surrogate representation of a biological process such as a disease itself, treatment susceptibility or response, or a measure of safety or toxicity. Biomarkers are often referred to as *fit for purpose*, meaning that the biomarker assay, the data readout, and the interpretation must be "fit" (competent) to serve a particular "purpose," i.e., to answer a specific question. As such, repurposing a biomarker might require alterations in, or at least revalidation of, the assay used to measure the biomarker, its performance characteristics, or interpretation. Traditional biomarkers such as blood pressure, electrocardiograms, and clinical chemistry and hematology tests have been validated for specific purposes as well as for general measures of physiology or disease. Biomarkers often bear an uncertain relationship to the disease process, but they are typically fluid-based analytes or imaging tests that aim to interrogate or reflect that status of lesional tissue or some of the events that are believed to constitute the chain of causation for that disease. Tests can be *single analyte* (measuring one parameter) or *multi-analyte*, composed of many measurements; modern omics-based biomarker tests can generate millions or billions or more measurements. Because much of the data that omics techniques generate is extraneous, statistical models can then be used to derive and then link a weighted single variable score to a probability of clinical response or outcome. There is further discussion of single- vs. multi-analyte tests in the chapter on next-generation sequencing. Biomarkers serve multiple purposes, but a major role of biomarkers in biopharmaceutical development is in the quantification of risk at each stage in a drug development program: How do you know which patients will tolerate and respond to therapy? (. . .or after some time on treatment, have responded to therapy?) What difference has the therapy made to the patient's disease or their outcome? A drug can hit its target in the test tube, in a cell-based system, and in a possibly relevant animal model, but all stakeholders want to see if the drug hits its target when given to human patients—and what happens next can make or break a drug and its backers. If, for example, a pathway-specific drug is administered at a dose in a phase 2B trial that leads to predicted changes in targeted pathway biomarkers, but the patients are not showing clinical improvement, then you know that the therapeutic hypothesis has been disproved: either the incorrect patient population was enrolled in the trial, or no such patients exist, and the project should be abandoned.

Valid biomarkers are of great interest to all stakeholders seeking new disease treatments. Unfortunately, they are difficult to come by. Two crucial, often untested assumptions underlie many searches for soluble biomarkers: (1) that they exist in fluids when they might only exist in lesional tissue; and (2) that they are true surrogates of the process in question, such as disease or therapeutic response. Often times, these assumptions can be tested rigorously as part of a drug development program through investigational studies of diseased tissue from relevant preclinical (animal) studies, as well as by using archival, or prospectively collected, human tissues. For example, diseased tissues can be profiled to identify molecular species that might be liberated as part of the disease process and detectable in body fluids as biomarkers. The opposite strategy is far more common: fluids (e.g., blood, urine, saliva) from diseased patients are profiled with high-content omics technologies, with the assumption that molecular species will be present and discovered, representing the underlying disease process—an approach that assumes biomarkers of disease exist in fluids when they might remain restricted to lesional tissue, or be ephemeral once liberated from tissue.

Genetic biomarkers (DNA sequence mutations or variants) are a special case and are of two broad types: constitutional or acquired. Constitutional genomic biomarkers, such as the host's germ line DNA sequence, are present in most to all cells of the body and are either inherited from parents or arise de novo during gametogenesis or embryogenesis. Acquired mutations are typically present only in lesional tissue. Neoplasia is probably the most prominent example of multiple acquired mutations as ultimate and proximate causes of disease. But, this concept is also relevant to other categories of disease, including autosomal dominant polycystic kidney disease, caused by second-hit somatic mutations in cyst-suppressor genes [35]; and autoimmune disease lesions enriched for autoreactive lymphocytes harboring unique, pathogenic immunoglobulin or T lymphocyte receptor gene rearrangements. With the right technology and an adequately sized collection of patient samples, a hypothesis-free (discovery-based) approach is feasible. For example, the genome-wide association study (GWAS) identifies genomic regions or loci associated with variance in a particular trait (such as a particular disease), and massively parallel sequencing can identify infrequent but more devastating gene mutations that contribute to risk of disease or cause disease. The challenge comes with follow-on studies: linking identified genetic mutations to disease diagnosis and pathogenesis and then determining in which pathways the altered genes normally function, so the mutant molecules or altered pathways can be manipulated for therapeutic advantage.

14 Companion Diagnostics

Biomarker tests that interrogate lesional cells or their products will increasingly be used to distinguish health from disease, select patients for specific therapies, and monitor safety and efficacy of therapies. Tests whose results determine eligibility for, or contraindicate, a particular therapy—a so-called "companion diagnostic" (CDX) test—carry risks to the patient commensurate with the therapeutic decisions they inform. Development and implementation of such tests require that they meet strict regulatory requirements, including specific performance criteria, and be interpreted by competent personnel in an accountable laboratory environment.

Among the most widely used tissue-based CDX tests are those that determine whether a breast cancer carries large quantities of active Her2 kinase and is thus highly likely to be responsive to anti-Her2 therapy, such as trastuzumab. The Her2 story, while an early case study in personalized medicine, illustrates many of the pitfalls and challenges in translating an observation initially made in research laboratories into a robust clinical test. In the early years of implementation, some Her2 determinations were made in error, either exposing patients to the risks of taking a drug from which they derived no benefit or denying them a potentially life-prolonging treatment. These issues have led to several revisions of Her2 testing guidelines [36]. More broadly, the FDA in 2011 released a draft guidance on CDX development, affirming the need for a well-defined regulatory pathway for tests that determine whether a given drug, with its risk/benefit profile, can be prescribed in a particular disease. In 2014, FDA released a draft guidance on laboratory-developed tests (LDTs), a far more common class of test than CDXs, that are used to diagnose or monitor disease but are typically performed in a single laboratory. LDTs and CDX tests will increasingly influence how diseases are defined as well as providing key data to inform interventions and therapies. As many CDX tests involve analysis and interpretation of lesional tissue, pathologists and scientists who understand the complexity of diseased tissue are uniquely prepared to create, develop, and implement CDX tests.

15 Summary and Concluding Thoughts

Light microscopy of fixed, stained tissue shed critical light on disease diagnosis, classification, and progression as histopathology became central to medical care in the late nineteenth and early twentieth centuries. Since then, ancillary molecular technologies that either keep tissue intact (IHC, ISH) or destroy it (ELISA, PCR) have become commonplace in anatomic pathology practice, extending and complementing H&E-based microscopic diagnosis. The past two decades have witnessed the development of several omics technologies that interrogate a specific type of chemical complexity in a diseased tissue or fluid sample. These technologies generate very large datasets, whose interpretation can provide new insights into disease classification, but when viewed in isolation often bear unclear relationships to the histopathological features of the disease in question. While histopathologic diagnosis suffers from a variety of pitfalls, including lack of standardization of laboratory techniques, controversial diagnostic criteria, and sometimes substantial inter-pathologist variability, it serves as a framework—a canvas and a landscape—to understand the interrelationships between events occurring in diseased tissues, including relevant cell types, and how their alterations promote disease.

One unifying theme of this book is that histopathology, instead of falling into disuse, will become increasingly important for personalized medicine. One major reason drugs fail in human clinical trials is that the animal models used to validate the target and develop the therapy do not replicate the chain of causative events of disease in the specified human populations. Experiments on appropriately banked, curated, and consented human blood or tissue specimens can be used to rigorously test hypotheses generated via animal models as well as generate new hypotheses based on natural genetic experiments in human populations [37]. While tissues are routinely evaluated by histopathology in preclinical safety and efficacy studies, only rarely is pathologic interpretation of diseased human tissue central to a clinical trial, a situation we anticipate will change as greater certainty in deciphering disease entities and designing effective therapies is demanded by all stakeholders. Depending on how it is calculated, the dollar value of a human body has been estimated to vary over seven orders of magnitude, from a mere $4.50 for its constituent elements to $45 million for tissues, but the diagnostic value of human tissue, particularly diseased tissue, is priceless. To understand disease, tissue is clearly "*where the money is*," so we must choose language that describes its myriad abnormalities in a manner that does not impede our search for truth.

Acknowledgements

Duojia Pan, Diego Castrillon, Maureen Miller, Vito Sasseville, Dave Young, Chikwendu Ibebunjo, and the editors are acknowledged for critique and comments. Portions of this chapter were written by the author as part of the medical school pathology course at UT Southwestern Medical School. The opinions expressed herein are those of the author and do not necessarily represent those of Novartis AG and Novartis affiliates.

References

1. Warman ML, Cormier-Daire V, Hall C, Krakow D, Lachman R, LeMerrer M, Mortier G, Mundlos S, Nishimura G, Rimoin DL, Robertson S, Savarirayan R, Sillence D, Spranger J, Unger S, Zabel B, Superti-Furga A (2011) Nosology and classification of genetic skeletal disorders: 2010 revision. Am J Med Genet A 155A(5):943–968. doi:10.1002/ajmg.a.33909
2. Rodin AE (1962) A historical survey of disease concepts. Can Med Assoc J 87:124–128
3. Veith I (1969) Historical reflections on the changing concepts of disease. Calif Med 110(6): 501–506
4. Bennett JP (1989) Man's conception of disease–a historical journey. Iowa Med 79 (4):175–179
5. Jack CR Jr, Albert MS, Knopman DS, McKhann GM, Sperling RA, Carrillo MC, Thies B, Phelps CH (2011) Introduction to the recommendations from the National

Institute on Aging-Alzheimer's Association workgroups on diagnostic guidelines for Alzheimer's disease. Alzheimers Dement 7 (3):257–262. doi:10.1016/j.jalz.2011.03.004

6. Loscalzo J (2011) Systems biology and personalized medicine: a network approach to human disease. Proc Am Thorac Soc 8 (2):196–198. doi:10.1513/pats.201006-041MS
7. Reynolds JR (1866) A system of medicine, vol 1. MacMillan, London
8. Zuzek C (2014) Obesity is a disease. Tex Med 110(1):45–48
9. Greenberg G (2013) The creation of disease. The New Yorker. http://www.newyorker.com/tech/elements/the-creation-of-disease
10. Johnson RL, Rothman AL, Xie J, Goodrich LV, Bare JW, Bonifas JM, Quinn AG, Myers RM, Cox DR, Epstein EH Jr, Scott MP (1996) Human homolog of patched, a candidate gene for the basal cell nevus syndrome. Science 272 (5268):1668–1671
11. Dork T, Dworniczak B, Aulehla-Scholz C, Wieczorek D, Bohm I, Mayerova A, Seydewitz HH, Nieschlag E, Meschede D, Horst J, Pander HJ, Sperling H, Ratjen F, Passarge E, Schmidtke J, Stuhrmann M (1997) Distinct spectrum of CFTR gene mutations in congenital absence of vas deferens. Hum Genet 100 (3–4):365–377
12. Keightley PD (2012) Rates and fitness consequences of new mutations in humans. Genetics 190(2):295–304. doi:10.1534/genetics.111.134668
13. Xue Y, Chen Y, Ayub Q, Huang N, Ball EV, Mort M, Phillips AD, Shaw K, Stenson PD, Cooper DN, Tyler-Smith C, 1000 Genomes Project Consortium (2012) Deleterious- and disease-allele prevalence in healthy individuals: insights from current predictions, mutation databases, and population-scale resequencing. Am J Hum Genet 91(6):1022–1032. doi:10.1016/j.ajhg.2012.10.015
14. Kher N, Marsh JD (2004) Pathobiology of atherosclerosis–a brief review. Semin Thromb Hemost 30(6):665–672. doi:10.1055/s-2004-861509
15. Heard E, Martienssen RA (2014) Transgenerational epigenetic inheritance: myths and mechanisms. Cell 157(1):95–109. doi:10.1016/j.cell.2014.02.045
16. Hochhaus A (2004) Imatinib mesylate (Glivec, Gleevec) in the treatment of chronic myelogenous leukemia (CML) and gastrointestinal stromal tumors (GIST). Ann Hematol 83 (Suppl 1):S65–S66. doi:10.1007/s00277-004-0850-2
17. Roschewski M, Staudt LM, Wilson WH (2014) Diffuse large B-cell lymphoma-treatment approaches in the molecular era. Nat Rev Clin Oncol 11(1):12–23. doi:10.1038/nrclinonc.2013.197
18. Kim JB (2014) Channelopathies. Korean J Pediatr 57(1):1–18. doi:10.3345/kjp.2014.57.1.1
19. Babcock JJ, Li M (2013) hERG channel function: beyond long QT. Acta Pharmacol Sin 34 (3):329–335. doi:10.1038/aps.2013.6
20. Cook JR, Ramirez F (2014) Clinical, diagnostic, and therapeutic aspects of the Marfan syndrome. Adv Exp Med Biol 802:77–94. doi:10.1007/978-94-007-7893-1_6
21. Virchow R (1858) Cellular pathology (trans: Chance F). Edwards Brothers, Ann Arbor, MI
22. Arnedo J, Svrakic DM, Del Val C, Romero-Zaliz R, Hernández-Cuervo H, Molecular Genetics of Schizophrenia Consortium, Fanous AH, Pato MT, Pato CN, de Erausquin GA, Cloninger CR, Zwir I (2014) Uncovering the hidden risk architecture of the schizophrenias: confirmation in three independent genome-wide association studies. Am J Psychiatry. doi:10.1176/appi.ajp.2014.14040435
23. Flint J, Munafo M (2014) Schizophrenia: genesis of a complex disease. Nature 511 (7510):412–413. doi:10.1038/nature13645
24. Friedberg EC, Castrillon DH, Galindo RL, Wharton KA Jr (2012) New-Opathies: an emerging molecular classification of human disease. World Scientific, Singapore
25. Capell BC, Collins FS (2006) Human laminopathies: nuclei gone genetically awry. Nat Rev Genet 7(12):940–952. doi:10.1038/nrg1906
26. Armanios M, Blackburn EH (2012) The telomere syndromes. Nat Rev Genet 13 (10):693–704. doi:10.1038/nrg3246
27. Prat A, Perou CM (2011) Deconstructing the molecular portraits of breast cancer. Mol Oncol 5(1):5–23. doi:10.1016/j.molonc.2010.11.003
28. Kool M, Korshunov A, Remke M, Jones DT, Schlanstein M, Northcott PA, Cho YJ, Koster J, Schouten-van Meeteren A, van Vuurden D, Clifford SC, Pietsch T, von Bueren AO, Rutkowski S, McCabe M, Collins VP, Backlund ML, Haberler C, Bourdeaut F, Delattre O, Doz F, Ellison DW, Gilbertson RJ, Pomeroy SL, Taylor MD, Lichter P, Pfister SM (2012) Molecular subgroups of medulloblastoma: an international meta-analysis of transcriptome,

genetic aberrations, and clinical data of WNT, SHH, Group 3, and Group 4 medulloblastomas. Acta Neuropathol 123(4):473–484. doi:10.1007/s00401-012-0958-8

29. Paul SM, Mytelka DS, Dunwiddie CT, Persinger CC, Munos BH, Lindborg SR, Schacht AL (2010) How to improve R&D productivity: the pharmaceutical industry's grand challenge. Nat Rev Drug Discov 9(3):203–214. doi:10.1038/nrd3078
30. Ciociola AA, Cohen LB, Kulkarni P, FDA-Related Matters Committee of the American College of Gastroenterology (2014) How drugs are developed and approved by the FDA: current process and future directions. Am J Gastroenterol 109(5):620–623. doi:10.1038/ajg.2013.407
31. Hay M, Thomas DW, Craighead JL, Economides C, Rosenthal J (2014) Clinical development success rates for investigational drugs. Nat Biotechnol 32(1):40–51. doi:10.1038/nbt.2786
32. Cook D, Brown D, Alexander R, March R, Morgan P, Satterthwaite G, Pangalos MN (2014) Lessons learned from the fate of AstraZeneca's drug pipeline: a five-dimensional framework. Nat Rev Drug Discov 13 (6):419–431. doi:10.1038/nrd4309
33. Akkoc N, van der Linden S, Khan MA (2006) Ankylosing spondylitis and symptom-modifying vs disease-modifying therapy. Best Pract Res Clin Rheumatol 20(3):539–557. doi:10.1016/j.berh.2006.03.003
34. Fishman MC (2013) Power of rare diseases: found in translation. Science Transl Med 5 (201): 201ps211. doi: 10.1126/scitranslmed.3006800
35. Harris PC (1999) Autosomal dominant polycystic kidney disease: clues to pathogenesis. Hum Mol Genet 8(10):1861–1866
36. Wolff AC, Hammond ME, Hicks DG, Dowsett M, McShane LM, Allison KH, Allred DC, Bartlett JM, Bilous M, Fitzgibbons P, Hanna W, Jenkins RB, Mangu PB, Paik S, Perez EA, Press MF, Spears PA, Vance GH, Viale G, Hayes DF, American Society of Clinical Oncology; College of American Pathologists (2013) Recommendations for human epidermal growth factor receptor 2 testing in breast cancer: American Society of Clinical Oncology/College of American Pathologists clinical practice guideline update. J Clin Oncol 31 (31):3997–4013. doi:10.1200/JCO.2013.50.9984
37. Plenge RM, Scolnick EM, Altshuler D (2013) Validating therapeutic targets through human genetics. Nat Rev Drug Discov 12 (8):581–594. doi:10.1038/nrd4051

Methods in Pharmacology and Toxicology (2015): 27–41
DOI 10.1007/7653_2014_16

Published online: 26 September 2014

A Field Guide to *Homo morphologicus* for Biomedical Scientists, Or How to Convey an Understanding of Pathology to Scientists in a Biopharma Enterprise

Humphrey Gardner

Abstract

Humphrey has spent about equal time as a pathologist and as a biologist, and about equal time in academia and industry. He has run clinical and anatomical pathology labs including knockout service facilities for mice, histomorphology laboratories doing immunohistochemistry (IHC) and immunofluorescence (IF) with image analysis, and CAP/CLA-approved laboratories doing IHC for oncology clinical trials. What follows is a collection of his observations of potential misunderstandings between scientists and pathologists that can arise in the analysis of tissue for clinical trials, with a particular emphasis on oncology.

Key words Immunohistochemistry, Histopathology, Pathology, Fixation, Image analysis, Veterinary pathology, Pharmaceutical industry

1 Introduction

Anatomic pathology is a far older discipline than most used in modern drug development. It is a highly empiric discipline. In essence, for a hundred or more years, tissue samples have been prepared, cut, and examined microscopically in a fairly consistent way. Appearances of tissues associated with various diseases have been documented, descriptively and extensively, long before good photography was available. These appearances have been associated with outcomes of disease, and have been given standard names that are, within limits, globally agreed upon and understood. The training of a pathologist includes the teaching of these associations and appearances.

The empiric nature of the discipline is often a source of confusion to scientists with reductionist, "bottom up" molecular biology training and world view. It can be frustrating for a biologist to ask a pathologist, "Why are you calling this the *tall cell variant of papillary carcinoma of the thyroid*," and be told, "Because that is what it looks like." As a biologist, you feel like Alice in Wonderland. If the pathologist were to say, "This histologic appearance, with these distinctive aspects of cellular morphology, has been shown in a sample size of many thousand to be associated with a worse outcome and slightly different demographics than regular papillary carcinoma. Although the genetics are not fully understood, Trk kinase mutations are found in a subset, and the odd morphology may be explained by altered Rac GTPase signalling...," the biologist might feel better. On the flip side, pathologists are bewildered by biologists, for whom the world is turned on its head by a new molecular hypothesis. Having witnessed a morphology-outcome correlation where the statistical sample is most of humanity, no pathologist is surprised or counter-surprised by results from experiments with a sample size of a few tens or hundreds.

The areas of confusion described below are the most common the author has encountered in his career in pharma. They all spring from mutual misunderstandings about the role and knowledge of diverse participants in the pharma enterprise.

2 Collecting and Managing Samples

2.1 So It'll Be Ready Tomorrow?

"The patient was biopsied this morning, so given that fixation takes 6 h and the IHC run on the autostainer takes three, we should have the results this evening, right?"

As often as I need to explain pathology procedures to my colleagues, I find they always forget the steps that bore them. Shipping of the sample via courier (who will—hopefully—neither freeze nor boil the samples)? Boring. Tissue processing in a big machine? Boring. So here is a quick summary: The biopsy is fixed in formalin, and transferred to 70 % ethanol at the end of the day. On day 2 it is shipped and 24 h later it arrives at a pathology lab. It is logged and processed, sections are taken, and an H&E stained slide is examined (day 3). Relevant IHC tests are batch stained on the automated immunostainer, coverslipped, and examined by the pathologist (day 4). On *working* day 5, barring any hitches, you can expect a report.

2.2 The Not So Splittable Biopsy

"We'll get a 2 mm needle biopsy, do a touch prep, freeze, then section longitudinally to get tumor content, and then take half for RNA in −70 and the other half for immunohistochemistry."

Many scientists indulge in magical thinking about the clinical world: that it is populated with surgeons with time to spare, who happen to have access to an expert histotechnologist (who cares about the labile nature of RNA), as well as a pathologist and a cytologist—all available instantly when a patient appears. Simplicity is key to sample collection. With a little experience, it is very easy to gut-check feasibility. A well-equipped clinical trial site will have the tools for sample collection, but they are probably far less well arrayed than a biopharma lab doing pharmacokinetic studies. In a typical clinical trial there are multiple, variably equipped sites, run by diverse, variably equipped personnel. The frozen section room is not where the patient is, but it is adjacent the operating room and it is in constant use. The cytologist is not in the frozen section room; she is in her office. The various fixatives and temperatures needed are often not in one place. A protocol so designed, when executed, will produce fragmented samples, material lost to shipping and handling problems, and a patchwork of incomprehensible data for the patient set. Therefore, a specimen triage protocol that minimizes real-time decisions and greatly simplifies handling—preferably with the initial step using standard solutions such as formalin—will usually work best. With the example above, clearly it would be desirable to have all the IHC assays—as well as any—omics profiling tests—working on formalin fixed paraffin embedded (FFPE) tissue. In this way, the biopsy can simply go into formalin, with no additional steps required other than ensuring an appropriate fixation time. The work to generate robust analytical techniques that work on FFPE tissue may seem annoying at first, but the 6 months to develop them is a sliver of time compared to the length of most trials. In the relatively rare event that something absolutely has to be done using a preservative other than formalin, the process of sample triage has to be thought through very carefully, with particular attention to the availability of internal controls for specimen adequacy.

"I've found this new fixative way better than formalin."
This issue comes up irregularly, seemingly every few years. Its occurrence must follow the path of some comet I don't know about, or the same fad clock such as those influencing McKinsey & Company's oscillatory opinions of outsourcing, cubicle design, or the ideal number of layers in an organization. Yes, there might be fixatives better than formalin. They might smell better, cost less, preserve this or that analyte better, or they might magically freeze living tissue into a state where it can be recovered living 200 years later. In the meantime, the brutal truth is that it does not matter if there are fixatives better than formalin. The vast majority of the data we have generated on the seven billion people on the planet and their ancestors is based on tissue fixed in formalin and embedded in paraffin wax—be it the artifacts we interpret as morphology or the relatively finely diced nucleic acid fragments that are still useful for a great many experiments. Over the last few years many manufacturers of nucleic acid diagnostic platforms have vastly improved their offerings to use FFPE tissues as starting material, and increasingly providers are able to analyze some proteins from FFPE tissues by mass spectrometry. I see no reason to look for another fixative during my lifetime, and I'll be happy to be fixed therein at the appropriate time myself. "...but Dr. Dusty says he always fixes in a mixture of Mercury and Plutonium"

A short coda to the discussion about formalin merits special attention. Dr. Dusty in this example is usually a charismatic pathologist from the old school, who imagines that crisp cytological definition on the H&E stained slide is going to enable superior morphological classification, and that the DNA in the specimen, cleaved into oligonucleotides by the fixative, is (still, in 2014) of entirely secondary value in diagnosing and prognosing the patient. Such characters are usually eloquent and persuasive dealing with their nonclinical, biologist colleagues, and it can take years to undo their evangelistic damage. In these instances one can only hope for a mandatory retirement age.

3 Reading and Interpreting a Slide

3.1 Discovery or Artifact?

When I look at slides with non-pathologists, I probably spend more time explaining artifacts than anything else. It is hard to think of my pathologist-self at the beginning of residency, but I remember artifacts as a great source of confusion. Here are a memorable few:

Knife marks: These may pull the tissue, compressing it in a long line through the sample, which, due to its increased thickness, makes every stain look dark.

Variation in thickness: A gradient in one direction across a sample is a good clue to its presence, but this can be distracting if the wedge of tissue only has tumor or lesion at one end or the other

Folds: To the untrained eye, a small fold can look like an intense region of staining with highly malignant cells.

Holes: "There's a hole here!" says the biologist. "What?" says the pathologist, seeing fat, because we are used to white space (meaning fat) in our paraffin embedded samples.

Blown out nuclei and other fixation artifacts: These are many and raise all kinds of specters to the imaginative biologist, such as viral infections or new kinds of stromal malignancy.

Planes of section: Particularly in skin, a slightly tangential section can leave islands of basal epidermis surrounded by dermis, often mistaken for a malignant invasive process. Obliquely sectioned blood vessels can cause similar confusion.

When the pathologist reviews a slide with a non-pathologist, he or she has to remember that with pathology training we learn to see things, but we equally learn to not see things that a neophyte observer would.

"What do you mean, the patient doesn't have cancer?"
This can be a source of anxiety in oncology clinical trials. The usual problem, of course, is that the wrong tissue blocks were sent from the hospital to the trial sponsor. It is not so uncommon for a random block to be selected by a hurried pathology team at the patients' sites of origin, and there is often a bias to "not waste the good blocks"... and so instead of tumor we see something less

helpful, like the "resection margin, 10 cm distal." Once, in a total of many thousands of cases, my colleagues and I were unable to confirm there existed any tumor in the diagnostic block, an incidence of misdiagnosis lower than found in daily clinical life. "This tumor is screamingly positive and you missed it!"

...says the cancer scientist who sees lots of brown stain on the IHC slide. In this circumstance, the painful education begins with what is tumor and what is not tumor. (Oh, and did I ever mention? Abandon all ego ye pathologists who enter pharma.) This usually happens when a biologist observer sees something so very dark brown that he or she is distracted from the actual question of where the tumor is. The arguments can then get very heated: "How do you KNOW that's not tumor"? etc., and even if you are a celebrated Fellow of the Royal College of Extremely Distinguished Pathologists, you will find yourself pulling out histology books to demonstrate to the enraged biologist that you really do know what you are talking about. If you just bark or stonewall you will be distrusted, and the biologist will develop an increasingly bizarre set of hypotheses, without looking for further input from anyone who actually understands tissue morphology.

"There's nothing in the literature about staining there!—OR—You mean that protein is expressed in normal tissue?"
It often comes as a shock to cancer biologists that the "onco" proteins they work on are also essential constituents of normal tissue. Once we pathologists have gone through the appropriate steps to validate the specificity of an antibody, we will also, in a calm and unsurprised fashion, document the locations of expression in the tissues we see. In cancer specimens, we will take cognizance of the expression in nearby normal tissue structures as a possible internal control for staining adequacy of the specimen. Many signaling proteins show useful and distinctive patterns in structures such as the epidermis and hair follicles. The problem is that they often haven't been published—certainly not in journals that cancer scientists read. So they get a little outraged when I say "I don't trust this biopsy result because the adnexa aren't stained right." A nice example of this is the staining of the phosphorylated forms of ribosomal protein S6 in skin, now a well-used marker for activation of the mTor pathway. When I first used it in skin biopsies it was of no particular surprise to me that the bulk of epidermal staining is in the granular cell layer, and that there was good modulation after drug treatment. Cancer biologists, however, used to equating "mitotically active" with "everything else active," assumed that staining "should" be strongest in the basal layer and taper to the upper layers. I could only deal with this by calmly and repeatedly showing the data and the many controls that had been employed to a variety of people. It wasn't until 2 years later that a paper from the appropriate academic "god" was published, vindicating our results [1].

3.2 Proportion of Tumor in a Specimen Is a "Guess-timate"

"This section is 60 % tumor so I can expect a maximum allele fraction of a heterozygous mutation of 30 % right? So why is it 10 %?"

It is tough to explain that even though at a particular location on the slide we appear to be looking at malignant tumor, there are still large numbers of cells that are simply not tumor: endothelial and stromal cells and sundry immune cells that may be practically invisible to the incompletely attentive viewer, which, even when visible, are hard to enumerate. If I were estimating tumor content these days I would base it on the allele fraction of the highest frequency somatic mutation from next-generation sequencing (NGS) analysis of the tumor. The estimate obtained from histology is reassuring if it correlates with expectation, but it is certainly not quantitative.

"The phospho-(insert your favorite protein here) increased after treatment with my new inhibitor from that biotech company we bought. You must have swapped the specimens!"

There is a black box in the middle of pharmacodynamic (PD) analysis of tumor samples that generally goes unmentioned in the literature, which is the problem of sampling. Anyone used to looking at completely resected tumor specimens will recognize the vast morphologic heterogeneity between hypoxic regions, well-vascularized areas, regions at the margins with features of epithelial to mesenchymal transition, and so forth. In a tissue PD study, which investigates the effect of drug on the tumor, the physician takes two (or more) tiny samples of a tumor that is still in the patient before and after a short period of therapy. The physician taking the post-treatment biopsy walks a line between trying to sample a similar geographic location as the first biopsy while trying to avoid areas of healing or other artifacts generated by the first biopsy. Depending on the tumor type and the ease of the procedure, this may not be possible, and the second sample might be taken from a more or less aggressive-looking region (i.e., a region of tumor with higher grade of malignancy or different growth pattern) than the first. For me this has two implications. Firstly, I only take the results of a PD marker seriously if I see a similar effect across a population of patients, not over-interpreting outliers. Secondly, in each patient, I make sure to comment on the level of morphologic similarity between the pre- and posttreatment tumor biopsies.

4 Generating (Semi)-quantitative Data

"*You called this* "Slightly Positive" and Dr. Blow called it "Moderately Positive"

Indeed, and neither of us knew what it meant! The fact remains that until a defined level of analyte (usually protein) abundance in a sample is associated with an outcome or a treatment, there is no motivation to standardize antibody, method, or interpretation of IHC stains. If I were a scientist faced with this kind of report, I would come back to the pathologist with questions along the lines of "did the expression appear to be greater than in control tissue X that from other analytical methods I know to express a lot"... and if you are met with a blank stare, you know that "slightly positive" and "moderately positive" are equally meaningless.

Immunohistochemistry can be reasonably quantitative, but only if the work is put in to make it so. This is discussed below. Why can't you quantitate this more accurately?

We pathologists invented IHC, and a certain generation of us grew up with it as a new and improved version of special (i.e., histochemical) stains. With IHC we could tell who was whom in the four-micron drama we visualized as the disease process—a play in still life acted out in front of us. IHC found particular favor with hematopathologists because the entities they wished to differentiate had relatively little morphologic variety, due largely to their minimal interaction with surrounding matrix and connective tissue. It also enabled cross-correlation with flow cytometric studies, now the linchpin of most hematopathology diagnoses. For the non-hematologic tumors, IHC in large part provided clues about the origin of poorly differentiated tumors, sometimes with dramatic significance for therapy. That led to a near reflex use of IHC for particular markers or their combination, not because it was necessarily needed every time in the view of the examining pathologist, but for the implications—often of legal rather than medical import—of being wrong. The types of markers being used tended to be lineage markers specific to one cell type or another, and the goal when developing the IHC test was to make it specific and sensitive, but with little or no dynamic range, so that the readout would be as close to a Boolean yes/no as possible. IHC only lends itself to quantitation with a wide dynamic range if a great deal of effort is put into assay design. For this reason, quantitation of an antigen by IHC—as opposed to techniques such as a protein ELISA or mass spectrometry that destroy tissue—is no one's first choice. However, it has happened, partly out of desperation and partly for the need to investigate the distribution of proteins that are in no way unique to the cell of interest. The myriad challenges to delivering a quantitative result include:

Inconsistent tissue fixation = many fold variation in antigen availability. In particular, labile phospho-epitopes of some proteins in tissue can change in level very rapidly, and in different directions from one another, with small delays in fixation.

Variation in tissue thickness (typically a 5 μm section is between 4 and 6 μm thick = 20 % variation in available antigen).

Variation in staining time in several linear but steep processes = at least a tenfold variation in intensity. These processes include antigen "retrieval" (methods like heating and exposure to proteases or pH change to denature and expose buried epitopes that allow detection by the antibody in use) as well as the steps by which a chromogen is deposited at the sites of antibody binding.

For these reasons, a great deal of effort needs to go into assay standardization. Fixation protocols need to be designed that are easy to perform. Antigen retrieval and immunostaining protocols are best performed using automation, so that all the reactions performed, however nonlinear or steep they may be, are at least performed identically on each slide. Control tissue sets need to be developed, which often include knockout and transgenic cell lines with levels of expression characterized in parallel by methods such as mass spectrometry or Western blot. Ideally, a control set in the form of a small array of cell lines of known expression level is placed on every slide, but this is rare.

As a profession we tend not to emphasize how inaccurate visual quantitation can be. Part of the reason is that in the same way our human eyes can adapt over an enormous dynamic range, we become very good at deriving a "result" from a slide by using internal cues. We read a slide in the context of other slides, but we also use all the internal cues present on the slide that we can discern. So long as we can read a stain ratiometrically, i.e., comparing staining intensity in a cell type of interest to some surrounding normal control tissue, we feel we can give a meaningful result with some confidence. In other words, we tend to care about precision rather than accuracy. It is probably fair to say that the presence of internal controls, like knowledge of the staining pattern in nearby normal tissue, encourages us to interpret samples and give results in instances where a clinical laboratory director might reject an ELISA result as out of range. We are willing to do this because tissue is precious and a repeat assay is not necessarily possible, and we owe the patient the most reasonable interpretation of the material available. While data like this can help guide an early clinical program, it is at best semiquantitative and its interpretation can rapidly mislead. Which brings me to...

"You had H scores of 20, 30, and 40 in these patients. I plotted the data and there's an exact inverse correlation with age!"
The H score is a semiquantitative method for expressing the staining density in tissue. It is a weighted sum, relying on the pathologist to categorize areas of staining into 4 levels: none, low, medium, and high (often expressed as 0,1+,2+,3+). The formula for the score is simple: %low + 2 × %medium + 3 × % high, and it first found published use in studies of the estrogen receptor [2].

I curse and proselytize the H score in equal measure. I do believe that capturing the components of the H score is the most efficient way for the pathologist looking in the microscope to capture information about staining density across a target tissue in as unbiased a way as possible. If there is more than one subcellular compartment to be considered I simply open the score to three compartments—e.g., nuclear, cytoplasmic, and membrane, and apply that to as many cell types as are relevant. You could easily imagine scoring in 18 columns for a membrane-associated coreceptor that shuttles to the nucleus upon activation, and that functions in both tumor and endothelial cells. Thus I score % high, medium, and low, in three compartments, in two cell types. Of course, once the biology is understood well, such an analysis can collapse down to a few salient observations, as have been made for Her2 immunostaining. However, if a pathologist makes the mistake of prematurely "understanding" the biology by noting a few patterns early on and not capturing any more data, then there is no room to create new boxes when something new comes along, and we as a profession are justly accused of being inconsistent and mysterious.

The beauty of the H score components is that, once they are understood to a statistician, reasonably good association studies can be made with little intervention on the part of the pathologist until they are done, and then the opportunity to reinterpret and reexamine arises. It is sadly true that most slide sets are sorely under-evaluated to arrive at truly useful and novel morpho-immunohistologic parameters of diagnosis, but at least doing the H score components and the association studies lay the groundwork for that to happen. This is an area where whole slide scanning and image analysis will soon find greatly extended use.

The horror of the H score components is that they are numbers. I find that within moments of numbers being released from the laboratory, someone will have "done" something with them. I am all for wide dissemination of data and freedom of analysis. But it adds unnecessary work when someone tells you that "there is a highly significant U shaped association between ratio of 0 and +1 intensity and outcome in my phase 1 with 30 patients." In the end, there is no substitute for sitting down with the investigator, discussing the results, and making sure that noise we have inadvertently generated does not lead to inaccurate reinterpretations of known biology.

"Wouldn't it be a lot more accurate if you used a camera to quantitate that brownness?"
Notwithstanding all the reasons for error in immunohistochemical quantitation I've discussed above, the one the biologist tends to focus on is the manual (via occipital lobe) assessment of staining intensity. These assessments are actually pretty reproducible, but for non-pathologists it seems as crude as holding a fluid-filled cuvette up to the light to measure concentrations instead of using a spectrophotometer. Any image analysis program suite can segment a scanned slide image and read integrated density very well, in regions of interest outlined by the operator. With current technology, the time spent manually selecting regions of interest in the scanned slide images, or training an image segmentation algorithm to find them, is far longer than the process of a quick, manual read.

In my experience the results from automated evaluation and manual evaluation rarely differ significantly, so the driver in favor of automation has to be the number of samples that will be evaluated. This leads us to a broader treatment of image analysis....

"Image analysis—Metamorph (or your favorite image analysis software) can replace a pathologist, right?"
One of the reasons image analysis—the use of software packages to analyze images of slides—is so attractive to basic scientists is that it promises predictability and repeatability in histological analysis. A two-step algorithm, one that first finds the areas of interest and then measures their integrated density (discussed above), can reduce drudgery as well as work in the hands of non-pathologist users. However, the pitfalls are large, and one has to do a lot of repeated work to justify training an algorithm. Indeed, algorithms are only as good as their training, and one area where all algorithms are poorly trained is in distinguishing desired staining from artifacts. Thus, it is easy for an algorithm to find objects of interest, but it does take significant additional knowledge to put those findings into context, and their results should always be crosschecked by a quick review of the marked up (post-segmentation) images.

There is no doubt that image analysis can dramatically change and improve the usefulness of a pathologist. Spending time manually counting islets in dozens to hundreds of sections of pancreas is not worthwhile beyond the Zen value of any mindless repetitive activity. Competent use of image analysis can lead to discoveries beyond the capability of the pathologist alone. This is not because image analysis can "see" things a pathologist can't, but because it is able to recognize and enumerate textures, patterns, or objects, without complaining or needing lunch, enabling the pathologist to do meta-analyses at an entirely higher level of sophistication. That does not obviate the need for a pathologist, but it certainly might change how hours are billed. If your pathologist is billing time for counting islets, rather than crosschecking algorithms for counting islets and doing something with the derived data, then you need to drag your pathologist from the Stone Age and into the Bronze Age, which is where most image segmentation algorithms are now, even in 2014.

"Can you please take a high-resolution photograph that captures everything that you say?"
Yes. Furthermore, I can capture that from practically any slide in the lab so long as it is the same tissue! In an era of many sources of hard analytical data, the pathologist's report is something of an outlier, insofar that much interpretation happens before any result whatsoever lands on a piece of paper or an entry in a spreadsheet. A single image is not the same data as the pathologists report. Imagine the following courtroom drama. "So, Dr. Pathologist, you said that the kidney was normal?" "Yes" "You also said that sclerotic glomeruli were found" "Yes" "But was the kidney thus not sclerotic?" "No,

normal kidneys can have sclerotic glomeruli because of a variety of accidents during their history."

"Can I borrow the slides?"
No. But you can copy the scan to your heart's content. I cannot stress how valuable a good slide scanner is, as well as a departmental policy to scan every slide. While the glass slide is a result of record, and can be stored very compactly and with less fuss than any other archive, the digital image and its associated database are the material of daily use, which may include manual reads, automated reads, markups, easy availability to pull images for presentations, consultation, and—the most valuable to me personally—the ability to fill a portable hard drive with a day's work and to read slides on the plane, train, or taxi (if you have a strong stomach). Coming back to the original question, my response was (and remains) always to give the requestors access to their images and the relevant software. There was usually a flurry of activity in the first week, rapidly followed by a subsidence...as they realized that looking at slides was less fun if the pathologist kept correcting your interpretations after the fact. There's no substitute for looking at slides together at the microscope or on a big screen, agreeing on the salient points, and taking representative images in real time. The advantages of digital slides are many, as you don't need to be in the same place to review them, and, better than a double-headed scope, the non-pathologists are not half in focus and closing one eyeball because they can't admit to not remembering how to focus the microscope's eyepiece.

5 Choosing the Right Pathologist

"If you don't know what you are looking for ask a DVM, not an MD"

About 90 % of "pathologists" in industry were first veterinarians, and then through one means or another decided to complete a fellowship in veterinary pathology, which includes organisms as diverse as fish, whales, and chimpanzees. Most of these pathologists serve a crucial role in drug development by examining the effect of potential drugs in animals, usually one rodent and non-rodent species as required by most health regulatory authorities. The remaining 10 % or so of pathologists in industry were trained in human medicine and usually carry the MD (or equivalent) degree. Many like myself have significant research experience, and feel that drug development is a worthy way to combine their love of biomedical science with the thrill of discovery. My friend and colleague Frank was gracious enough to hand over the keys to his kingdom to me and hang around for some time before he "retired." When we looked at tumors and IHC stains over the double-headed microscope, I sometimes wondered why he took so long to score a slide, though I never ever had a reason to doubt the score. "This looks normal," I'd say, "Let's find some tumor." "Normal?" he'd say, "What do you think of the pyknosis over there and the juiced up fibroblasts over here?" I'd look at him blankly and move the stage, shaking my head. But then after about a week I got it. My whole training as a human pathologist was to find recognizable features of danger to the patient, and I had learned those features directly and indirectly from generations of my ilk who had the advantage, by many studies, of knowing the association between appearance and outcome. In short, from my training I knew what 500 things looked like and didn't care about the rest. He, on the other hand, had the reciprocal training—of knowing normal in several species, and screening for unknown effects of unknown drugs. Instead of triaging through a list of abnormal findings, he had to distinguish between normal and not normal. Then, by looking at not normal, he at least had to arrive at a first level guess of the pathophysiology causing what he was looking at. As you can imagine, this was a harder job than mine because describing the unknown is a lot harder than identifying the absence of the known. Therefore I conclude (as the maker of knockout mice, creatures hitherto unknown to the planet), (1) if you want to know whether your mouse is abnormal, ask a DVM pathologist, and (2) if you want to know whether your mouse is a close phenotypic mimic of a human disease, ask an MD pathologist. And if the answer to 1 is yes, it is abnormal, and to 2 is no, it is not a human disease, then read a lot, because the only person who can help you is you!

6 The Challenges of Manually Scoring a Large Dataset

For one trial I set up a data capture macro in an Excel spreadsheet that locked data in the row containing the H scores, and timestamped it, at the moment the pathologist clicked a confirmatory button. I'm sure all kinds of software can do this. As I was looking at the inter-pathologist correlation data with a statistician, we suddenly thought what fun it would be to see how the timestamp data varied and whether it correlated with any of the other study variables. We had the H score data, and the time submitted. For one particularly soft IHC biomarker we found ourselves able to track the pathologist as though he were on a secret camera: "5 min interval...4 min interval.... 2 min interval...last 3 cases 1 min apart.... Hmm, looks like all the variability in the score has regressed to a mean...wait—here's a fifteen minute gap...aha! 5 min interval again. Guess he's refreshed now!" Clearly the pathologist was going to great effort to see some difference in staining between the cases for this immunohistochemical test, and took the time to mark them accordingly. As the cases progressed he could recognize the pattern of apparent differences wasn't really panning out, and his increasing ennui was detectable by the more rapid scoring of each successive slide. After a break, he redoubled his efforts to see differences between cases that were really blandly similar. A marker like this, with no dynamic range, tends to be dropped from the portfolio after a few tries.

Another more serious challenge along similar lines is a readout where you know (or think you know) the incidence of the biomarker finding, such as PTEN or Her2 immunohistochemistry. For example, among all patients with breast cancer approximately 15–25 % have Her2 3+ score, making them eligible for trastuzumab (Herceptin) therapy. Any slide reader begins to sweat when more than a few cases in a row appear to be the same, if the population incidence of marker positivity is more than a fifth and less than four fifths. Many times I found myself racing through a tray of stained slides, which I knew were stained on sequential slots on the autostainer, to find a case with the opposite result, just to reassure myself that all was right with the stains. There is nothing intrinsically discomforting about this, but it certainly does speak to the desirability of supervised automation in reading immunohistochemical datasets.

7 Conclusions

Interactions between pathologists and bioscientists in pharma can be enormously rewarding for both parties. The pathologist's understanding of normal variation and the pitfalls of tissue analysis, along with input from adept statisticians, can ensure that the biologists asking questions design appropriately scaled, executed, and analyzed in vivo experiments. Pathologists—and statisticians—need to be involved early in experimental design, be it for a preclinical study or for a pivotal clinical study with patient selection. As a demographic, pathologists can be meek when interacting with their veterinary or medically trained colleagues, and sometimes give in too easily. They can also, in the interest of a "can-do" spirit, tend to underestimate the resources necessary to do their work. The potential cost of these traits is a suboptimal dataset with a large bill. As a profession we have to be forthright and proactive to make sure useful work is planned, and done.

Discovery biologists in pharma often ping-pong between worshipping the pathologist for her all knowing aura and resenting her inscrutability. As a profession we integrate more data on the fly than many others in order to generate an "output" from a test. When you look hard under the hood of apparently Boolean data, like mutational analysis, you will find a great deal of interpretation goes on, but this is hidden in a black box. When a pathologist and a biologist look at an image of tissue as viewed under the microscope, they are seeing the same raw data, but the chasm between the two observers is immediately evident. For this reason, I try to offer a very detailed, transparent commentary as I look at a slide with a colleague, saying what I see, what I am sure about, what I do not understand, and what I will get back to later. This is a lot more useful to the inquiring mind than a grunt appended by a comment about how the Boston Red Sox are doing this year, and a diagnosis, which is the output many surgeons prefer. Communication is everything.

8 The Future

Pathology has been two distinct disciplines for a long time. Pathology in preclinical drug development incorporates new techniques to explore new avenues of pathophysiology and mechanism of action of new compounds, whereas Pathology in support of human patient care tends to cling to tried and tested histomorphologic examination of samples as the underpinning of the result, and of the profession.

These two disciplines are merging before our eyes, as the practice of traditional diagnostic anatomic pathology is being significantly shaped by the first wave of (currently fairly primitive) sequencing methodologies, while clinical chemistry and hematology, also branches of pathology, are bringing a new emphasis to the cutting edge methodologies and range of tests. Newly minted pathologists are more likely to see histomorphology as an adjunct to a full understanding of the patient, rather than seeing other tests as an adjunct to histopathological diagnosis.

Even the "bread and butter" of tumor pathology—e.g., knowledge of tumor margins in surgically resected specimens—may become increasingly irrelevant. Natural histories of cancer progression are becoming better understood, alternative methods for understanding tumor dissemination in the patient such as circulating tumor cell and DNA analysis are maturing, and therapies that are more closely attuned to individual tumor biology are coming to the forefront. A significant proportion of common solid tumors may turn out to be disorders of cell lineage differentiation that manifest in particular tissues but require systemic treatment, as exemplified by some recent findings in lung cancer [3].

In the meantime, our computing power, still following Moore's law, continues to ensure that multidimensional modeling of pathophysiological processes, or even machine interpretation of morphology, is limited by our imagination, not by our tools. A significant problem in machine-based morphologic analysis is the lack of standardization in approaches, and it is likely that the field will only attain wide clinical usage when a set of open-source computational modules is developed by a wider community using common standards, and when commercial organizations are rewarded for validating such modules in the clinic, rather than for locking in intellectual property and attempting abortive moves at market domination. It would not be surprising if a few cross-trained radiologists, with deep expertise in analysis of in vivo imaging data, take the field by storm.

Consistent with recent experience, trends in hematopathology might predict where solid organ pathology will be in a decade, with the majority of entities diagnosed by molecular and flow cytometric methods, and an increasing number treated with targeted therapies whose efficacy is monitored by nucleic acid tests. Histomorphology is unlikely to ever be entirely supplanted, however, as it remains cheap, efficient, and a remarkably effective route to the initial diagnosis of a multitude of disease states.

References

1. Tabernero J, Rojo F, Calvo E, Burris H, Judson I, Hazell K, Martinelli E, Ramon y Cajal S, Jones S, Vidal L, Shand N, Macarulla T, Ramos FJ, Dimitrijevic S, Zoellner U, Tang P, Stumm M, Lane HA, Lebwohl D, Baselga J. Dose- and schedule-dependent inhibition of the mammalian target of rapamycin pathway with everolimus: a phase I tumor pharmacodynamic study in patients with advanced solid tumors. J Clin Oncol. 2008;26(10):1603–1610. doi: 10.1200/JCO.2007.14.5482. Epub 2008 Mar 10
2. van Netten JP, Thornton IG, Carlyle SJ, Brigden ML, Coy P, Goodchild NL, Gallagher S, George EJ (1987) Multiple microsample analysis of intratumor estrogen receptor distribution in breast cancers by a combined biochemical/immunohistochemical method. Eur J Cancer Clin Oncol 23(9):1337–1342
3. Katz RL, He W, Khanna A, Fernandez RL, Zaidi TM, Krebs M, Caraway NP, Zhang HZ, Jiang F, Spitz MR, Blowers DP, Jimenez CA, Mehran RJ, Swisher SG, Roth JA, Morris JS, Etzel CJ, El-Zein R (2010) Genetically abnormal circulating cells in lung cancerpatients: an antigen-independent fluorescence in situ hybridization-based case-control study. Clin Cancer Res 16:3976–3987

Methods in Pharmacology and Toxicology (2015): 43–63
DOI 10.1007/7653_2014_28

Published online: 03 December 2014

Outsourcing Tissue Histopathology Investigations in Support of Clinical Trials for Novel Therapeutics: Considerations and Perspectives

Keith A. Wharton Jr., Benjamin H. Lee, Pierre Moulin, Dale Mongeon, Rainer Hillenbrand, Arkady Gusev, Bin Ye, and Xiaoyu Jiang

Abstract

Tissue histopathology investigations are central to discovery and preclinical development of novel therapeutics and for routine medical care, but their variable use in clinical trials represents a missed opportunity to improve our understanding of disease and the effects of various therapies on disease. Here we discuss major considerations and propose a question-based framework for implementation of tissue histopathology biomarker investigations in clinical trials for novel therapeutics.

Keywords Outsourcing, Tissue biomarkers, Clinical trials, Histopathology, Endpoints, Operations, Compliance, Workflow, CRO, Academic laboratory

1 Introduction

Rapidly emerging knowledge about diseases has created unprecedented opportunities to develop novel therapies for patients. Innovation, increased throughput, and reduced costs of "omics" technologies (such as next generation sequencing, mass spectrometry, etc.) are revealing the molecular underpinnings of many diseases and fueling optimism that cures are imminent. However, insights critical for the development of novel therapies arise in diverse and unpredictable ways—from serendipitous convergence of disparate basic science studies to targeted clinical studies of diseased patients. The path from scientific insight to new therapy—particularly for novel molecular targets, and complex, heterogeneous, and still mysterious diseases—can be tortuous. Overall, drug development is time consuming, expensive, and risky. The majority of drugs fail during human clinical testing, with only ~5–10 % of therapies tested in man in recent years progressing to the marketplace. There are diverse reasons for failure, but they are generally related to biology of the target and consequences of its engagement in diseased tissue, efficacy and/or safety in tested patients, marketplace considerations, and the incentives faced by multidisciplinary teams [1]. Stakeholders including patients, health care providers, taxpayers and foundations who fund research, insurance companies and other payers who reimburse health care costs, and investors in biopharma companies will no longer tolerate such a high failure rate, especially if no progress is made to inform subsequent investments. Increasing costs and stricter regulatory standards, coupled to declining reimbursements, suggest that industry-wide biopharma R & D investment practices are unsustainable.

One approach to improve the efficiency of drug development has been movement away from the shotgun-like "blockbuster" approach, predominant in the twentieth century, toward rifle-like "proof of concept" (POC) strategies that aim to create a highly impactful therapy for a precisely defined, often devastating disease for which few to no effective treatment options exist [2]. Notable recent examples in oncology include imatinib for chronic myeloid leukemia, crizotinib for lung cancers with ALK gene translocations, and vemurafenib for tumors with B-RAF mutations [3]. Advantages of the POC approach include increased confidence and impact of efficacy on the disease, early assessment of safety in man, multiple accelerated regulatory approval pathways, and (to date) substantial reimbursements from a relatively small population of affected patients that justify the high costs of development and clinical trial activities. A subset of these highly targeted therapies will be found to be beneficial to broader categories of patients with more prevalent but heterogeneous diseases.

However, further improvements in efficiency will require derisking each step in the drug development pipeline. The authors believe that routine incorporation of tissue-based translational pathology investigations in clinical trials will improve drug development by enhancing fundamental knowledge about human disease, by revealing common features of diseased patients as well as unique attributes of subsets of patients, and by illuminating possible differences between the disease in human patients and the preclinical disease models used to validate candidate drugs. For a variety of reasons that will become clear, such investigations often require coordination of in-house activities with outsourcing some or all of the hands-on work to well-equipped partners. Our goal in this chapter is to share, within a question-based framework, a biopharma sponsor perspective in the design and execution of outsourced tissue histopathology investigations.

2 The Biomarker Paradigm

Biomarkers are defined as measured characteristics of a biological state, condition, or effect [4]. A validated biomarker can serve as the foundation of a surrogate for a more difficult to measure but desirable endpoint, such as rate of disease progression, functional status, or survival. Biomarker measurements can be descriptive and qualitative, or multivariable and quantitative; ideally what is measured relates to the underlying biology, but we refer here to a biomarker "paradigm" because as long as the biomarker measurement has a specific utility, then its relationship to biology need not be understood. For example, earlobe creases in human are a qualitative observation implicated as a biomarker of underlying cardiovascular disease, blood pressure is a two variable quantitative biomarker of real-time cardiovascular function, troponin T is a muscle protein that can be measured in blood following its liberation from injured or dead cardiac muscle, and a radionucleotide angiography (MUGA) scan generates quantifiable images that depict cardiovascular dynamics in real time; each of these "biomarkers" reflects a distinct aspect of cardiovascular status and can be useful in specific contexts, but none of these biomarker assays require collection or analysis of diseased tissue. In practice, most biomarker assays involve measurement of specific molecular entities from relevant biological samples with a defined method and experimental environment, in order to achieve a specific purpose. Indeed, biomarkers are often described as "fit for purpose" because the phenomenon or analyte measured, the assay employed, and the results themselves, should only be interpreted—and any conclusions reached and actions taken—within a defined context [5]. Context is crucial—*ceteris paribus*, it is generally easier to validate a biomarker for a narrowly defined than a broadly defined context; consequently, a biomarker measurement out of context (e.g., a different species, disease, or treatment with a different drug) can be misinterpreted. For example, the growth of breast and gastric adenocarcinoma can be driven by increased Her2 kinase and thus be susceptible to anti-Her2 therapy (e.g., trastuzumab), but the relationships between tests of Her2 abundance and recommended therapy are subtly distinct for each disease [6].

For the development of novel therapies, a key context important for each biomarker measurement is the underlying disease, or at least how the disease is defined—the topic of an earlier chapter in this book. Matching the disease to the biomarker measurement and its context (e.g., prognostic, predictive, safety vs. efficacy) is crucial, requiring close, iterative collaboration between clinicians with expertise in diagnosis and management of the disease, and scientists expert in various biomarker technologies and assays. Disease diagnosis is central to the practice of medicine, increasingly requiring integration of diverse datasets—from clinical observations and traditional laboratory tests to cutting-edge omics investigations. Because tissue histopathology represents a "gold standard" for diagnosis of many diseases, pathologists are well equipped to bridge gaps between clinicians and scientists in relating biomarker measurements to disease, and can advocate for histopathology to serve as a framework to integrate tissue-based with non-tissue based biomarker investigations.

We propose that including collection of human tissue samples in selected clinical trials, with proper analysis by qualified pathologists, is one of the most efficient ways to advance knowledge about human disease and the effects of new therapies. An underlying assumption of many biomarker investigations is that a soluble analyte or analytes such as cytokines, detectable in a fluid sample with sensitive assays, reflects the disease state or treatment effect in the patient. Such analytes might be produced by diseased tissue and distribute into the fluid sample, or might be produced elsewhere in the body (e.g., other tissues or circulating cells) in response to disease. Body fluids such as blood or urine are generally easier to collect and process than lesional (i.e., diseased or target) tissue; but depending on the behavior of the biomarker (e.g., rates and sites of synthesis, stability, and metabolism in vivo), the type of disease, or effect of the disease on biomarker kinetics, such samples might not—under any circumstances—be representative of the desired aspect of the disease. Studies can be designed to test a specific hypothesis (typically with a well-established technology platform such as ELISA or LC-MS/MS) or be exploratory in nature, in order to investigate whether an emerging technology can discover or identify biomarkers that correlate with disease, prognosis, or treatment. Depending on study design, one risk of investigations in which fluids are the only biological samples collected is that if the data do not support the hypothesis that underlies biomarker measurements, then insights to inform subsequent investigations might be lacking. For exploratory studies, failure to identify a valid soluble biomarker reinforces their often well-earned reputation as "fishing expeditions"—if there are no fish in the lake, you will catch no fish there. We believe a better approach is to design the overall investigation to test a limited number of hypotheses of interest while poising samples for future exploratory studies, even if the right technology is not available (or even invented) at the time of sample collection. In this way, knowledge can be advanced in case the primary biomarker hypothesis is not supported by trial data. Moreover, by including analysis of tissue samples, particularly lesional tissues, investigations of specific soluble biomarkers can be related to the abundance or quality of the biomarker in lesional tissue (assuming lesional tissue is its site of synthesis or accumulation, which might not be the case, for example, for diseases involving endocrine axes). Tissue-based biomarkers might be more readily preserved from samples, better conserved between preclinical toxicology or disease model species and human, and amenable to post-study data-driven analytics (i.e., correlation with clinical or other biomarker endpoints). For diseases in which valid biomarkers do not yet exist, investigations on adequately sized tissue collections might be the only path to discover, measure, and exploit the biomarker paradigm to advance knowledge about disease and develop rational diagnostic and therapeutic strategies.

Proper deployment of tissue histopathology in human clinical trials requires consideration of how tissue pathology is typically performed in clinical, research, and regulatory contexts. Histopathology-based diagnosis is widespread in human medical care, and can provide crucial insight into research investigations, e.g., by determining whether animal models recapitulate aspects of the disease they are intended to model. Histopathology as part of (human) medical practice is performed in (USA—CLIA/CAP; EU—ISO) approved laboratories by practitioners, usually led by MD or DO-degreed (in the USA) pathologists whose licenses are issued by state medical boards and whose certifications are granted by national specialty boards. Histopathology in animal research is often performed by veterinary (DVM or equivalent) pathologists as part of research projects that are approved by institutional animal care and use committees (IACUCs), who increasingly scrutinize proposals to ensure judicious and humane use of animals. Research activities involving human tissue require assurances to funding agencies and regulatory bodies that appropriate protections for humans and their tissues are in place, including data privacy and protection, control of fate and use of material, and informed consent. In biopharmaceutical development, significant histopathology is practiced by veterinary pathologists in support of so-called "non-clinical" (which usually means "non-human", as distinct from "clinical" that describes clinical trial-related activities) toxicology studies, ultimately under Good Laboratory Practice (GLP) guidelines that ensure to regulatory agencies that specific individuals are accountable for results and that systems used to track data allow, if necessary, its reconstruction. Depending on the intended use of the data, tissue biomarker investigations in support of novel therapeutic development in humans require consideration of each of these contexts in order to devise a proper experimental plan and quality management strategy, as well as comply with appropriate regulations.

The value of tissue sampling (and resampling) varies with the disease. In oncology practice it is becoming crucial. Recent investigations of tumor heterogeneity and evolution indicate that serial access to tumor tissue, particularly with resistance to targeted therapy, enables selection of the next appropriate therapy (e.g., [7]). Despite recent setbacks, there remains interest in relating properties of circulating tumor cells (CTCs) or other tumor products in blood samples, to tumor qualities including progression, prognosis, and susceptibility to therapy [8]. Potential limitations of tissue biopsy include accessibility and representativeness of the sampled tissue to the disease. Skin, oral cavity, gut, and bone marrow are generally accessible and can be sampled in limited quantities, but access to internal organ or CNS tissue carries markedly increased risks, and is based on medical necessity. For some systemic diseases, single or even multiple biopsies might not be representative of disease progression or therapeutic effect, as is the case with some diseases of the neuromuscular or skeletal systems where only a small fraction of the total system can be sampled, if at all. Tissue investigations can provide critical data, but like biomarkers themselves must be interpreted by qualified practitioners in the context of disease and treatment effect.

Recent articles highlight key issues associated with outsourcing activities in support of drug development, particularly biomarker activities (e.g., [9]), but we focus here on special issues that concern analysis and interpretation of biomarkers involving pathologic interpretation of tissue—a skill typically ascribed to the anatomic pathology lab but which increasingly integrate molecular pathology investigations (e.g., in situ hybridization, qPCR, transcript profiling) and/or "omics" techniques such as microarray profiling or massively parallel sequencing. Reasons to outsource generally fall into five categories—capability, innovation, cost, capacity, and time. Options for outsourcing single analyte brightfield immunohistochemical (IHC) assays are numerous because they are widely used in current human clinical practice, but novel multianalyte assays with image quantitation (or other "bells and whistles") might only be performed competently in-house by big pharma companies or by a few cutting-edge laboratories. Generally, the more innovative or technically demanding the assay is, the more expensive the setup and technical acumen required to validate the assay, driving up costs. Pressure to deliver definitive results as quickly as possible thus gives established outside laboratories with high throughput capacity and quick turnaround times an advantage over in-house facilities with no specific assay experience. Vetting and accurate interpretation of data often requires engagement of key opinion leaders (KOLs), whose reputation for expertise and influence in a particular discipline can help gauge the credibility of any findings and their implications.

Depending on capabilities and needs, investigations typically require coordination between in-house activities and outsourcing to contract research organizations (CRO), private hospital or academic research labs, or platform-based biotech companies. Each entity has unique cultures, operational modes, procedures, and incentives for success that must be considered. Outsourcing introduces complexity, cost, and logistical challenges to a clinical trial, but by bringing the best expertise to bear on precious human samples, it is ultimately aimed at maximizing access to technologies, managing capacity, and reducing costs—all to extract maximum value from a limited precious resource.

This chapter is intended to provide a biopharma sponsor perspective on outsourcing tissue biomarker/pathology investigations for clinical trials. However, for some projects we find that outsourcing per se is only a minor component; indeed, if the work is performed in-house, outsourcing is eliminated, but similar project planning, organization, and management issues need to be addressed. The chapter is roughly organized in a chronological fashion, addressing issues in the order they might arise as a potential therapy progresses through the drug development pipeline. For the planning stages, we address major questions for consideration (the why's, who's, what's, etc.) and conclude with trial monitoring and follow on activities. But such linear, predictable scenarios are the exception rather than the rule. Drug development occurs between the dynamic world of biomedical discovery and a volatile health care marketplace, and opportunities to interrogate tissue can arise from either end of the spectrum. For example, a newly reported insight into a particular disease might justify a phase 3 trial modification that allows sampling tissues from patients in order to develop a potential (companion) diagnostic test that predicts response to the therapy, thus enriching patient selection in subsequent trials. Flexibility and innovation are key. Our intended audience is anyone who participates in planning and operation of clinical trials, with little to no familiarity with pathology in research or clinical practice.

3 Trial Design and Initiation

As with any project, thoughtful planning pays. It is crucial at the outset of the project, and periodically thereafter, to query all relevant sources of information, including literature and scientific meeting abstracts, competitive intelligence, as well as key internal program contributors to date. The decision to procure tissue from anyone, let alone a patient in a clinical trial, is not to be taken lightly: tissues are scarce sources of information about each individual patient, their disease, and the potential effect of therapy, and should be treated as such. Obtaining tissue from healthy volunteers in phase 1 trials is feasible as long as safety is ensured, and benefit outweighs risk. For example, buccal swabs and small skin biopsies are relatively safe with minimal risk to the patient, whereas the markedly increased risk associated with biopsy of internal organs such as liver may only be justified if such tissue is needed for an essential endpoint, and there is no surrogate biomarker. Effort and cost of tissue biomarker studies can be substantial, so the experimental plan should be devised to fulfill multiple goals: (1) ensure that an adequate number of samples will be available to test trial hypotheses accurately and on time, (2) prepare any additional matched or remaining tissue for follow-on experiments that might not be defined at the outset of the trial, and (3) enable integrative analysis with other study endpoints, including other clinical endpoints and biomarker assays. Methods chosen should thus be linked to answer key questions the trial is designed to ask, as well as exploit any unique investigative opportunities that arise by creating a curated tissue collection, to include traditional histopathological investigations as well as to compare emerging with existing technologies. How tissue pathology is handled also depends on whether the measurement is a primary or secondary vs. exploratory objective of the trial; although all uses require special attention to issues related to handling human tissues and managing protected health information, the former place more strict requirements on regulatory quality, data and system controls, and timelines—and thus resources including cost and employee time—that the latter generally do not.

4 Why Is Tissue Needed for the Trial?

For practical reasons, many later stage clinical trials in diseased patients use externally measured endpoints related to disease progression, such as time to walk a certain distance in multiple sclerosis, or the maximum dimension of a tumor measured by imaging techniques in cancer. While such endpoints can be highly relevant for patients and are generally acceptable to regulatory authorities, they can be somewhat subjective and often bear an indirect relationship to the disease process itself. Clinical trial teams will consider incorporating tissue analysis when it is required for enrollment (e.g., to confirm a diagnosis or investigate a sample of a patient's tumor), is available for other reasons (e.g., a renal biopsy to ascertain a particular disease), or is necessary to assess a particular endpoint in vivo (e.g., a lymph node sample at primary tumor resection, a tumor sample following neoadjuvant therapy, or a skin sample to assess pathway-specific pharmacodynamics measurements in systemic disease). Tissue is not needed for germline genome sequence analysis, as starting material is readily obtained from blood, cheek swab, or sputum. Fluids such as blood, urine, or CSF are more readily available than tissue, but the analytes of interest might not be present in them, or if they are measurable, substantial variation in their abundance not attributable to disease or treatment might prevent definition of useful reference ranges and cut-points that distinguish normal from abnormal, or responder from nonresponder. Depending on the disease, tissue samples might be useful (1) in small scale, early stage trials to establish safety and possibly efficacy at various dose levels (thus de-risking the trial, engendering confidence that the therapy is having the desired pharmacological consequences), (2) to investigate biomarkers as a basis for possible companion diagnostics, (3) to generate or test hypotheses about pathophysiology and heterogeneity of disease and the impact of therapy, or (4) to investigate mechanisms of drug resistance as a foundation to discover more effective follow-on or combination therapies, or dosing regimens.

Collection of tissue in later stage trials is currently limited to some oncology trials, owing to the fact that cancer is nearly as diverse as the millions of patients that develop it each year, and patients usually undergo surgery to remove a primary tumor, but the need for tissue depends on the hypothesis of the trial as well as the data (and hence the patient samples) required to test the hypothesis, or to make the next discovery. Designating biopsies as "optional" in a clinical trial protocol might compromise the diversity of samples and preclude meaningful statistical analysis across the trial, although complex datasets from small numbers of samples can help to generate new hypotheses. Conversely, designating biopsies as "mandatory" might limit trial enrollment if alternative treatments exist. Justification for, and use of, the biopsy in any trial—whether for clinical decision making, biomarker stratification, or purely research purposes—should be fully disclosed in plain terms in the informed consent, and be directly linked to the scientific questions the trial aims to answer [10].

5 What Tissue Is Needed?

Will archival tissue suffice, or is a fresh tissue sample necessary? Anatomic pathology practice is noteworthy for its lack of experimental standards, so depending on whether the measurement of interest is qualitative (i.e., present or absent), semiquantitative (grading scale), or quantitative, how the tissue was originally handled/processed, and how it has been stored, might influence whether an accurate answer can be obtained from an archival specimen. If a fresh specimen is necessary, it must be handled and processed to generate accurate information in time to influence possible decision making (e.g., to diagnose a specific disease and thus determine whether the patient is eligible for trial enrollment, dose escalation, or extension of therapy). Pre-analytic variables (i.e., everything that happens before an assay is run), as further elaborated in chapters about IHC and NGS in this book, must be understood and controlled where possible. As tissue biopsy is only a "snapshot in time" of tissue qualities, consideration of timing relative to dosing or anticipated therapeutic response, as well as the possibility and consequences of serial sampling, must be considered. Must diseased tissue be sampled, or will non-diseased tissue suffice? For example, pathologic intracellular inclusions of Parkinson's disease—the alpha synuclein positive Lewy body—are characteristically present in affected neurons of the midbrain and cerebral cortex, but similar structures have been identified in peripheral nerves and ganglia of far more accessible tissues such as the colon [11]. For core intercellular signaling and functional pathways that can be active in many tissues, sampling one tissue might provide some confidence of pharmacologic effect in other tissues—such as sampling plucked hair follicles to investigate modulation of the specific pathways (www.epistem.co.uk)—but proof of concept studies in appropriate preclinical animal species increase the likelihood the experimental method and biological context will translate to human. Method development has to be uniquely matched to the tissue obtained and answers sought; any histology lab can process, and at some level, interpret gut, skin, muscle, or brain tissue, but specialized expertise with the tissue of interest might be required to answer key questions. Uniform tissue collection, procurement, and handling methods across a multisite clinical trial will help to limit preanalytic variation and might aid in development of a potential companion diagnostic, whether the diagnostic is developed in parallel with the drug, or whether tissue is banked and then later used for retrospective analysis and bridging/validation studies after outcomes are unblinded.

6 Who Does the Work? Where Is the Work Done?

This brings us to the ostensible purpose of this chapter—outsourcing. Although "outsourcing" is a growing activity across industries and economies, and is driven by a variety of factors, we focus here on its role in generating accurate, timely, and cost-effective data. Historically, tissue pathology/biomarker studies were typically performed in-house: either in research-focused labs by personnel experienced in handling animal tissues for preclinical studies, or outsourced to a local academic hospital or university research lab. Increasing costs, regulatory concerns, operational flexibility, and need for human subject protections and quality management on par with medical diagnostic (clinical) labs can make partnering with external providers preferable to supporting such activities in-house. When addressing who does the work, and where, consider the entire process, from sample procurement to data readout, to ensure that each player in the process has built-in means to ensure accountability for their contribution, including quality checks in case the process breaks down or unexpected results are obtained. For example, for IHC studies it can be inadvisable to have tissue sections cut in one lab and IHC stains performed in a different lab, because if the stains do not yield interpretable data it might be difficult to pinpoint which step in the process or lab is responsible. Nevertheless, output from one outsourced partner (e.g., stained slides) might need to be transmitted internally or to a different partner for subsequent analysis (e.g., image analysis), in which case all parties must be aware of the overall goals of the study and agree to how quality will be checked at each stage of the workflow.

Scientific, operational, and business factors should be considered in selecting with which laboratory to partner. Scientific factors include technical competence and experience of key personnel, laboratory reputation and capabilities, and any assay-specific issues such as familiarity with the disease or readout, specific markers, techniques, or equipment. Operational considerations include the obvious and less obvious: How well are samples handled and tracked, how are metrics kept, and how are any discrepancies and deviations from procedures captured and reconciled? Are SOPs (standard operating procedures/protocols) properly developed, implemented, and updated? Does the organization have experience with other biopharma partners and clinical trial samples/data? A site visit and technical assessment should engender confidence that experiment will be performed in a way that is highly likely to achieve study objectives on budget and on time. Business considerations include professional and expedient communications, the ability to define deliverables and milestones in order to compose a competitive bid, business continuity planning, and operational transparency as well as quality management systems and procedures. Personal information about patients and data privacy also needs to be secured according to HIPAA or other guidelines, depending on country of origin. Each type of laboratory—including academic research labs, hospital diagnostic labs, private diagnostic labs of various sizes, or small platform biotechnology companies—typically has characteristic strengths and weaknesses in each of these areas, which must be considered when sourcing work to a given partner and environment. For many sponsors, a potential vendor must first be qualified to perform the work through an assessment that scrutinizes the organization according to predefined criteria.

Here we define the role of the "central" lab: for a single site trial often a nearby lab that will handle, process, and analyze trial samples, but for some single site and nearly all multisite trials often part of a large contract research organization with infrastructure to receive, track, process, distribute, and (if necessary) properly archive or dispose of samples. Some central labs have the capabilities of pathology laboratories, whereas others are best positioned to serve as hubs to manage sample chain of custody and shipments to include distribution of tissue samples to a specialty pathology or research lab. A central lab plays an important role in a multisite trial involving pathology endpoints because many small company or academic labs do not have the infrastructure and resources in place to responsibly handle and process specimens sent from multiple trial sites. This process is aided by automated LIMS (laboratory information management systems), but it is important that the central lab LIMS communicates in a manner compatible with the pathology lab's data tracking system, and if not, that the sponsor's data and sample handling requirements be defined and met prior to the onset of the study. Vertical integration of sample collections via a central lab thus eases communications and logistics of the outsourced activity while enabling coordination with the sponsor's internal activities.

The sponsor organization should take the lead in defining roles and responsibilities, while remaining flexible to suggestions from potential outsourcing partners. Within the sponsor organization it is important to have project responsibilities clearly defined by individual as well as by line function, including the scientific lead, operational outsourcing manager, biosample coordinator, and quality assurance resource. Partial redundancy to include primary and backup responsibilities in case of a lost or discrepant sample, employee absence or turnover, missing contact information, etc., is also important. For example, a scientific lead might develop the overall plan, but must also engage several internal stakeholders including the research, development, and clinical teams, legal, business/contracts, sample management, data analysis and statistics, etc., and must be prepared to coordinate communications among these line functions with the potential vendor so that communications follow the "three Cs" of consistency, conciseness, and clarity. Geographic factors impact cost and speed as well: shipping samples across international boundaries can introduce shipping delays due to customs, additional costs, and administrative burden, possibly increasing the likelihood of specimen loss or destruction.

7 How to do the Work?

Next we consider method development, or adaptation of an established method, to answer trial-relevant questions. Technical and operational factors must be jointly considered as the study plan is developed, with every step in the process thought out – from sample procurement at the trial site (including informed consents), shipments, tracking, and chain of custody, to experimental methods and interpretation, and sample disposition and results reporting. At the trial site, particularly in a multisite trial, it is important to keep the tissue collection and handling process as simple and uniform as possible, while transferring the tissue in a manner that ensures it is properly identified, handled, processed (if at all), and shipped to the appropriate lab. Close collaboration with the central lab is important for robust sample transfer. The primary informed consent the patient signs before trial enrollment should describe, in layperson's terms, the use, scope of analysis, and disposition of tissue. While it is possible to use fresh tissue in an assay performed at or near a trial site, shipping delays limit (but do not eliminate) the practical use of fresh tissue in clinical trials. Plus, most histological techniques involve tissue preservation.

A major question is thus how to stabilize or fix the tissue—by freezing, by immersion in a cross linking fixative such as formalin, or in a solution designed to preserve specific analytes, such as

RNAlater® to preserve RNA (www.lifetechologies.com). Heterogeneity of freezing medium (liquid N_2 vs. solid CO_2 vs. chilled isopentane), as well as technique, from person to person and site to site, can lead to variable preservation of labile analytes (e.g., signal activation-dependent phosphoproteins) or to histological artifacts (e.g., ice crystals, prominent in muscle biopsies) that can render collected tissues unsuitable for the intended analyses. Freezing tissue for techniques intended to destroy tissue architecture is distinct from freezing tissue for histologic interpretation, the latter usually requiring controlled freezing in cooled isopentane with optimal cooling temperature (OCT) or other support medium. Custom specialized liquid immersion media designed to preserve specific analytes might not preserve tissue morphology for robust microscopic interpretation [12]; depending on specimen number and geometry, it is generally desirable either to split specimens in a defined manner at time of harvest, or to collect multiple equivalent specimens for each type of assay. It is also important to avoid subdividing a very small sample such as a punch biopsy prior to fixation, because the method used to divide the sample might prevent morphologic interpretation of even the fixed sample. A method that stabilizes analytes of interest and can be performed identically across different trial sites is usually the best option, but for formalin fixation, care should be taken to limit fixation time if follow-on assays such as IHC, ISH, or other profiling assays that depend on reversing formalin-induced cross links impact analyte visualization or recovery. Since optimal tissue fixation and processing times can vary with tissue type, size, and geometry, it is important to define these variables, as well as the relationship of the pathologic lesion (if present) to the overall sample, as precisely as possible in the onsite laboratory manual. We remain optimistic that "one size fits all" methods of tissue fixation and processing will emerge to satisfy the pathologist's demand for high quality tissue morphology as well as preserve all analytes for follow-on studies, but at present none exist (e.g., [13]).

Interpretation of some tissues requires proper processing and orientation prior to sectioning, with some tissue-specific techniques such as those required to process renal or muscle biopsies only performed to standards acceptable to specialists by certain research or clinical labs. For nearly all studies, including those that destroy tissue, consider gating performance of additional experiments on an assessment of tissue adequacy and cellular composition for each specimen via H & E (hematoxylin and eosin) staining and microscopic evaluation by a pathologist. Data derived from incorrectly identified or handled (or even autolyzed) tissue can be uninformative, irrelevant, or misleading, biasing or confounding statistical analyses and interpretation. If staining methods such as IHC or ISH are done, automated platforms that generate highly reproducible stains should be used. For such instruments, the frequency of batch staining runs should balance the need for rapid data generation with minimizing assay-associated inter-run and inter-sample variability and cost. Regardless of the experimental procedure, methods need to be described in detail, performance should be documented for each sample run, and should be executed by trained and qualified personnel.

Whether a qualitative, semiquantitative, or quantitative readout is desired will influence the tissue sectioning scheme. A qualitative readout—e.g., whether a particular disease process is present or absent, and what type it is—is generally assessed by viewing H & E sections but might require additional stains depending on the tissue and specialist interpreting them. Many pathologic conditions and IHC assays are graded according to progressive histopathologic features or staining criteria, a semiquantitative or "binned variable" approach. Prominent examples include the WHO classification of lupus nephritis [14], and breast cancer IHC for the kinase Her2 (Herceptest; http://www.dako.com/us/29036_herceptest_interpretation_manual-breast_ihc_us.pdf). If classification of tissue biopsy by these criteria are crucial to patient enrollment or data analysis following the trial, then it is important that the criteria used to grade the tissue are clear and widely accepted by pathologists interpreting the stains, or if a distinct scoring system is used, that it is intended to test a novel or trial-specific hypothesis. Having multiple pathologists interpret or grade slides enables study transparency and enhances peer review, but introduces ever present inter-pathologist variability that can plague some diagnoses or grading systems [15]. Discrepant diagnoses between primary and peer review pathologists can be adjudicated by an expert panel, as is routinely done in interpretation of pathological lesions in preclinical toxicology studies [16]. Increasingly, fully quantitative pathologic endpoints are desired, but they place additional requirements on assay development, experimental design, expertise of personnel, cost, and completion time. In lieu of manual (and possibly biased) counting of cells on a slide (e.g., with a manual cell counter), quantitative readouts increasingly require whole slide scanning and microscopic image segmentation and quantitation—techniques that are only now being adopted by some clinical labs. Depending on the endpoint desired, it is important to consider: (1) tissue sampling frequency, orientation, specimen geometry, and size and distribution of the entity or lesion to be quantitated in light of the principles of systematic uniform random sampling [17]; (2) controls, including replicate staining runs, to ensure each value generated is representative of the tissue sample [18]; and (3) efforts to determine whether specimen quality is adequate for slide scanning and segmentation by the image analysis algorithm are in place through the entire workflow. The diversity of variables that must be considered in such studies are generally greater than required for development of some bioanalytical assays, but an experimental design that both limits inter-sample variation and allows recognition of, and correction for, extraneous or confounding variables, increases the likelihood of success. Data transfer through defined protocols and secure portals—not just excel files emailed between individuals—is critical, and should be agreed upon prior to the onset of the study. Quantitative data also places additional requirements on a statistical analysis plan, requiring involvement of professional statisticians and not simply pathologists (claiming to be) adept at statistical software packages.

A statistical analysis plan should be drafted prior to the onset of the study, and should be negotiated before unblinding data or conducting any analysis, in order to avoid biases. A plan usually starts with the study objectives and endpoints for quantification and analysis. The clinical questions are then "translated" into statistical questions, for which the corresponding statistical methods or models need to be specified. For example, to compare the means of two groups, a two-sample *t*-test may be used, but the assumptions that underlie the *t*-test, such as whether the variances of the two groups are equal or not, need to be checked. When a more complex model such as a linear model is proposed, all components in the model, including the dependent variable, the independent variable of interest, and any covariates, should be specified. The intended use and interpretation of statistical analysis and results should be documented in enough detail to be comprehensible to trial team members and critically judged by peer statisticians. If applicable, power analysis for determining minimal number of samples should be included, with expected effect and estimated variability in data estimated from prior knowledge. On the other hand, tissue collections from small clinical studies that are underpowered from a statistical standpoint can still be useful to generate rich multidimensional datasets (e.g., samples with matched clinical, pathological, and molecular data), discern possible trends, and generate novel hypotheses, but the results from such studies cannot be interpreted with a pretense of statistical significance. Lastly, analysis results and visualization should also be discussed in the analysis plan, and vetted by the clinical team, to ensure consistent, focused messages are generated by study data. A thorough treatment of statistical analysis of histopathology data is beyond the scope of this chapter, but other chapters in this book provide representative examples.

8 When Is the Data Needed?

In the competitive marketplace for development of new therapies, the goal of any tissue-based biomarker study should be to generate accurate data according to timelines. Histological readouts as a primary (as opposed to exploratory) trial endpoint and/or a desire to publish data coincident with other trials or their milestones can place the analysis on the "critical path" to achieving program or company goals. For these reasons, it is important at the outset to define realistic expectations for time to assay completion with internal stakeholders as well as expectations and milestones for progress among potential outsourcing partners, including backup plans in case an outsourcing partner is no longer able to perform the work. Corporate partners with experience will generally be adept in negotiating a study plan and scope of work, but depending on the expertise required, specific academic labs or core institutional labs might provide the best environment to perform the work. CROs and clinical labs are generally fee-for-service operations, and while they share responsibility for their contributions, they generally do not desire to own the data, and there generally is not an expectation of publication coauthorship or shared intellectual property. Academic investigators, while they too need to pay bills, are generally motivated by advancing knowledge, gaining peer recognition through publications, and—especially their host institutions—intellectual property. Negotiating legal and financial aspects of contracts can occupy what seems to be inordinate amounts of time, so it is important that all parties be responsive in the back and forth negotiations required to put contracts and work orders in place in a timely fashion. Given the up-front activities to establish relationships with a service laboratory, working with an established network of preferred and specifically qualified providers can increase efficiencies over the longer term. Negotiating sequential milestones in the study plan helps to align expectations at the beginning of the project and aids in monitoring progress through the project. Finally, the priority of activities can change dramatically depending on the trial progress and unforeseen external factors, so it is important for both sponsor and outsourced partner to remain flexible to renegotiate timelines or scope of work as the study progresses, including the possibility of interim analyses to give the sponsor feedback on the progress of the trial as well as the likelihood the trial will meet its endpoints.

9 Trial Execution

Prior to the trial going "live," usually designated by the FPFV (first patient, first visit), it is important to test run each portion of the study, including sample delivery to the laboratory, all experimental procedures (preferably with similar but non-trial or nonessential specimens), and any follow-on activities such as slide scanning, quantitative and statistical analysis. Monitoring of the trial study involves establishing clear responsibilities of key individuals on both sponsor and CRO side (with backups to account for vacations or employee turnover), chain of custody communications, periodic scheduled meetings to discuss progress and any issues arising, and rapid identification of root causes for any problems. As with the biomarker itself, the scrutiny with which these activities are monitored and tracked should be "fit for purpose"—highest for pivotal registration studies or companion diagnostic qualification studies as compared to exploratory studies. While ensuring data quality is crucial, it is equally important for both sponsor and partners to be able to track study progress and data accumulation throughout the study, e.g., via a periodically updated file or data output that allows quick data visualization and assessment, such as graphic output, color coded values representing different assays or groups, data within or out of range, of different quality (e.g., number vs. text), or missing. As the study progresses, the sponsor might find an onsite visit beneficial, to ensure business and operational continuity as well as observe maintenance of technical capability, particularly if the lab ownership or key personnel changes occur during the study.

As a part of study monitoring, it is crucial to follow each specimen closely through each stage of the process—from trial sites to the service provider lab, through experimental procedures, and to interpretation and any test data analysis. Individual steps and components of the study might not automatically integrate, particularly if departments, companies, or people that do not normally communicate with each other must routinely do so as part of the trial. All parties should be prepared to make adjustments to quickly fix any problems that arise, and prevent them in the future. Ideally, the central lab and/or the lab performing the analysis (if different from the central lab) periodically (e.g., weekly) provides the sponsor with a running summary of overall progress—again, in a format that makes assessment of progress simple and transparent to all parties. The goal is to establish high standards up front for experimental quality and performance monitoring, and then to maintain or improve these metrics throughout the study. If dry run (non-trial) samples are not available, then initial clinical samples must be handled and processed to assess variation and minimize it through the rest of the study. Regularly scheduled teleconferences or videoconferences with key personnel are a good way to discuss results and resolve any issues that arise. In most cases, labs will remain blinded to treatment groups throughout the trial. The sponsor should also perform periodic monitoring, including, but not limited to, confirming maintenance of accreditation, proficiency testing, or site visits, the latter particularly useful if data reconstruction is necessary. If interim data analysis is anticipated, the sponsor should clearly articulate any organization-specific processes and deadlines well in advance of any deliverables. As the trial nears closure, any data errors or discrepancies must be reconciled, including any repeat analyses needed due to equivocal, unusual, or out-of-range results.

10 Post-trial Activities

After trial closure, it is imperative to finalize the data as quickly as is reasonable, according to established timelines. The laboratory should have stored and handled the data in a qualified and secure LIMS and periodically reviewed summary data with the sponsor, so there are no surprises at the end. The data is first "cleaned"—each data point is assessed for correct format, identity, and range—in order to ensure quality data is analyzed. According to a statistical analysis plan, once the data is locked it cannot be adjusted, and such locking is required prior to un-blinding and final statistical analysis. Disposition of materials, both used (slides, blocks, etc.) and unused (extra tissues, unstained slides, etc.), should be negotiated among all parties; they are typically specified in the informed consent to which patients agree, and should be understood by all parties. Remaining processed and curated samples from a clinical trial can be extremely valuable as starting points for follow-on primary or secondary research – investigations that directly follow or diverge from the primary trial hypothesis, respectively. Contract labs will generally have little interest in storing samples beyond the time necessary, but academic research labs might have a vested or competitive interest in retaining specimens for further study. However, additional factors must be considered prior to use of such specimens. Current informed consent practices, for better or worse, place time limitations on use of tissue. Long-term storage of tissues in formalin, done in some tissue banks, continue the chemical cross linking process far beyond the initial minimum fixation time of ~8–48 h in formalin for a typical tissue biopsy, introducing variability and thus limiting the utility of tissue for certain assays or investigations. With expense, both frozen and formalin fixed paraffin embedded (FFPE) tissues can be stored indefinitely, and depending on the assay, might be useful for several years to decades after the trial. Whether stored tissues are useful indefinitely is another matter, as many factors including oxidation, moisture, and bacterial or chemical contamination could affect or even eliminate analytes of interest, even in FFPE tissue. Innovations and improvements in technology, particularly those optimized for use of archival FFPE specimens, have the potential to reveal insights not possible at the time of trial closure. Some of these improvements are described in the chapter in this book on NGS pre-analytic factors.

The sponsor should assess and reflect on all aspects of the outsourcing activity, document lessons learned, and consider them when deciding whether to continue partnering with the lab or seek alternate partners for future activities; establishing relationships with multiple potential outsourcing partners promotes healthy competition and keeps downward pressure on costs, but it is important to factor in all costs—including sponsor employee time (e.g., establishing and maintaining the relationship and interface) and quality and compliance of deliverables, not just contract-related financial costs—when sourcing subsequent studies. Unlike many soluble analyte assays on liquid samples, tissue pathology biomarker analysis is rarely standardized across sites, labs, or countries; each lab or pathologist will have a preferred way to process, cut, stain, and interpret tissue, necessitating that variables impacting workflow and data analysis are specified and agreed upon prior to each trial.

11 Conclusions and Future Directions

Tissues from patients with disease that are administered novel therapies are unique, essential resources to investigate the molecular basis of disease, discover and validate biomarkers, and visualize the consequences of therapy. The effort, time, and cost required to collect and process human tissue, particularly for histopathology endpoints, are substantial. Investigations should be custom-tailored to answer questions of immediate importance to trial objectives and program development as well as poise remaining or matching tissue for future investigations that advance knowledge of disease, mechanisms of therapy resistance and candidate targets to overcome resistance, and future trial design—irrespective of trial outcome. The optimal environment to perform different types of studies remains in flux: key reagents and assays are typically developed in academic or biopharma research laboratories, but the diversity and quality of commercially available reagents increases daily. Increased assay standardization, quality controls, and documentation requirements necessitate that many early trial-based investigations be performed to more rigorous standards than research studies when performed in-house, or by certified specialty clinical or research labs; whereas multisite national labs are often the best equipped to perform a well-established, robust assay in support of large, phase 3 trials. In order to improve reproducibility of data, the qualitative or semiquantitative assessments historically done by pathologists on stained tissue sections will eventually be replaced by whole slide scanning, image segmentation and analysis algorithms to generate fully quantitative data. But, the biological significance of such assessments is often not appreciated when assays are developed, and so the added value of quantitative analysis to each trial must be demonstrated on a case by case basis, especially when considering diagnostic standards for the disease in question as well as whether there is medically justified rationale to convert a trial-based tissue assay into a companion diagnostic test. Finally, a holistic picture of disease progression and response to therapy will emerge by integrating histopathology-based assessment of disease with insights gleaned from diverse technologies.

Acknowledgements

We acknowledge colleagues within Preclinical Safety, Biomarker Development, and Translational Medicine who support our efforts at Novartis/NIBRI. Additional thanks to Steve Potts, Dave Eberhard, Stephen Johnson, and Irina Koroleva for helpful conversations and comments on the chapter.

References

1. Cook D, Brown D, Alexander R, March R, Morgan P, Satterthwaite G, Pangalos MN (2014) Lessons learned from the fate of AstraZeneca's drug pipeline: a five-dimensional framework. Nat Rev Drug Discov 13 (6):419–431. doi:10.1038/nrd4309
2. Fishman MC (2013) Power of rare diseases: found in translation. Sci Transl Med 5 (201):201ps211. doi:10.1126/scitranslmed.3006800
3. Jackson SE, Chester JD (2014) Personalised cancer medicine. Int J Cancer. doi:10.1002/ijc.28940
4. Aronson JK (2005) Biomarkers and surrogate endpoints. Br J Clin Pharmacol 59(5):491–494. doi:10.1111/j.1365-2125.2005.02435.x
5. Lee JW, Devanarayan V, Barrett YC, Weiner R, Allinson J, Fountain S, Keller S, Weinryb I, Green M, Duan L, Rogers JA, Millham R, O'Brien PJ, Sailstad J, Khan M, Ray C, Wagner JA (2006) Fit-for-purpose method development and validation for successful biomarker measurement. Pharm Res 23(2):312–328. doi:10.1007/s11095-005-9045-3
6. Hofmann M, Stoss O, Shi D, Buttner R, van de Vijver M, Kim W, Ochiai A, Ruschoff J, Henkel T (2008) Assessment of a HER2 scoring system for gastric cancer: results from a validation study. Histopathology 52(7):797–805. doi:10.1111/j.1365-2559.2008.03028.x
7. Sequist LV, Waltman BA, Dias-Santagata D, Digumarthy S, Turke AB, Fidias P, Bergethon K, Shaw AT, Gettinger S, Cosper AK, Akhavanfard S, Heist RS, Temel J, Christensen JG, Wain JC, Lynch TJ, Vernovsky K, Mark EJ, Lanuti M, Iafrate AJ, Mino-Kenudson M, Engelman JA (2011) Genotypic and histological evolution of lung cancers acquiring resistance to EGFR inhibitors. Sci Transl Med 3 (75):75ra26. doi:10.1126/scitranslmed.3002003
8. De Mattos-Arruda L, Cortes J, Santarpia L, Vivancos A, Tabernero J, Reis-Filho JS, Seoane J (2013) Circulating tumour cells and cell-free DNA as tools for managing breast cancer. Nat Rev Clin Oncol 10(7):377–389. doi:10.1038/nrclinonc.2013.80
9. Lane RF, Friedman LG, Keith C, Braithwaite SP, Frearson JA, Lowe DA, Longo FM, Refolo LM, Watterson DM, Tsaioun K, Shineman DW, Fillit HM (2013) Optimizing the use of CROs by academia and small companies. Nat Rev Drug Discov 12(7):487–488. doi:10.1038/nrd4057
10. Peppercorn J, Shapira I, Collyar D, Deshields T, Lin N, Krop I, Grunwald H, Friedman P, Partridge AH, Schilsky RL, Bertagnolli MM (2010) Ethics of mandatory research biopsy for correlative end points within clinical trials in oncology. J Clin Oncol 28(15):2635–2640. doi:10.1200/JCO.2009.27.2443
11. Beach TG, Adler CH, Sue LI, Vedders L, Lue L, White Iii CL, Akiyama H, Caviness JN, Shill HA, Sabbagh MN, Walker DG, Arizona Parkinson's Disease Consortium (2010) Multiorgan distribution of phosphorylated alpha-synuclein histopathology in subjects with Lewy body disorders. Acta Neuropathol 119 (6):689–702. doi:10.1007/s00401-010-0664-3
12. Staff S, Kujala P, Karhu R, Rokman A, Ilvesaro J, Kares S, Isola J (2013) Preservation of nucleic acids and tissue morphology in paraffin-embedded clinical samples: comparison of five molecular fixatives. J Clin Pathol 66(9):807–810. doi:10.1136/jclinpath-2012-201283
13. Belloni B, Lambertini C, Nuciforo P, Phillips J, Bruening E, Wong S, Dummer R (2013) Will PAXgene substitute formalin? A morphological and molecular comparative study using a new fixative system. J Clin Pathol 66(2):124–135. doi:10.1136/jclinpath-2012-200983
14. Markowitz GS, D'Agati VD (2009) Classification of lupus nephritis. Curr Opin Nephrol Hypertens 18(3):220–225
15. Haratake J, Horie A, Tokudome S, Era S, Fujii H, Kawachi J, Miyamoto Y, Suko S, Tokunaga M, Tsuji K et al (1987) Inter- and intrapathologist variability in histologic diagnoses

of lung cancer. Acta Pathol Jpn 37 (7):1053–1060

16. Morton D, Sellers RS, Barale-Thomas E, Bolon B, George C, Hardisty JF, Irizarry A, McKay JS, Odin M, Teranishi M (2010) Recommendations for pathology peer review. Toxicol Pathol 38(7):1118–1127. doi:10.1177/0192623310383991
17. Boyce RW, Dorph-Petersen KA, Lyck L, Gundersen HJ (2010) Design-based stereology: introduction to basic concepts and practical approaches for estimation of cell number. Toxicol Pathol 38(7):1011–1025. doi:10.1177/0192623310385140
18. Dunstan RW, Wharton KA Jr, Quigley C, Lowe A (2011) The use of immunohistochemistry for biomarker assessment–can it compete with other technologies? Toxicol Pathol 39 (6):988–1002. doi:10.1177/0192623311419163

Methods in Pharmacology and Toxicology (2015): 65–78
DOI 10.1007/7653_2014_20

Published online: 01 October 2014

Histopathology in Mouse Models of Rheumatoid Arthritis

Patrick Caplazi and Lauri Diehl

Abstract

Murine models of rheumatoid arthritis are widely used for mechanistic studies and for the validation of therapeutic targets. Many models exist and can be classified into induced and spontaneous types. Even though models vary considerably in the pathogenesis of lesions, overlapping spectra of morphological features render the commonly used models suitable for standard histopathological examination and scoring strategies.

Key words Mouse models, Autoimmunity, Rheumatoid arthritis, Histopathology

1 Introduction

Rheumatoid arthritis is a complex autoimmune disease with a pathogenesis that involves numerous genetic and environmental factors [1]. The disease manifests predominantly as chronic synovitis of diarthrodial joints, particularly of the hands and feet. Key genetic factors include HLA variants, notably those that encourage antibody reactions towards citrullinated antigens [2, 3]. Rat and mouse models that recapitulate aspects of the human disease, including immunity to citrullinated antigens [4–6], have been in use for decades [7–11]. In general, most of these models are useful to study early events in the joint pathology that may not be accessible for direct investigation in human tissue samples. Examples include intervention studies (preclinical efficacy studies) or studies in genetically engineered mice for elucidation of disease mechanisms or pathways.

For the purpose of this review, the discussion is limited to a selection of commonly used models that rely on systemically induced autoimmunity and often utilize histopathology of the fore and hind paws as one of the main end points. Models that target larger joints, for instance those induced by intra-articular injection of antigen [12] or cytokines [13], are excluded, as are more complex models, e.g., those that rely on adoptive cell transfer [14, 15].

Histomorphologically, lesions in the commonly examined paw joints are fairly stereotypical regardless of the model. Lesions are characterized by periarticular edema, immune cell infiltration (predominantly neutrophilic), intra-articular exudate, cartilage destruction, and bone remodeling. The bone marrow near arthritic joints may be densely populated by neutrophils and their precursors [8]. Indeed, this lesion may precede the occurrence of frank arthritis. Often, periarticular synovial structures (e.g., tendon sheaths) are also affected and may contribute significantly to the soft tissue ("joint") swelling observed grossly in vivo. In fact, tenosynovitis may precede arthritis in some models [16, 17].

Inflammation coincides with periarticular fibroplasia that may peak and persist after the inflammation has started to regress. Fibroplasia in human RA [18, 19] as well as murine models [20] may in part depend on recruitment of bone marrow-derived fibrocytes into the affected joints. Similar to human disease, fibrovascular tissue may form a pannus (Latin: "cloth, rag") that, in concert with inflammatory cells, contributes to cartilage injury [21]. The classic pattern, however, where aggressive fibrovascular pannus grows from the capsular angle along the periosteal surface to extend an avascular aggregate of fibroblasts and macrophages onto and into the articular cartilage surface [22, 23] is usually indistinct in arthritic rodent paws. Rather, aggregates of spindle cells occupy areas of peri- and subcartilaginous osteolysis, thereby obliterating the junction of bone and cartilage. In this process, cartilage is undermined, eroded, lifted, and ultimately lost (Fig. 1). Early subchondral osteolysis has been described in human RA [24], but may be a more prominent feature in RA models due to the relatively

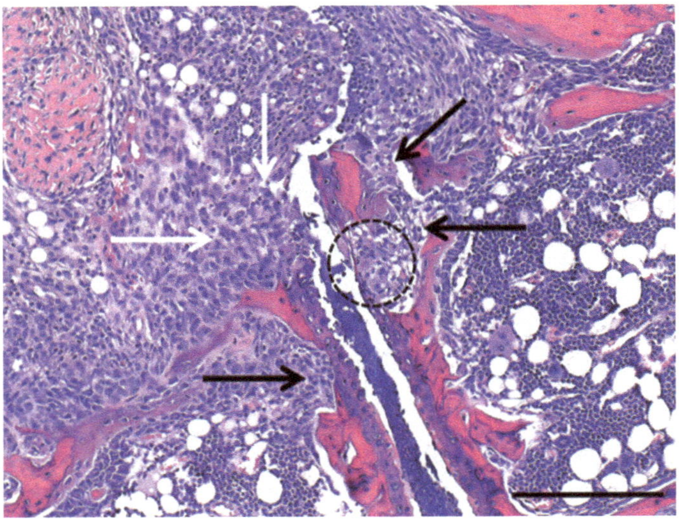

Fig. 1 CAIA, day 14, tarso-metatarsal joint, sagittal, H + E. Morphology of pannus in mouse models of arthritis. Fibroplasia ("non-classical pannus," *black arrows*) approaches the basal surface of the mineralized articular cartilage in foci of subchondral osteolysis resulting in segmental loss of cartilage (*dashed circle*). More classical pannus (*white arrows*), encroaching on the cartilage laterally from the capsular angle, is also seen

delicate subchondral bone in rodents and the proximity of capsular junction and joint surfaces. Cartilage injury occurs regularly in more severely affected joints. Depending on the duration of lesions, severity ranges from superficial erosion, to segmental or complete loss. If segmental or complete loss occurs, and lesions are allowed to mature, partial or complete ankylosis may result. Bone remodeling begins very early in the progression of lesions [13], typically coinciding with spindle cell proliferation and preceding cartilage injury. More substantial osteolysis is typically first visible in the vicinity of cortical blood vessels distal of the epiphyses and then extends along periosteal and endosteal surfaces. Osteolysis is accompanied by predominantly periosteal new bone formation, including formation of cartilaginous or osseous exostoses.

2 Models

2.1 Collagen-Induced Arthritis (CIA)

Collagen-induced polyarthritis (CIA) is one of the most commonly used animal models of rheumatoid arthritis. Murine variations of CIA were developed following the description of the model in rats, which developed erosive polyarthritis after immunization with the cartilage-specific collagen II (but not other collagens) in complete Freund's adjuvant [25]. Type II collagen is the major constituent collagen of articular cartilage and immunoreactivity to type II collagen can be shown in rheumatoid arthritis patients [26, 27]. CIA shares many important features with human RA, including the main morphological and serological features including presence of rheumatoid factor (RF) and anti-citrullinated peptide antibody (ACPA) [28]. Th17 response is essential for the production of lesions in this model [29], while effects of IFN-γ are divergent [28] with enhancement of some aspects of the pathogenesis and suppression of others, resulting in a net modulating effect of endogenous IFN-γ in the model as outlined here [28, 30].

Practical aspects of collagen-induced arthritis have been previously reviewed [31]. The most commonly used variants of the model rely on the use of heterologous collagen II to produce an acute to subacute, monophasic polyarthritis. A generic induction protocol starts by immunizing mice from a susceptible strain (MHC haplotype q, H-2q; e.g., DBA/1) with bovine type II collagen in complete Freund's adjuvant, followed by a booster of type II collagen in incomplete Freund's adjuvant at day 21. Effects of experimental treatment may be studied using either a prophylactic design (treatment initiated before onset of clinical signs at around day 20–30) or a therapeutic design (treatment initiated after onset of clinical signs on animals selected for presence of clinical signs and randomized into treatment groups). Treatment effects can be studied longitudinally in vivo (e.g., caliper measurements of ankle thickness) and terminally (e.g., micro-CT, MRI, histology). In a study with therapeutic design, tissue collection may occur around day 70. For histology, fore and hind paws are collected, fixed in formalin, decalcified, and processed to sagittal hemisections representing larger (e.g., tarsal, carpal) as well as smaller digital joints [8]. Although there is considerable individual variability with regard to onset and incidence, lesions in this model are usually progressive and quite severe with extensive cartilage injury and bone remodeling [32].

2.2 Proteoglycan-Induced

Proteoglycan-induced arthritis (PGIA) is induced by intraperitoneal immunization of H-2d haplotype (e.g., BALB/c) mice with heterologous cartilage-derived proteoglycan and complete Freund's adjuvant. PGIA is a progressive and destructive polyarthritis, characterized by synovitis, cartilage injury, and bone remodeling [33, 34]. Disease onset is delayed compared to CIA, with peak disease occurring around day 50. In contrast to CIA, PGIA is Th1-mediated and requires the presence of IFN-γ [35].

2.3 Antibody Transfer-Induced

Arthritogenic antibodies are produced in a number of RA models (e.g., CIA, GP6PI-induced, K/BxN) and can be exploited for disease induction [36]. A widely used application is collagen antibody-induced arthritis (CAIA) [37, 38]. CAIA as a derivative of CIA has clinical and histologic features similar to CIA [39, 40] and thus replicates features of human rheumatoid arthritis including inflammatory synovitis, pannus formation, cartilage degradation, and bone remodeling. The CAIA model is used when there is need for a rapid/high-throughput model, a need for greater synchronicity in disease onset, or a desire to evaluate arthritis in normally CIA-resistant mouse strains.

CAIA can be induced by serum transfer from arthritic mice [41] or human RA patients [42] but is most commonly induced using a cocktail of monoclonal antibodies which target various regions of type II collagen [37, 43]. The commercial availability of a monoclonal antibody cocktail for disease induction makes this model feasible in most laboratories [40], even though some variability in disease induction and robustness of lesions may occur even with the use of commercial reagents. Administration of lipopolysaccharide (LPS) can be used to reduce the amount of monoclonal antibody required to induce arthritis [44]. Disease has a rapid onset with maximum disease severity reached by approximately day 7–8 following induction depending on mouse strain, age, and sex [37, 40]. Typically, clinical arthritis subsides within about 1 month [38]. Likewise, histological lesions regress depending on their extent and character at peak disease.

The speed of the CAIA model allows for an abbreviated treatment period, which can be very useful when testing expensive or difficult to produce therapeutic molecules; however, the rapid time course is also one of the major caveats of this model. CAIA is an acute arthritis model which relies on immune complex formation and complement activation [45]. CAIA is B- and T-cell independent and does not fully represent the complexity of immune and tissue remodeling responses, which occur in a chronic disease such as human rheumatoid arthritis. Scientific questions to be addressed should be carefully considered as some immunologic pathways and cell types are not represented in this model. Also, the spontaneous remission that occurs in CAIA limits the potential study duration.

2.4 Chemically Induced, Including Adjuvant-Induced

Autoimmune disease, including polyarthritis, may evolve in rodents in the context of chronic inflammation. This principle is exploited by use of oily adjuvants in many of the arthritis models. Pristane is a component of mineral oils and an example of a substance that by itself may induce arthritis in the context of chronic systemic inflammation. The discovery of pristane-induced arthritis (PIA) in mice was made serendipitously during the study of genetic susceptibility of mouse strains to formation of plasmacytomas after intraperitoneal injection of pristane [46]. PIA shares a number of features with human RA, including the presence of autoantibodies. Most important drawbacks include long latency (>50 days) to onset of overt arthritis and the relatively modest severity of lesions. An important advantage is the reported poor response to anti-TNFα therapy, which may render the model valuable for examinations of targets relevant to human anti-TNFα nonresponders [47].

Classical adjuvant-induced arthritis (AIA) by administration of complete Freund's adjuvant is mainly a rat model [48], although variants of this model are also used in mice [49, 50].

2.5 Selected Genetic/ Spontaneous Models

2.5.1 IL-1RA Knock-Out Mouse

IL-1, which includes both IL-1α and IL-1β, is a proinflammatory cytokine that is important in the control and resolution of infection [51]. IL-1 is expressed by a variety of cell types including activated monocytes, macrophages, fibroblasts, and synovial cells which are integral to rheumatoid arthritis-associated inflammation and joint destruction [52].

The presence of naturally occurring IL-1 inhibitors, including IL-1R antagonist (IL-1RA), is critical to avoid excessive inflammatory responses. IL-1RA KO mice backcrossed onto a BALB/cA background spontaneously develop arthritis characterized by swelling and redness in multiple joints, especially tarsal joints [53]. Histologic lesions are characterized by proliferation of synovial lining cells and formation of an inflamed pannus, which results in significant bone and cartilage erosion. These lesions are reported to occur as early as 5 weeks of age with all mice showing lesions by 13 weeks of age. No arthritis was present at 16 weeks of age in KO mice on the C57BL/6J strain, although occasional mice developed disease at more advanced ages [53].

The IL-1RA KO model has a number of interesting features in addition to its relatively early disease onset and phenotypic similarity to human RA. IL-1RA KO mice have rheumatoid factor and develop autoantibodies to double-stranded DNA and type II collagen, although the antibody titers do not always correlate with disease severity [53]. T-cells are important in the pathogenesis of human RA [54–57] and are also central to the pathogenesis of arthritis in IL-1RA KO mice. T-cell deficient IL-1RA KO mice do not develop arthritis, and transferring T-cells from IL-1RA mice to nude mice induces arthritis [58, 59]. Expression of IL-6 and, to a lesser extent, TNF-α is increased in arthritic joints, and IL-17 is required for arthritis development [58–60].

There is considerable evidence that IL-1 plays a role in the pathogenesis of RA [61–64] and the phenotype observed in IL-1RA KO mice recapitulates many of clinical features of human disease. However, while treatment of RA patients with recombinant human IL-1Ra, anakinra, is effective, it has been generally inferior to TNF-α blocking agents [65]. These differences in efficacy may be due to pharmacokinetic challenges [66] or they may reflect other aspects of disease biology. While IL-1 is clearly an important player in RA pathogenesis, it is acting in a complex inflammatory cytokine milieu. Also, the dominant cytokine profile in RA may shift over time, which adds considerably to the complexity of the human disease being modeled [67]. The IL-1RA KO model can be very useful to dissect specific scientific questions, but it may not model the range of inflammatory processes occurring in RA patients.

2.5.2 T-Cell Transgenic Models

F1 offspring of NOD mice crossed with KRN T-cell receptor (TCR) transgenic mice spontaneously develop severe arthritis [68]. The TCR in these K/BxN mice recognizes a peptide of glucose-6-phophoisomerase (G6PI), in the context of MHC-II A^{g7}. Owing to preferential T-cell help to B-cells producing antibodies towards G6PI, K/BxN mice produce large amounts of anti-G6PI antibody, which is arthritogenic, even after transfer to non-transgenic C57/Bl6 recipients [69, 70]. K/BxN arthritis is a severe, destructive arthritis with onset at 4–5 weeks of age. In the antibody-transfer variant, arthritis occurs rapidly within 2 days after induction, peaks within 7–14 days, and wanes by day 21 although residual lesions (e.g., ankylosis) may persist. Anti-GPI antibodies, neutrophils, macrophages, TNF-α, IL-1, and complement contribute to the pathogenesis in K/BxN arthritis [36, 71].

2.5.3 TNF-α Driven Models

Proinflammatory cytokines play an important role in the pathogenesis of rheumatoid arthritis. TNF-α has emerged as a key cytokine in RA as demonstrated by the impact of TNF-inhibitors on disease progression and quality of life measurements [72].

Two murine models have been described that focus on the role of TNF-α in arthritis development—TNF-α transgenic mice and the TNFΔ^{ARE} mouse.

Human TNF-α transgenic (hTNF-tg) mouse strains develop a spontaneous, chronic, and progressive inflammatory polyarthritis that can be prevented by treatment with antibodies against human TNF-α [73]. hTNF-tg mice develop synovial hyperplasia with lymphocytic and neutrophilic infiltrates leading to pannus formation and cartilage destruction. Even though this model differs from human RA due to the initiating role of TNF-α and the absence of detectable rheumatoid factor, it does incorporate a number of features relevant to human rheumatoid arthritis [74]. Despite the lack of rheumatoid factor, hTNF-tg mice develop disease-associated B cell changes, and B cell depletion is protective [75]. Further, hTNF-tg mice develop a complex proinflammatory cytokine response with upregulation of interleukin-1 (IL-1) and interleukin-6 (IL-6) that occurs in conjunction with the development of early lesions [16]. Another caveat of this model is that human TNFα binds only to mouse tumor necrosis factor receptor 1 (TNFR1) and not to tumor necrosis factor receptor 2 (TNFR2) [76], and therefore, the arthritis observed in hTNF-tg mice is mediated exclusively through TNFR1.

Spontaneous arthritis driven by murine TNFα occurs in the murine TNFΔ^{ARE} model. The TNFΔ^{ARE} mouse was created by deletion of the TNF AU-rich elements (ARE), which mediate rapid degradation of TNFα transcripts unless the transcripts are stabilized during cell activation [77]. Because TNF-α is primarily regulated at the posttranscriptional level, ARE loss results in transcript stabilization and chronic TNFα elevation [78].

TNFΔ^{ARE} mice develop chronic inflammatory arthritis, a Crohn's disease-like transmural small intestinal inflammation, and growth retardation [78]. Disease onset is very rapid in homozygous mice, but more protracted in heterozygous animals, which develop overt arthritis at age 6–8 weeks. At week 16, heterozygous wt/TNFΔ^{ARE} mice have polyarthritis involving nearly all paw joints typically examined. Lesions at this stage are chronic and characterized by extensive fibroplasia, cartilage injury, and osteolysis. In contrast to other models, there is only minimal formation of new bone, resulting in the absence of ankylosis in spite of the otherwise severe and destructive character of lesions. Relative lack of new bone formation is likely attributable to the altered Ca^{++} metabolism in these mice, which may also account for the growth retardation [78].

Important features shared between RA and commonly used mouse models are summarized in Table 1.

Table 1
Summary of features seen in RA and reported for the commonly used mouse models [8, 28]

Feature	RA	CIA	CAIA	PGIA	KBxN
Polyarticular	+	+	+	+	+
Symmetrical	+	–	–	–	+ (Distal)
Systemic lesions	+	+ [28, 83]	n/r	n/r	n/r
Progressive	Years	Weeks	–	Weeks	Weeks
Genetic predisposition	MHC-II	H-2	–[a]	H-2	+[b]
Rheumatoid factor	+	+	n/r	n/r	n/r
Anti-collagen II antibody	+	+[b]	+[b]	+ [11]	n/r
Anti-citrullinated peptide antibody	+	+ [84]	–[c]	n/r	n/r

n/r not reported
[a]MHC-independent, but strain and sex differences in susceptibility/severity exist [36, 40]
[b]By definition of the model
[c]Addition of ACCP antibodies aggravates disease [84]

3 Histopathological Evaluation and Scoring

3.1 Technical Considerations

Bolon et al. [8] provide a thorough review of considerations for the selection of models and end points. They also provide detailed technical instructions for the collection and preparation of histological samples. For practical reasons, routine histological examination is most often limited to the paws. Since lesion distribution is often random and nonsymmetrical, there is individual animal variability in number and location of arthritic joints. This is especially true for efficacy studies that contain groups with partially effective treatment. As a consequence, confidence in the assessment of number of affected joints and amplitude of severity is increased or decreased by the number of joints examined. For mouse models, sagittal hemisections of all fore and hind paws collected proximal of the carpal or tarsal joints, respectively, can easily be presented on two standard histological sections, and sections will typically include not only the major joints, but also at least some of the smaller digital joints. Examination of all paws from each animal may not always be possible due to logistic limitations, or because paws are used to collect end points other than histopathology (e.g., RNA, cytokines, FACS). If the latter is the case, care should be taken when interpreting correlations between histopathology and other end points. For example, in our hands up to 30 % of animals in CIA studies may have marked differences in lesion distribution and severity—including

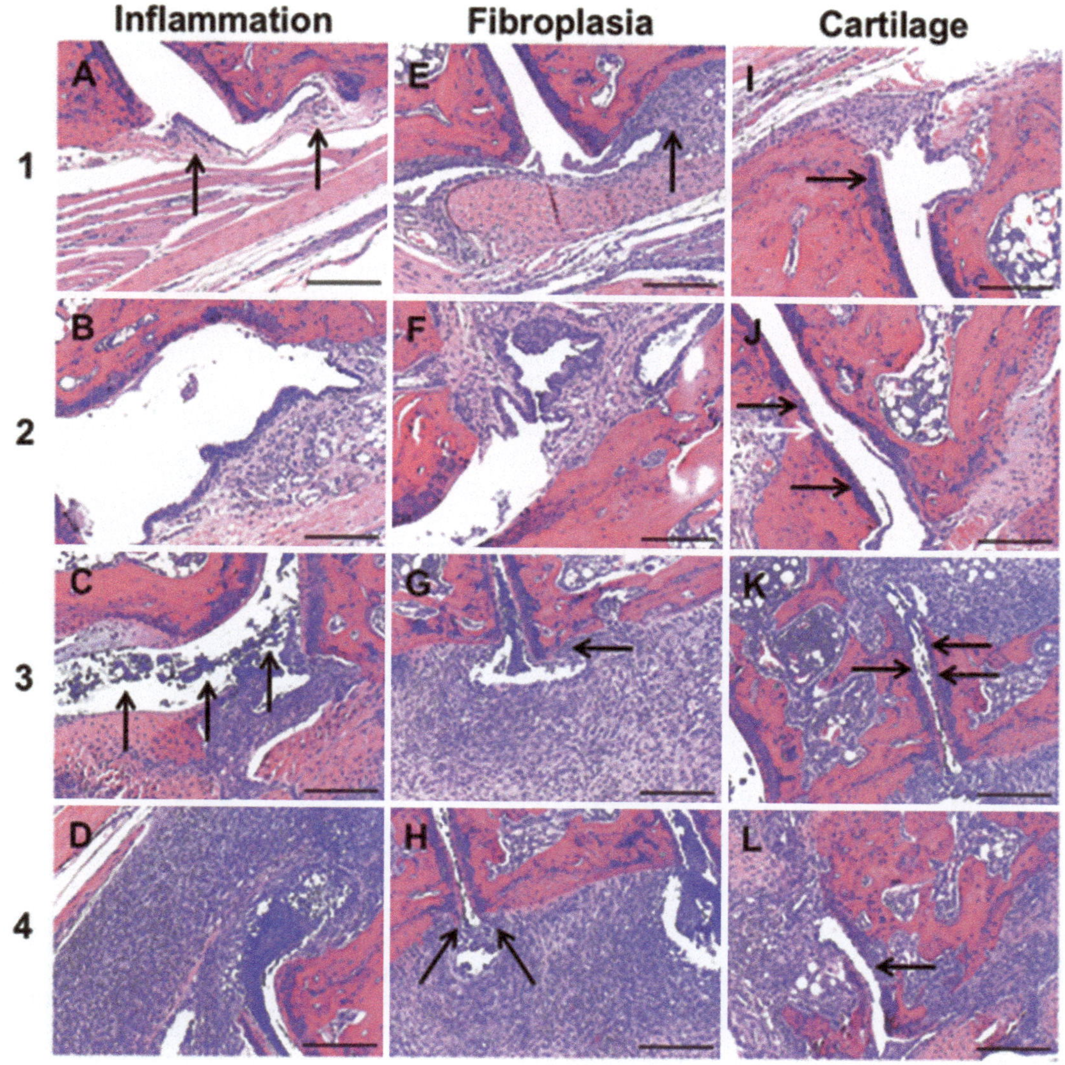

Fig. 2 CAIA, day 14. Hind Paw, sagittal, H + E. Examples of lesion scores 1–4 for inflammation, fibroplasia, and cartilage injury. Bar = 200 μm. (**a**) Inflammation, score 1. Sparse periarticular infiltration with inflammatory cells (*arrows*). The overlying synovial lining has slight hyperplasia and hypertrophy. (**b**) Inflammation, score 2. Sparse periarticular infiltration with inflammatory cells, more extensive than in **a** and with more pronounced synovial hyperplasia, but without formation of intra-articular exudate as in **c**. (**c**) Inflammation, score 3. Focally extensive, densely cellular periarticular infiltration. Overall synovial architecture is preserved, but there is intra-articular exudate and cellular debris. (**d**) Inflammation, score 4. Massive periarticular infiltration, abundant intra-articular exudate. In contrast to **c**, the architecture of the synovium is perturbed. (**e**) Fibroplasia, score 1. Mild focal expansion of the periarticular soft tissue by fibroplasia (*arrow*). (**f**) Fibroplasia, score 2. Mild expansion of the periarticular soft tissue by fibroplasia that is more extensive than in **e** and spans the entire width of the affected joint. There is no pannus. (**g**) Fibroplasia, score 3. There is extensive periarticular fibroplasia. Formation of pannus (*arrow*) is minimal. Overall synovial architecture is only mildly perturbed. (**h**) Fibroplasia, score 4. Extensive periarticular fibroplasia that spans multiple joints. Pannus (*arrows*) and osteolysis are beginning to undermine the articular cartilage. (**i**) Cartilage injury, score 1. Discrete area of chondrocyte degeneration and attenuation of staining restricted to the non-mineralized superficial cartilage. Cartilaginous surfaces are smooth and show no evidence of erosion. (**j**) Cartilage injury, score 2. Lesions as in **i** but multifocal and more extensive. Incipient erosion in the center of the lesion (*white arrow*). (**k**) Cartilage injury, score 3. Focal erosion of superficial non-mineralized cartilage (*arrows*) on opposing portions of articular cartilage. (**l**) Cartilage injury, score 4. A large central segment of cartilage is lost (*arrow*), exposing an area of subchondral osteolysis and fibroplasia

Table 2
Example scoring scheme for mouse RA models

Criterion	Score	Definition
Polyarthritis	0	No lesions in 2 sections representing 1 paw
	1	≤3 joints in 2 sections; restricted to phalangeal joints
	2	≤3 joints in 2 sections; at least one large joint[a]
	3	≤5 joints in 2 sections; at least one large joint
	4	>5 joints in 2 sections; at least two large joints
	5	All joints/2 sections
Soft tissue inflammation[b]	0	None detected in 2 sections representing 1 paw
	1	Sparse focal perisynovial infiltration, with or without hypertrophy of synovial lining
	2	Sparse focally extensive perisynovial infiltration with hypertrophy of synovial lining
	3	Focally/multifocally extensive (peri-)synovial infiltration with synovial hypertrophy and intra-articular exudate
	4	Dense focally/multifocally extensive (peri-)synovial infiltration with disruption of synovial lining; intra-articular exudate
	5	Dense locally extensive and destructive (peri-)synovial infiltration; intra-articular exudate
Fibroplasia[b]	0	None detected in 2 sections representing 1 paw
	1	Sparse focal perisynovial fibroplasia
	2	Sparse focally extensive perisynovial fibroplasia without distortion of local architecture
	3	Sparse focally extensive perisynovial fibroplasia with mild distortion of local architecture; no or minimal pannus
	4	Robust focally extensive perisynovial infiltration spanning multiple joints; pannus, including subcartilaginous, present
	5	Focally extensive, locally destructive fibroplasia
Cartilage injury[b]	0	None detected in 2 sections representing 1 paw
	1	Small area with attenuation of staining; slight superficial roughening
	2	As 1, but more extensive or multifocal
	3	Few erosions, but not beyond tideline
	4	Extensive erosion, segmental loss, lifting
	5	Complete loss
Bone remodeling[b]	0	None detected in 2 sections representing 1 paw
	1	Minimal periosteal osteolysis in areas of fibroplasia at capsular insertion and near cortical blood vessels
	2	As 1, but extending to subchondral bone
	3	Focally extensive periosteal or endosteal osteolysis; at least mild reactive bone formation
	4	Marked osteolysis in areas of fibroplasia distant from the capsular junction and subchondral areas; marked periosteal reactive bone formation
	5	(Incipient) ankylosis

[a]Large joints: tibiotarsal/-talar, tarsal, tarsometatarsal; radiocarpal, carpal, carpometacarpal. Carpal and tarsal joints counted as 1 joint

[b]Score obtained from most severely affected joint

unilateral distribution—between hind paws. Examination is performed on sections stained with hematoxylin and eosin, though other histological stains may give superior contrast for the evaluation of cartilage (e.g., Toluidine blue, Safranin-0).

3.2 Histopathological Features and Scores

In addition to estimating the extent of polyarthritis by counting affected joints per section, histopathological features to evaluate include infiltration with inflammatory cells, fibroplasia and pannus, cartilage injury, and bone remodeling (Fig. 2). While it is common to integrate these features and report them in aggregate arbitrary severity scores (e.g., [31]), such aggregate severity scores do not provide substantially more information than clinical scores (e.g., based on caliper measurements of joint thickness). In addition, results are difficult to replicate between observers. By contrast, reporting of scores for individual features of interest can reduce interobserver discrepancies and provide detail on type and pathogenesis of lesions. The latter can be further augmented by combination with quantitative data based on histomorphometry [79] or provided by imaging modalities such as micro-CT or MRI [80–82]. If overall lesion severity scores are required in efficacy studies to evaluate group differences of treatment effects, such scores can easily be derived by combining individual feature scores into overall lesion scores. An example of a detailed scoring scheme that is applicable for a broad range of lesion severities is given in Table 2. Severity scores for the major scored features are illustrated in Fig. 2.

References

1. Imboden JB (2009) The immunopathogenesis of rheumatoid arthritis. Ann Rev Pathol 4:417–434
2. Eyre S, Bowes J, Diogo D, Lee A, Barton A, Martin P et al (2012) High-density genetic mapping identifies new susceptibility loci for rheumatoid arthritis. Nat Genet 44:1336–1340
3. Raychaudhuri S, Sandor C, Stahl EA, Freudenberg J, Lee H-S, Jia X et al (2012) Five amino acids in three HLA proteins explain most of the association between MHC and seropositive rheumatoid arthritis. Nat Genet 44:291–296
4. Wegner N, Lundberg K, Kinloch A, Fisher B, Malmström V, Feldmann M et al (2010) Autoimmunity to specific citrullinated proteins gives the first clues to the etiology of rheumatoid arthritis. Immuol Rev 233:34–54
5. Cordova KN, Willis VC, Haskins K, Holers VM (2013) A citrullinated fibrinogen-specific T cell line enhances autoimmune arthritis in a mouse model of rheumatoid arthritis. J Immunol 190(4):1457–1465
6. Cantaert T, Teitsma C, Tak PP, Baeten D (2012) Presence and role of anti-citrullinated protein antibodies in experimental arthritis models. Arthritis Rheum. doi: 10.1002/art.37839
7. Sakaguchi S, Sakaguchi N (2005) Animal models of arthritis caused by systemic alteration of the immune system. Curr Opin Immunol 17:589–594
8. Bolon B, Stolina M, King C, Middleton S, Gasser J, Zack D et al (2011) Rodent preclinical models for developing novel antiarthritic molecules: comparative biology and preferred methods for evaluating efficacy. J Biomed Biotechnol 2011:569068
9. Kannan K, Ortmann RA, Kimpel D (2005) Animal models of rheumatoid arthritis and their relevance to human disease. Pathophysiology 12:167–181
10. Lindqvist A (2002) Mouse models for rheumatoid arthritis. Trends Genet 18:S7–S13
11. Joe B, Wilder RL (1999) Animal models of rheumatoid arthritis. Mol Med Today 5:367–369

12. van den Berg WB, Joosten LAB, van Lent PLEM (2007) Murine antigen-induced arthritis. Methods Mol Med 136:243–253
13. Joosten LAB, Netea MG, Kim S-H, Yoon D-Y, Oppers-Walgreen B, Radstake TRD et al (2006) IL-32, a proinflammatory cytokine in rheumatoid arthritis. Proc Natl Acad Sci U S A 103:3298–3303
14. Bárdos T, Mikecz K, Finnegan A, Zhang J, Glant TT (2002) T and B cell recovery in arthritis adoptively transferred to SCID mice: antigen-specific activation is required for restoration of autopathogenic CD4+ Th1 cells in a syngeneic system. J Immunol 168:6013–6021
15. Kis-Toth K, Radacs M, Olasz K, van Eden W, Mikecz K, Glant TT (2012) Arthritogenic T cells drive the recovery of autoantibody-producing B cell homeostasis and the adoptive transfer of arthritis in SCID mice. Int Immunol 24:507–517
16. Hayer S, Redlich K, Korb A, Hermann S, Smolen J, Schett G (2007) Tenosynovitis and osteoclast formation as the initial preclinical changes in a murine model of inflammatory arthritis. Arthritis Rheum 56:79–88
17. Reilkoff RA, Bucala R, Herzog EL (2011) Fibrocytes: emerging effector cells in chronic inflammation. Nat Rev Immunol 11:427–435
18. Galligan CL, Siminovitch KA, Keystone EC, Bykerk V, Perez OD, Fish EN (2010) Fibrocyte activation in rheumatoid arthritis. Rheumatology 49:640–651
19. Galligan CL, Fish EN (2013) The role of circulating fibrocytes in inflammation and autoimmunity. J Leukoc Biol 93:45–50
20. Galligan CL, Fish EN (2012) Circulating fibrocytes contribute to the pathogenesis of collagen antibody-induced arthritis. Arthritis Rheum 64:3583–3593
21. Firestein GS (2003) Evolving concepts of rheumatoid arthritis. Nature 423:356–361
22. Shiozawa S, Tsumiyama K, Yoshida K, Hashiramoto A (2011) Pathogenesis of joint destruction in rheumatoid arthritis. Arch Immunol Exp Ther 59:89–95
23. Kobayashi I, Ziff M (1975) Electron microscopic studies of the cartilage-pannus junction in rheumatoid arthritis. Arthritis Rheum 18:475–483
24. Goldring SR (2002) Pathogenesis of bone erosions in rheumatoid arthritis. Curr Opin Rheumatol 14:406–410
25. Trentham DE, Townes AS, Kang AH (1977) Autoimmunity to type II collagen an experimental model of arthritis. J Exp Med 146:857–868
26. Cook AD, Rowley MJ, Stockman A, Muirden KD, Mackay IR (1994) Specificity of antibodies to type II collagen in early rheumatoid arthritis. J Rheumatol 21:1186–1191
27. Tarkowski A, Klareskog L, Carlsten H, Herberts P, Koopman WJ (1989) Secretion of antibodies to types I and II collagen by synovial tissue cells in patients with rheumatoid arthritis. Arthritis Rheum 32:1087–1092
28. Schurgers E, Billiau A, Matthys P (2011) Collagen-induced arthritis as an animal model for rheumatoid arthritis: focus on interferon-γ. J Interferon Cytokine Res 31:917–926
29. Park MJ, Park HS, Oh HJ, Lim JY, Yoon BY, Kim HY et al (2012) IL-17-deficient allogeneic bone marrow transplantation prevents the induction of collagen-induced arthritis in DBA/1J mice. Exp Mol Med 44:694–705
30. Kelchtermans H, Billiau A, Matthys P (2008) How interferon-gamma keeps autoimmune diseases in check. Trends Immunol 29:479–486
31. Williams RO (2007) Collagen-induced arthritis in mice. Methods Mol Med 136:191–199
32. Wooley PH, Luthra HS, Stuart JM, David CS (1981) Type II collagen-induced arthritis in mice. I. Major histocompatibility complex (I region) linkage and antibody correlates. J Exp Med 154:688–700
33. Glant TT, Mikecz K, Arzoumanian A, Poole AR (1987) Proteoglycan-induced arthritis in BALB/c mice. Clinical features and histopathology. Arthritis Rheum 30:201–212
34. Glant TT, Mikecz K (2004) Proteoglycan aggrecan-induced arthritis: a murine autoimmune model of rheumatoid arthritis. Methods Mol Med 102:313–338
35. Cao Y, Doodes PD, Glant TT, Finnegan A (2008) IL-27 induces a Th1 immune response and susceptibility to experimental arthritis. J Immunol 180:922–930
36. Nandakumar KS, Holmdahl R (2006) Antibody-induced arthritis: disease mechanisms and genes involved at the effector phase of arthritis. Arthritis Res Ther 8:223
37. Nandakumar KS, Svensson L, Holmdahl R (2003) Collagen type II-specific monoclonal antibody-induced arthritis in mice. Description of the disease and the influence of age, sex, and genes. Am J Pathol 163:1827–1837
38. Nandakumar KS, Holmdahl R (2007) Collagen antibody induced arthritis. Methods Mol Med 136:215–223
39. Holmdahl R, Mo J, Nordling C, Larsson P, Jansson L, Goldschmidt T et al (1989) Collagen induced arthritis: an experimental model for rheumatoid arthritis with involvement of

both DTH and immune complex mediated mechanisms. Clin Exp Rheumatol 7:S51–S55
40. Khachigian LM (2006) Collagen antibody-induced arthritis. Nature Protocols 1:2512–2516
41. Stuart JM, Dixon FJ (1983) Serum transfer of collagen-induced arthritis in mice. J Exp Med 158:378–392
42. Wooley PH, Luthra HS, Singh SK, Huse AR, Stuart JM, David CS (1984) Passive transfer of arthritis to mice by injection of human anti-type II collagen antibody. Mayo Clinic Proc 59:737–743
43. Terato K, Hasty KA, Reife RA, Cremer MA, Kang AH, Stuart JM (1992) Induction of arthritis with monoclonal antibodies to collagen. J Immunol 148:2103–2108
44. Terato K, Harper DS, Griffiths MM, Hasty DL, Ye XJ, Cremer MA et al (1995) Collagen-induced arthritis in mice: synergistic effect of E. coli lipopolysaccharide bypasses epitope specificity in the induction of arthritis with monoclonal antibodies to type II collagen. Autoimmunity 22:137–147
45. Banda NK, Hyatt S, Antonioli AH, White JT, Glogowska M, Takahashi K et al (2012) Role of C3a receptors, C5a receptors, and complement protein C6 deficiency in collagen antibody-induced arthritis in mice. J Immunol 188:1469–1478
46. Potter M, Wax JS (1981) Genetics of susceptibility to pristane-induced plasmacytomas in BALB/cAn: reduced susceptibility in BALB/cJ with a brief description of pristane-induced arthritis. J Immunol 127:1591–1595
47. Patten C, Bush K, Rioja I, Morgan R, Wooley P, Trill J et al (2004) Characterization of pristane-induced arthritis, a murine model of chronic disease: Response to antirheumatic agents, expression of joint cytokines, and immunopathology. Arthritis Rheum 50:3334–3345
48. Waksman BH (2002) Immune regulation in adjuvant disease and other arthritis models: relevance to pathogenesis of chronic arthritis. Scand J Immunol 56:12–34
49. Gauldie SD, McQueen DS, Clarke CJ, Chessell IP (2004) A robust model of adjuvant-induced chronic unilateral arthritis in two mouse strains. J Neurosci Methods 139:281–291
50. Parvathy SS, Masocha W (2013) Gait analysis of C57BL/6 mice with complete Freund's adjuvant-induced arthritis using the CatWalk system. BMC Musculoskelet Disord 14:14
51. Miller LS, O'Connell RM, Gutierrez MA, Pietras EM, Shahangian A, Gross CE et al (2006) MyD88 mediates neutrophil recruitment initiated by IL-1R but not TLR2 activation in immunity against staphylococcus aureus. Immunity 24:79–91
52. Dayer J-M, Bresnihan B (2002) Targeting Interleukin-1 in treatment of rheumatoid arthrtitis. Arthritis Rheum 46:574–578
53. Horai R, Saijo S, Tanioka H, Nakae S, Sudo K, Okahara A et al (2000) Development of chronic inflammatory arthropathy resembling rheumatoid arthritis in Interleukin 1 receptor antagonist-deficient mice. J Exp Med 191:313–320
54. Leipe J, Skapenko A, Lipsky PE, Schulze-Koops H (2005) Regulatory T cells in rheumatoid arthritis. Arthritis Res Ther 7:93–99
55. Svendsen P, Andersen CB, Willcox N, Coyle AJ, Holmdahl R, Kamradt T et al (2004) Tracking of proinflammatory collagen-specific T cells in early and late collagen-induced arthritis in humanized mice. J Immunol 173:7037–7045
56. Cope AP (2008) T cells in rheumatoid arthritis. Arthritis Res Ther. 10
57. Samson M, Audia S, Janikashvili N, Ciudad M, Trad M, Fraszczak J et al (2012) Inhibition of Interleukin-6 function corrects Th17/Treg cell imbalance in patients with rheumatoid arthritis. Arthritis Rheum 64:2499–2503
58. Horai R, Nakajima A, Habiro K, Kotani M, Nakae S, Matsuki T et al (2004) TNFa is crucial for the development of autoimmune arthritis in IL-1 receptor antagonist-deficient mice. J Clin Invest 114:1603–1611
59. Nakae S, Saijo S, Horai R, Sudo K, Mori S, Iwakura Y (2003) IL-17 production from activated T cells is required for the spontaneous development of destructive arthritis in mice deficient in IL-1 receptor antagonist. Proc Natl Acad Sci 100:5896–5990
60. Koenders MI, Devesa I, Marijnissen RJ, Abdollahi-Roodsaz S, Boots AMH, Walgreen B et al (2008) Interleukin-1 drives pathogenic Th17 cells during spontaneous arthritis in Interleukin-1 receptor antagonist-deficient mice. Arthritis Rheum 58:3461–3470
61. Corvaisier M, Delneste Y, Jeanvoine H, Preisser L, Blanchard S, Garo E et al (2012) IL-26 is overexpressed in rheumatoid arthritis and induces proinflammatory cytokine production and Th17 cell generation. PLoS Biol 10
62. Kay J, Calabrese L (2004) The role of interleukin-1 in the pathogenesis of rheumatoid arthritis. Rheumatology 43:iii2–iii9
63. Dayer J-M (2003) The pivotal role of interleukin-1 in the clinical manifestations of rheumatoid arthritis. Rheumatology 42: ii3–ii10

64. Feldmann M, Brennan FM, Maini RN (1996) Role of cytokines in rheumatoid arthritis. Ann Rev Immunol 14:397–440
65. Kalliolias GD, Liossis S-NC (2008) The future of the IL-1 receptor antagonist anakinra: from rheumatoid arthritis to adult-onset Still's disease and systemic-onset juvenile idiopathic arthritis. Expert Opin Investig Drugs 17:349–359
66. Liu M, Huang Y, Hu L, Liu G, Hu X, Liu D et al (2012) Selective delivery of interleukine-1 receptor antagonist to inflamed joint by albumin fusion. BMC Biotechnol 12:68
67. Ferraccioli G, Bracci-Laudiero L, Alivernini S, Gremese E, Tolusso B, De Benedetti F (2010) Interleukin-1b and Interleukin-6 in arthritis animal models: roles in early phase of transition from acute to chronic inflammation and relevance for human rheumatoid arthritis. Mol Med 16:552–557
68. Monach P, Hattori K, Huang H, Hyatt E, Morse J, Nguyen L et al (2007) The K/BxN mouse model of inflammatory arthritis: theory and practice. Methods Mol Med 136:269–282
69. Kouskoff V, Korganow AS, Duchatelle V, Degott C, Benoist C, Mathis D (1996) Organ-specific disease provoked by systemic autoimmunity. Cell 87:811–822
70. Maccioni M, Zeder-Lutz G, Huang H, Ebel C, Gerber P, Hergueux J et al (2002) Arthritogenic monoclonal antibodies from K/BxN mice. J Exp Med 195:1071–1077
71. Sakaguchi S, Benham H, Cope AP, Thomas R (2012) T-cell receptor signaling and the pathogenesis of autoimmune arthritis: insights from mouse and man. Immunol Cell Biol 90:277–287
72. Lekander I, Borgstrom F, Lysholm J, van Vollenhoven RF, Lindblad S, Geborek P et al. The cost-effectiveness of TNF-inhibitors for the treatment of rheumatoid arthritis in Swedish clinical practice. Eur J Health Econ 2012
73. Keffer J, Probert L, Cazlaris H, Georgopoulos S, Kaslaris E, Kioussis D et al (1991) Transgenic mice expressing human tumour necrosis factor: a predictive genetic model of arthritis. EMBO J 10:4025–4031
74. Li P, Schwarz EM (2003) The TNF-a transgenic mouse model of inflammatory arthritis. Springer Semin Immunopathol 25:19–33
75. Li J, Kuzin I, Moshkani S, Proulx ST, Xing L, Skrombolas D et al (2010) Expanded CD23+/CD21hi B cells in inflamed lymph nodes are associated with the onset of inflammatory-erosive arthritis in TNF-transgenic mice and are targets of anti-CD20 therapy. J Immunol 184:6142–6150
76. Lewis M, Tartaglia LA, Lee A, Bennett GL, Rice GC, Wong GH et al (1991) Cloning and expression of cDNAs for two distinct murine tumor necrosis factor receptors demonstrate one receptor is species specific. Proc Natl Acad Sci U S A 88:2830–2834
77. Carballo E, Lai WS, Blackshear PJ (1998) Feedback inhibition of macrophage tumor necrosis factor-alpha production by tristetraprolin. Science 281:1001–1005
78. Kontoyiannis D, Pasparakis M, Pizarro TT, Cominelli F, Kollias G (1999) Impaired on/off regulation of TNF biosynthesis in mice lacking TNF AU-rich elements: implications for joint and gut-associated immunopathologies. Immunity 10:387–398
79. Buttgereit F, Zhou H, Kalak R, Gaber T, Spies CM, Huscher D et al (2009) Transgenic disruption of glucocorticoid signaling in mature osteoblasts and osteocytes attenuates K/BxN mouse serum-induced arthritis in vivo. Arthritis Rheum 60:1998–2007
80. Nishida S, Tsurukami H, Sakai A, Sakata T, Ikeda S, Tanaka M et al (2002) Stage-dependent changes in trabecular bone turnover and osteogenic capacity of marrow cells during development of type II collagen-induced arthritis in mice. Bone 30:872–879
81. Barck KH, Lee WP, Diehl LJ, Ross J, Gribling P, Zhang Y et al (2004) Quantification of cortical bone loss and repair for therapeutic evaluation in collagen-induced arthritis, by micro-computed tomography and automated image analysis. Arthritis Rheum 50:3377–3386
82. Lee S-W, Greve JM, Leaffer D, Lollini L, Bailey P, Gold GE et al (2008) Early findings of small-animal MRI and small-animal computed tomography correlate with histological changes in a rat model of rheumatoid arthritis. NMR Biomed 21:527–536
83. Schurgers E, Mertens F, Vanoirbeek JAJ, Put S, Mitera T, Langhe ED et al (2012) Pulmonary inflammation in mice with collagen-induced arthritis is conditioned by complete Freund's adjuvant and regulated by endogenous IFN-γ. Cellular Immune Response 42:3223–3234
84. Kuhn KA, Kulik L, Tomooka B, Braschler KJ, Arend WP, Robinson WH et al (2006) Antibodies against citrullinated proteins enhance tissue injury in experimental autoimmune arthritis. J Clin Invest 116:961–973

Methods in Pharmacology and Toxicology (2015): 79–85
DOI 10.1007/7653_2014_32

Published online: 02 December 2014

Markers Used for Visualization and Quantification of Blood and Lymphatic Vessels

Mohamed E. Salama, David A. Eberhard, and Steven J. Potts

Abstract

In order to study angiogenesis and vascular networks in histology samples, an appropriate stain for endothelial cell surfaces is required. This chapter reviews the various possible antibodies used to identify blood and lymph vessel endothelium, and other vascular markers in both immunohistochemical and immunofluorescent studies. Two categories of human endothelial cell-specific antibodies are commonly used: the pan-endothelial cell markers and antibodies that bind selectively to activated or proliferating endothelium. The chapter concludes with a review and discussion of potential clinical uses of angiogenesis quantification.

Key words Angiogenesis, Microvessel density, CD31, CD34, Vascular, Digital pathology, Tissue image analysis

1 Introduction

Tumor microvessel density (MVD) has been reported to be a significant predictor of overall and relapse-free survival in various tumors [1, 2]. However, other studies reported contrasting result where MVD failed to predict outcome [3, 4]. In some tumors such as clear cell renal cell carcinoma, two categories of tumor vessels correlate with patient prognosis in contrasting ways [5]. Although these differences could be due to distinct tumor biologic behavior, it also may reflect variation in investigators methods to detect and quantify these vessels. Thus a comprehensive understanding of the specificity and limitations of markers and techniques that are used in detection of vessels is crucial first step in the evaluation. In this chapter, the markers available for use in immunohistochemical and immunofluorescence evaluation of tissues are reviewed. In the following chapter, traditional as well as computer-based quantification strategies will be discussed in detail.

Immunohistochemical staining of endothelial cell-specific antibodies is the most widely used approaches to detect blood vessels and study angiogenesis in histologic samples, particularly in paraffin-embedded archived samples. Immunofluorescence is a very useful supplemental technique that allows simultaneous analysis and comparison of multiple markers in a single non-fixed tissue section.

Two categories of human endothelial cell-specific antibodies are commonly used (*see* Table 1): the pan-endothelial cell markers and antibodies that bind selectively to activated or proliferating endothelium.

Table 1
Commonly used endothelial cell-specific antibodies

		Fixation reactivity		
Antibody	**Expression**	**Frozen**	**Paraffin**	**References**
Anti-CD31: (PECAM-1, platelet endothelial cell adhesion molecule)	Expressed on the surface of endothelial cells, but is also weakly expressed on many peripheral leukocytes (macrophages, granulocytes, lymphocytes) and platelets Blood vessels stain strongly Low expression on lymphatic vessels Expressed on plasma cells, alveoli and alveolar ducts, glomerular cells	+	+	[6–8]
CD34	Expressed on blood vascular endothelial cells Depending on its glycosylation, it can serve as a ligand for L-selectin Expressed on hematopoietic precursors and stromal cells	+	+	[9]
Anti-FVIII-RAg	Variable staining of large vessels and capillaries Monoclonal antibody more specific Stromal staining with polyclonal antibody	+	+	[10]
VE-cadherin (CD144): endothelial cell-specific	Stains blood vessels as well as lymphatic vessels	+	+	[8, 11]
MECA-32	Expressed by most blood vascular endothelial cells	+	+	[12, 13]
CD105	Small intra-tumoral vessels positive Large inter-tumoral vessels negative Weak normal tissue vessel staining	+	Fixation depended	[14, 15]
D2-40 (podoplanin)	Selective for lymphatic vessels	+	+	[16, 17]

2 Endothelial Cell Markers

The pan-endothelial markers show equal intensity of staining for small and large vessels and are reactive to both fresh and fixed specimens, whereas other antibodies. Bind selectively to activated or proliferating endothelium but show little or no staining of lymphatics and normal quiescent blood vessels. Thus, these markers can distinguish between tumor neovascularization and preexisting vessels. This distinction is very important in the assessment of tumor angiogenesis and could provide more accurate prognostic information. These markers mainly react with fresh or frozen tissues, while; their activity in paraffin-embedded samples is fixation dependent.

CD31 (PECAM-1) and CD34 are utilized as the pan-endothelial markers of choice for paraffin sections. Anti-CD31 may be preferred over anti-CD34 due to its more selective expression on blood vessels and not lymphatic vessels. However, CD31 can be expressed in various immune and inflammatory cell types as well as alveoli and glomerular cells [18]. In contrast, anti-CD34 will highlight stem cells and progenitor cells in tissues such as bone marrow [9]. Nonvascular cells or structures with positive staining can often be distinguished from endothelial cells on the basis of morphological differences. However, this distinction is not always obvious, and the nonspecific expression of these markers can result in false higher quantification of vessels if not taken into account depending on the method adopted for quantification. Approaches to identify false positive results will be discussed in details in the next chapter. Another common limitation of CD31 is frequent antigen loss due to fixatives that contain acetic acid [19], though adequate antigen retrieval may eliminate this problem. In conclusion, careful selection of the most suitable marker to be used should be planned depending on the tissue type and fixation status.

Other markers extensively reported in the literature and utilized for quantification of angiogenesis include FVIII-RA, VE-cadherin (CD144), MECA-32, UAE-1, and thrombomodulin. Recently, nestin expression in endothelial cells accompanying the process of angiogenesis has been reported [31]. Lymphatic vascular endothelial hyaluronan receptor-1 (LYVE-1), Podoplanin, and Prox1 are available lymphatic vascular markers that produce strong and specific signals in immunofluorescence on mouse tissue sections [20–24].

A number of markers have been reported to be upregulated in tumor-associated activated endothelium compared with normal tissues and include EN7/44, CD105 (endoglin), TEC-11, EN7/44, Integrin aV3, and endosialin. Some markers, such as CD105, a TGF-b-receptor, have shown correlation with outcome. CD105 antibodies have shown a greater specificity for tumor vasculature in comparison to pan-endothelial markers. In a study of 106 patients with breast carcinoma, the MVD was assessed using a pan-endothelial marker CD34 and a monoclonal antibody to CD105. In contrast to CD34, MVD values for CD105 expression showed statistically significant correlation with relapse-free survival and overall survival [25]. The superiority of the anti-CD105 antibody was confirmed on other studies; thus, CD105 immerged as a promising vascular target that can be used for tumor imaging, prognosis, and proposed to bear therapeutic potential in patients with solid tumors and other angiogenic processes [15].

3 Endothelial Cell Proliferation Fraction

In addition to MVD, the number of proliferating endothelial cells in the tissue can be quantified. To detect proliferating cells in tissue sections, staining for Bromodeoxyuridine (5-bromo-2-deoxyuridine, BrdU) or for Ki67 can be used. BrdU is a synthetic analogue of thymidine that is typically administered systemically to the experimental animal several hours before tissue acquisition. During this time, BrdU will be incorporated into newly synthesized DNA in replicating cells, such as proliferating endothelial cells, where it can later be detected by staining tissue sections with a BrdU-specific antibody.

The Ki67 (MIB-1) antigen is a nuclear protein that is expressed by proliferating cells in all phases of the active cell cycle, and therefore can be used as an endogenous marker to detect proliferating cells in the tissue. Double immunohistochemistry or immunofluorescence for antibodies to CD31 or CD34 (for endothelium) and BrdU or MIB-1 (as proliferation markers) are required to quantify endothelial cells proliferation index.

4 Potential Clinical Use of Angiogenesis Quantification

Methods to quantitatively assess angiogenesis in histologic samples will be discussed in detail in the next chapter. These methods apply a variety of techniques to assess different histomorphologic parameters that may be biologically informative, but all depend on the use of vessel markers described here in order to identify and visualize the particular vessels of interest. Each of these methods has a potential for clinical application in a range of diagnostic and research applications. Some of these methods have been reported with convincing data suggesting that the parameter and method could be implemented to provide prognostic information in clinically relevant settings [16, 26, 27]. Further work has aimed to make clinical implementation more practical by identifying angiogenesis-related features that can be easily assessed in routine surgical pathology specimens without using special markers. For example, the presence of fibrotic foci on standard hematoxylin-eosin stained slides can serve as a surrogate marker of high Chalkley count and prognostic factor in breast cancer [19]. In the research direction, our understanding of vessel biology will be deepened by the development of more refined methods to assess vessel subpopulations including lymphatic versus blood vessels [16], endothelial cell apoptosis or proliferative fractions, or multiple markers to assess structural components of vessels such as pericytes and smooth muscle cells as well as endothelial cells and the relationships of other microenvironment cells such as immune and inflammatory cells and tumor cells to vessels. Advances in double immunohistochemical staining and multiple immunofluorescences labeling in conjunction with the novel application of whole slide imaging and image analysis are emerging as promising tools that will expand the utility of these applications. Extending these applications beyond quantification into studying the spatial distribution and histotopography is expanding our understanding of the disease biology and sheds light on new therapeutic approaches.

Arrays of antiangiogenic inhibitors, either as monotherapy or in combination with other cytotoxic and chemotherapy drugs, have been approved or are in clinical trials for the treatment of cancer and other indications. However, the expense of these drugs and the potential for significant, even fatal, treatment-associated adverse events create a strong need for improved ways to select patients who are most likely to benefit and least likely to be harmed by their use. Furthermore, the problem of tumor resistance to therapy continues to be a challenge. Thus, despite the recent advances in knowledge of various antiangiogenic agents, and understanding the molecular mechanisms and potential targets by which these agents elicit their responses, there is continuous need of laboratory research and well-designed studies to improve on the existing agents and approaches for the use of antiangiogenic agents for cancer treatment.

Three major classes of antiangiogenic agents include antibodies that target and neutralize Vascular Endothelial Growth Factor (VEGF) such as bevacizumab and ranibizumab, VEGF decoy receptors such as aflibercept (also known as VEGF-Trap), and small molecule tyrosine kinase inhibitors of the receptors for VEGF, PDGF, or other angiogenic molecules (e.g., imatinib, sorafenib, sunitinib, axitinib, and pazopanib) [28]. Significant relationships have been found between tumor vessel measurements and the expression of angiogenic ligands and their receptors that represent drug targets, and to polymorphisms in the sequences of the receptors [16, 29]. However, investigations into the potential of tumor tissue MVD to serve as a predictive marker for response to antiangiogenic therapies have often not been fruitful [26]. It may be that specific vessel subsets, identified by particular immunostaining patterns, are more directly related to therapeutic response [30]. Further work is needed to establish the clinical utility of specific approaches to assess and quantify the various vascular subpopulations that can be identified in tissue sections using appropriate immunohistochemical markers.

References

1. Horak ER, Leek R, Klenk N, LeJeune S, Smith K, Stuart N, Greenall M, Stepniewska K, Harris AL (1992) Angiogenesis, assessed by platelet/endothelial cell adhesion molecule antibodies, as indicator of node metastases and survival in breast cancer. Lancet 340:1120–1124
2. Tanigawa N, Amaya H, Matsumara M, Shimomatsuya T (1997) Association of tumour vasculature with tumour progression and overall survival of patients with non-early gastric carcinomas. Br J Cancer 75:566–571
3. Page DL, Jensen RA (1995) Angiogenesis in human breast carcinoma: what is the question? Hum Pathol 26:1173–1174
4. Lindmark G, Gerdin B, Sundberg C, Påhlman L, Bergström R, Glimelius B (1996) Prognostic significance of the microvascular count in colorectal cancer. J Clin Oncol 14:461–466
5. Qian CN, Huang D, Wondergem B, Teh BT (2009) Complexity of tumor vasculature in clear cell renal cell carcinoma. Cancer 115(10 Suppl):2282–2289
6. Ebata N, Sawa Y, Nodasaka Y, Yamaoka Y, Yoshida S, Totsuka Y (2001) Immunoelectron microscopic study of PECAM-1 expression on lymphatic endothelium of the human tongue. Tissue Cell 33:211–218
7. Ilan N, Madri JA (2003) PECAM-1: old friend, new partners. Curr Opin Cell Biol 15:515–524
8. Baluk P, Fuxe J, Hashizume H, Romano T, Lashnits E, Butz S, Vestweber D, Corada M, Molendini C, Dejana E, McDonald DM (2007) Functionally specialized junctions between endothelial cells of lymphatic vessels. J Exp Med 204:2349–2362
9. Baumhueter S, Dybdal N, Kyle C, Lasky LA (1994) Global vascular expression of murine CD34, a sialomucin-like endothelial ligand for L-selectin. Blood 84:2554–2565
10. Augustin HG, Kozian DH, Johnson RC (1994) Differentiation of endothelial cells: analysis of the constitutive and activated endothelial cell phenotypes. Bioessays 16:901–906
11. Breier G, Breviario F, Caveda L, Berthier R, Schnürch H, Gotsch U, Vestweber D, Risau W, Dejana E (1996) Molecular cloning and expression of murine vascular endothelial-cadherin in early stage development of cardiovascular system. Blood 87(2):630–641
12. Halin C, Tobler NE, Vigl B, Brown LF, Detmar M (2007) VEGF-A produced by chronically inflamed tissue induces lymphangiogenesis in draining lymph nodes. Blood 110:3158–3167
13. Hallmann R, Mayer DN, Berg EL, Broermann R, Butcher EC (1995) Novel mouse endothelial cell surface marker is suppressed during differentiation of the blood–brain barrier. Dev Dyn 202:325–332
14. Hasan J, Byers R, Jayson GC (2002) Intratumoural microvessel density in human solid tumours. Br J Cancer 86:1566–1577
15. Duff SE, Li C, Garland JM, Kumar S (2003) CD105 is important for angiogenesis: evidence and potential applications. FASEB J 17 (9):984–992
16. Choi WW, Lewis MM, Lawson D, Yin-Goen Q, Birdsong GG, Cotsonis GA, Cohen C, Young AN (2005) Angiogenic and lymphangiogenic microvessel density in breast carcinoma: correlation with clinicopathologic parameters and VEGF-family gene expression. Mod Pathol 18(1):143–152
17. Kahn HJ, Bailey D, Marks A (2002) Monoclonal antibody D2-40, a new marker of lymphatic endothelium, reacts with Kaposi's sarcoma and a subset of angiosarcomas. Mod Pathol 15 (4):434–440
18. Fernàndez JE, Deaglio S, Donati D, Beusan IS, Corno F, Aranega A, Forni M, Falini B, Malavasi F (1998) Analysis of the distribution of human CD38 and of its ligand CD31 in normal tissues. J Biol Regul Homeost Agents 12(3):81–91
19. Vermeulen PB et al (2002) Second international consensus on the methodology and criteria of evaluation of angiogenesis quantification in solid human tumours. Eur J Cancer 38:1564–1579
20. Oliver G, Sosa-Pineda B, Geisendorf S, Spana EP, Doe CQ, Gruss P (1993) Prox 1, a prospero-related homeobox gene expressed during mouse development. Mech Dev 44:3–16
21. Wigle JT, Oliver G (1999) Proxl function is required for the development of the murine lymphatic system. Cell 98:769–778
22. Banerji S, Ni J, Wang SX, Clasper S, Su J, Tammi R, Jones M, Jackson DG (1999)

LYVE-1, a new homologue of the CD44 glycoprotein, is a lymph-specific receptor for hyaluronan. J Cell Biol 144:789–801

23. Breiteneder-Geleff S, Soleiman A, Kowalski H, Horvat R, Amann G, Kriehuber E, Diem K, Weninger W, Tschachler E, Alitalo K, Kerjaschki D (1999) Angiosarcomas express mixed endothelial phenotypes of blood and lymphatic capillaries: podoplanin as a specific marker for lymphatic endothelium. Am J Pathol 154:385–394
24. Schacht V, Ramirez MI, Hong YK, Hirakawa S, Feng D, Harvey N, Williams M, Dvorak AM, Dvorak HF, Oliver G, Detmar M (2003) T1alpha/podoplanin deficiency disrupts normal lymphatic vasculature formation and causes lymphedema. EMBO J 22:3546–3556
25. Kumar S, Ghellal A, Li C, Byrne G, Haboubi N, Wang JM, Bundred N (1999) Breast carcinoma: vascular density determined using CD105 antibody correlates with tumor prognosis. Cancer Res 59:856–861
26. Hlatky L, Hahnfeldt P, Folkman J (2002) Clinical application of antiangiogenic therapy: microvessel density, what it does and doesn't tell us. J Natl Cancer Inst 94(12):883–893
27. Uzzan B, Nicolas P, Cucherat M, Perret GY (2004) Microvessel density as a prognostic factor in women with breast cancer: a systematic review of the literature and meta-analysis. Cancer Res 64(9):2941–2955
28. Al-Husein B, Abdalla M, Trepte M, Deremer DL, Somanath PR (2012) Antiangiogenic therapy for cancer: an update. Pharmacotherapy 32(12):1095–1111
29. Glubb DM, Cerri E, Giese A, Zhang W, Mirza O, Thompson EE, Chen P, Das S, Jassem J, Rzyman W, Lingen MW, Salgia R, Hirsch FR, Dziadziuszko R, Ballmer-Hofer K, Innocenti F (2011) Novel functional germline variants in the VEGF receptor 2 gene and their effect on gene expression and microvessel density in lung cancer. Clin Cancer Res 17(16):5257–5267. doi:10.1158/1078-0432.CCR-11-0379, Epub 2011 Jun 28. PMID: 21712447
30. Zhao YY, Xue C, Jiang W, Zhao HY, Huang Y, Feenstra K, Resau JH, Qian CN, Zhang L (2012) Predictive value of intratumoral microvascular density in patients with advanced non-small cell lung cancer receiving chemotherapy plus bevacizumab. J Thorac Oncol 7(1):71–75. doi:10.1097/JTO.0b013e31823085f4, PMID: 22011670
31. Mokry J, Cizkova D, Filip S et al (2004) Nestin expression by newly formed human blood vessels. Stem Cells Dev 13:658–664

Methods in Pharmacology and Toxicology (2014): 87–100
DOI 10.1007/7653_2014_31

Published online: 02 December 2014

Practical Approaches to Microvessel Analysis: Hotspots, Microvessel Density, and Vessel Proximity

Steven J. Potts, David A. Eberhard, and Mohamed E. Salama

Abstract

This chapter reviews the approaches that have evolved over the last several decades to make quantitative measurements related to vascularity and angiogenesis in histology tissue sections. The early work focused on hotspot analysis, measuring antibody stained vessels in areas of intense vasculature on a tissue section. With the advent of whole slide scanning, newer algorithms could perform whole tissue section analysis, and quantify vascularity across an entire tumor section or other larger region of a microscope slide. The newest technique, one employed by the authors, uses vessel proximity as an approach to quantitation, where the percentage of a given cell type is quantified based on the degree of access to the vascular network. An example error analysis evaluation is provided, to assist in estimating the degree of error for a given quantitation approach. Challenges related to false positive detection of positive stained non-endothelial cells are discussed, and a further algorithm example discussed to address this area.

Key words Microvessel density, Hotspots, CD31, CD34, Vessel proximity, Angiogenesis, Tissue image analysis

1 Introduction

The previous chapter reviewed the use of various immunohistochemical and immunofluorescence markers in angiogenesis studies. This chapter discusses the authors' collective experience in the use of quantitative approaches to vascular hotspots, microvessel density (MVD), and a new technique using vessel proximity analysis. As discussed previously, while it has been established in hundreds of studies that microvessel density can be a useful prognostic factor, it is not necessarily a useful indicator of antiangiogenic therapy. A pharmaceutical researcher needs to well consider the clinical complexity of anti-angiogenesis therapies and its effect on tumor vasculature. Microvessel density by itself is not a measure of angiogenic activity; at its best it simply provides a two-dimensional snapshot of the number of existing vessels in a given tissue section.

The sheer volume of publications on both prognostic significance and techniques in histopathology angiogenesis can deceptively make it appear either that the problem is well solved or that anti-angiogenic therapy is well understood. One can simplify the area from a pharmaceutical drug development perspective by considering it as three categories of biomarkers, and three overlapping approaches to measuring vascularity.

Vascularity measurements in histology sections can be classified into three categories of biomarkers that collectively serve the prospect of personalized medicine:

(a) *Prognostic.* Will a given patient have a poorer prognosis because of higher levels of MVD? The previous chapter reviewed a large number of papers illustrating that poor prognosis is associated with high levels of vascularity.

(b) *Pharmacodynamic.* Is angiogenic therapy actually creating the desired angiogenic effect?

(c) *Response predictive.* Will a patient benefit from angiogenic therapy?

One can also classify the approaches to measuring vascularity into three categories.

(a) *Hotspots*: Measures the areas of a tumor sample with the highest level of vascularity. This approach has dominated the field since Wiedner's paper in 1991. Counts are made in several fields selected by the researcher as containing the most vascularity. Improvements to counting can be made by including a Chalkey counter. The details are further described in detail later in this review, and the most important thing to keep in mind is that the endpoint is a measurement on the most intense areas of vascularity in a given sample.

(b) *Whole tissue analysis*: Evaluates vascularity across the entire tissue section. Prior to the advent of whole slide scanning, this was done by selecting random fields of view and recording counts of vessels. With image analysis areas of vessels could be reported, and with whole slide scanning either the counts per area or vascular area per area of the entire tumor sample could be reported, excluding necrosis and other artifacts. Regardless of the specific quantitative approach used in this rapidly evolving technology, the endpoint is an average measure of vascularity across an entire tissue section.

(c) *Vessel proximity*: Evaluates the percentage of a given cell type that has access to the vascular network. This is a newer approach that asks not about the vascular network but about the percentage of cells that have access to that network. Known as proximity analysis, it asks a fundamentally different question.

To better understand each of these general approaches, the changes and advancements in the field to date are first reviewed.

2 The Evolution of Vascular Analysis in Histopathology

A manual technique to evaluate vascularity in tissue samples was proposed in 1972, with a subjective composite rating of 0–100 based on vasoproliferation, endothelial cell hyperplasia, and endothelial cytology [1].

3 Hotspots

Weidner developed an MVD method in 1991 [2] where the tissue was surveyed with a 4× objective, and the areas with the most vascularization, or "hotspots," were identified. In these areas, the field of view with the highest vascularization was counted for vessels with a 20× or 40× objective. The second highest field of view is counted, followed by the third, and up to ten fields of view are tabulated in this fashion. While studies have ranged from three to five fields of view or more, most studies utilize the average of the three most vascularized fields of view when reporting results. The counting method itself can be made slightly more objective after the hotspot regions of interest are selected by evaluating the fields of view using a Chalkey grid eyepiece [3, 4]. While the methodology overlap, and both critically depend on evaluating and selecting "hotspots," the Chalkey method can be considered a correlate for vessel area, while MVD is more of a correlate for vessel density.

The field and number of investigations using methodology related to this approach grew tremendously in the following 10 years. A review paper of MVD limited only to breast carcinoma studies illustrates this growth. In 2002, forty-three independent previous studies linking microvessel density to clinical outcome in breast cancer were reviewed, and the clinical utility of the method confirmed as a prognostic factor [5]. Factor VIII was used in 27 of these studies, CD31 in 10, and CD34 in 8. The majority (39 of 43) included measurement from the technique developed by Weidner, three studies included Chalkey methods, and seven studies utilized image analysis for an area-based method. The authors stress the need for better standardization in MVD, as there was high degree of variability in the number of fields observed, and the exact methodology of the counting technique.

The extent that a human observer is unreliable in estimating and identifying hotspots is well illustrated from a paper evaluating MVD in breast carcinomas [6]. In these studies, the manual technique from Weidner was followed, with the observer identifying and then counting vessels in order, for what was perceived as the ten vascularized highest microscope fields of view. The first field counted actually contained the greatest number of microvessels in only 20 % of the sections. In the apparent highest five fields identified by observer, the highest field of view was only found in these five fields 65 % of the time.

Multiple researchers have undertaken the approach to use image analysis for removing the observer variability introduced when attempting to identify hotspots. A semiautomated technique acquired all fields of view from a tissue section, and then identified hotspots based on the higher areas of positive endothelial staining [7]. The technique was improved upon with the introduction of image analysis morphology to use number of vessels per field rather than area when choosing hotspots [8].

Microvessel density, Chalkey counts, and image analysis methods were analyzed in depth and correlations between MVD and Chalkey counts, both relying on the "hot spot" approach, were compared [9]. The use of two versus three fields of view with Chalkey counts was evaluated, and the degree of correlation ($r = 0.93$) was considered high enough that only two fields were recommended as sufficient, although using the top two versus top three fields of view will generally produce slightly higher average values. The publication [9] evaluated the prognostic ability comparison between MVD and Chalkey in large cohorts of prostate, breast, bladder, and non-small cell lung carcinomas. It would be highly instructive to reevaluate these results using new whole slide imaging techniques, and rather than comparing two methods based on hotspots, compare with whole tumor vascularization or vessel proximity.

More recent studies from that era conceptualized image analysis-based total microvascular area (TVA) along with MVD counts [10]. For those thinking of measurements on a digital slide, the name total microvascular area might suggest the total area occupied by vasculature across an entire tumor section, but the methodology also predates digital pathology, and TVA, MVD, and Chalkey counts all use the same approach with selecting hotspots and several fields of view.

Microvessel density, vascular area, Chalkey counting, and stereological area of vascular profiles were compared in breast cancer. The authors found highest reproducibility in Chalkey counting and stereology, and recommended Chalkey counting overall [11]. In a previous study, microvessel density, Chalkey count, and area-based computer image analysis were compared in breast carcinomas, with the authors recommending Chalkey counts [12].

In the analyzed hotspots, some researchers have looked at other image analysis-based measurements of individual vessels. These have included major axis length, minor axis length, perimeter, compactness (perimeter/area), and more esoteric measurements like shape factor and Feret diameter [13, 14].

One researcher derived five classes of microvessel patterns in breast carcinomas and used these for stratification and prognostic outcome. The five classes were (a) increased, blood-filled capillaries with some clustering in the tumor; (b) small-sized capillaries in the tumor; (c) small-sized capillaries condensing at the periphery of the tumor (d); compressed delicate capillaries in the tumor; and (e) compressed delicate capillaries surrounding the tumor islands [15].

Researchers have noted that heterogeneity of vascularity, as measured by the coefficients of variation of microvessel density or area in randomly sampled regions, is lower in tumors compared to normal tissues in prostate [16, 17]. This may be explained by strongly increased levels of angiogenic factors that result in a saturation of the vascular bed. Vessel density may actually exceed metabolic requirements in tumors, and the result is uniform over-vascularization [18].

4 Whole Slide Analysis

With the introduction of digital pathology, the entire slide is available as a digital image for image analysis. This is a vastly different biological endpoint than the preceding hot spot analyses. The entire tumor section is potentially available for sampling, rather than only high areas of vascularity. The technology has been improved over the last several years by a number of software image analysis vendors. Area-based algorithms have been developed initially for ACIS (Chromovision), followed by object-based counting by Aperio [19], and more recently object-based counting by Definiens and Visiopharm. While many researchers have studied in detail the intra-technique differences within hotspots approaches (e.g., MVD versus Chalkey versus TVA) or more recently intra-technique differences within whole slide analysis approaches (vessel counting versus vessel areas), it is surprising how few if any studies have asked whether hotspots themselves are a better technique versus overall vascularity with whole slide analysis.

One should also return to the question of the end purpose when developing an analytical method. Is the goal to record how many vessels are in a given tissue, or is the goal to evaluate what percentage of the tumor or tissue is accessible to the vascular network? Authors have looked at the addition of a perimeter statistic on vessels, as a possible better correlate with oxygenation than number of vessels, but this requires a high degree of computer accuracy in identifying individual vessels. This assumes that large, non-oxygenating vessels are removed (an important software quality control technique we recommend) and also still suffers from the challenge of adequately assigning endothelial cells to vessel counts. When developing a new algorithm we reconsidered this question and introduced a technique that overcomes some of the theoretical and experimental challenges of MVD, but also is a closer representation of tumor hypoxia and oxygenation.

5 Vessel Proximity Analysis

In vessel proximity analysis (VPA), the central question is what percentage of the cells of interest are near a vessel, rather than the number of vessels themselves. The method is outlined below (Figs. 1 and 2), and has been applied successfully to evaluate percentages of myeloma cells near vessels [20].

6 Heterogeneity Approaches

Rather than either looking at hotspots, or considering the full tumor section, one could also utilize heterogeneity measurements of angiogenesis as a potential biomarker. This would be conducted similar to other approaches used in HER2 heterogeneity measurements in breast [21] and gastric cancer samples [22]

7 Example Error Analysis Evaluation

Regardless of the technique chosen, it is important to evaluate the error rates of the computer-based image analysis routine. As an example, the following is an error analysis performed to evaluate the accuracy of AngioMap™ (Flagship Biosciences, Westminster, CO) to correctly identify vascular areas across the whole slide images in multiple myeloma bone marrow samples. Because in the vessel proximity method a segmented vessel (e.g., one vessel being segmented into multiple vessels) does not contribute to error, only two types of errors were counted. A false positive was counted when the computer program identified a vessel where a vessel was not present, for example a positive staining non-endothelial cell. A false

Fig. 1 The "Loch Ness Monster" problem and the value of proximity to vessels. (Drawn by Aaron Young, Flagship Biosciences)

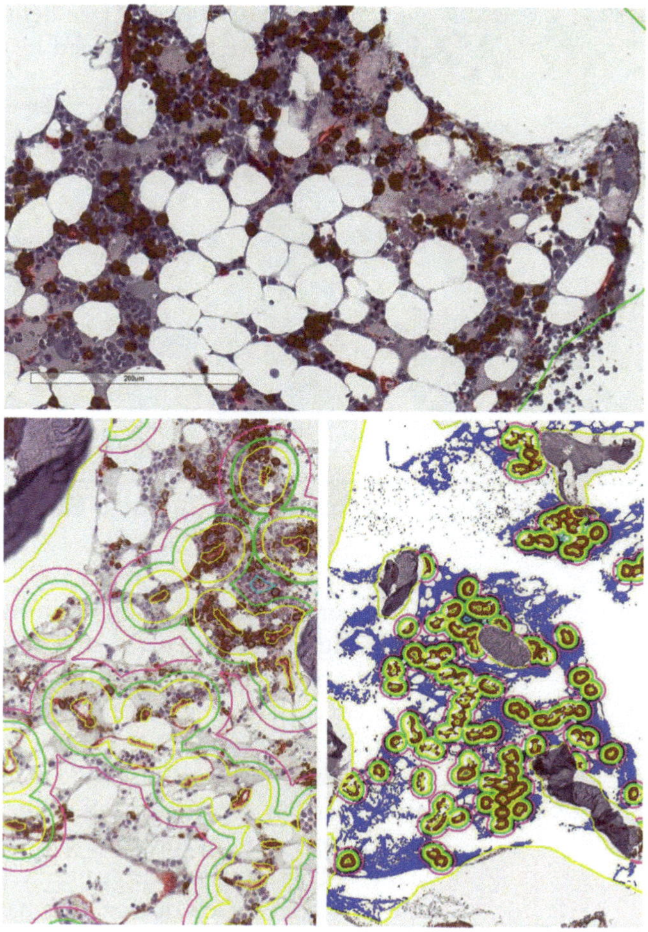

Fig. 2 Vessel Proximity Analysis. *Top*: Dual staining for vessels (CD34, *red*) and myeloma cells (CD138, *brown*). *Bottom left*: Software identifies each vessel, and then computes a given distance from the vessel. *Bottom right*: The number of vessels, geometric information on the vessels, and the percentage of myeloma cells within a given distance of vessels are calculated. Fat vacuoles and other tissue can be included or excluded in the calculation

negative was counted when vessel could be identified by a pathologist but the computer program did not identify it. Since the error analysis requires a pathologist to review every vessel, randomly selected slides and randomly selected regions on slides were chosen based on a methodology used previously [22]. In brief, five slides where chosen from the full slide set using systematic random sampling, and six regions from each of these slides were chosen using a GridMap™ (Flagship Biosciences) program that systematically random samples across the whole slide image. The results are shown in the table below, with error rate defined as the total of false positive and false negative errors divided by the number of vessels:

Slide	Total vessels	Total non-endothelial cells	False positives	False negatives	Error rate
A	70	32	0	4	5.7 %
B	92	48	2	5	7.6
C	186	13	5	4	4.8
D	65	17	0	1	1.5
E	42	67	1	1	4.8
TOTAL	455	177	8	15	5.0 %

8 Using Automated Image Analysis to Optimize Microvessel Detection Specificity and Sensitivity

The examples above describe various approaches for quantitatively analyzing microvessels in order to provide information about tumor angiogenesis. All of these methods rely on immunohistochemistry using an antibody to label the vessels of interest. However, the commonly used antibodies for detecting vessels can also react with certain "off-target," i.e., nonvascular, cell types. Furthermore, various pigments such as melanin, hemosiderin, lipofuscin, and carbon soot (anthracotic pigment) may be detected as positive signals. These types of nonspecific stained objects must be differentiated from vessels during automated image analysis. This is done by adjusting variable parameters such as shape (perimeter/area ratio), size, and RGB intensities in order to optimize the specificity and sensitivity for microvessel detection. As discussed in the previous section, the optimization process generally involves comparing the objects identified as microvessels by the analysis software to those identified manually by a pathologist (the gold standard) [23, 24].

Optimization can be challenging when the goal is to set a standard analysis approach to apply across a large set of samples, because the samples may have differences in biological, preanalytical, and even assay performance variables that may influence the presence, type, and amount of nonspecific staining (detected signal) as well as the intensity of vessel-specific immunostaining. Using tissue microarrays (TMAs) to study large series of cases allows all of the tissue samples in the TMA to be subjected to the IHC assay process at the same time so that assay conditions for the sample set are identical, but cannot address preanalytical or biological variation. Optimizing the image analysis parameters for TMAs could be approached in either of two ways, each with its own set of challenges. One approach is to individually optimize the analysis parameters for each sample; that is, adjusting the image analysis settings, running the analysis, manually inspecting the result, then readjusting the settings, rerunning the analysis, reinspecting, etc. on each tissue core in the TMA. This approach may seem the most accurate but could introduce subjective bias into the process, and is quite tedious both in the optimization process and in the subsequent analyses to be performed across the aggregate sample set. Another approach is to select a few "representative" samples in the TMA on which to optimize, and then apply those optimized settings across the entire TMA so that the entire sample set can be analyzed at once. While more rapid and uniform in the analysis, in this approach one must somehow decide which samples are "representative" for the analysis, again with the potential for subjective bias.

We have developed a simple method to evaluate signal intensity data and quickly identify samples in digital images of TMAs that show significant nonspecific staining, which facilitates establishing optimal vessel detection parameters for microvessel analysis in the TMA. This approach is based on the observation that CD31 staining intensity of endothelial cells often is different (more intense) than that of other cell types such as macrophages and granulocytes, and that the intensity of other pigments, such as carbon debris in the lung, may be higher than endothelial cell (CD31) IHC staining. Therefore, if a variety of stained objects in a sample image are analyzed at different signal detection sensitivity levels (DSLs), weakly stained objects will be identified only at the higher sensitivity settings, whereas strongly stained objects will be identified even at low sensitivity settings. Creating a scatterplot of the object counts of all the samples in TMA, analyzed over a range of DSLs, allows the easy identification of particular samples that show unusual trends across the DSL range that suggest different types of nonspecific staining (Fig. 3). In lung cancers, samples with very high object counts at high DSLs, but not at low DSLs, often contain significant inflammatory cell infiltrates. On the other hand, samples with high object counts at low DSLs often contain large amounts of carbon debris from smoking, dust, or air pollution (Fig. 4). Such "outlier" samples can be used to help establish the optimal DSL and, if necessary, can be excluded from subsequent analyses of the sample series due to unacceptable amounts of nonspecific staining.

9 Discussion

The main endpoint used in MVD has been the number of vessels per mm^2 of tissue section. There are both theoretical and experimental problems with this endpoint. When one considers microvessel density from a stereological viewpoint, recognizing that a two-dimensional

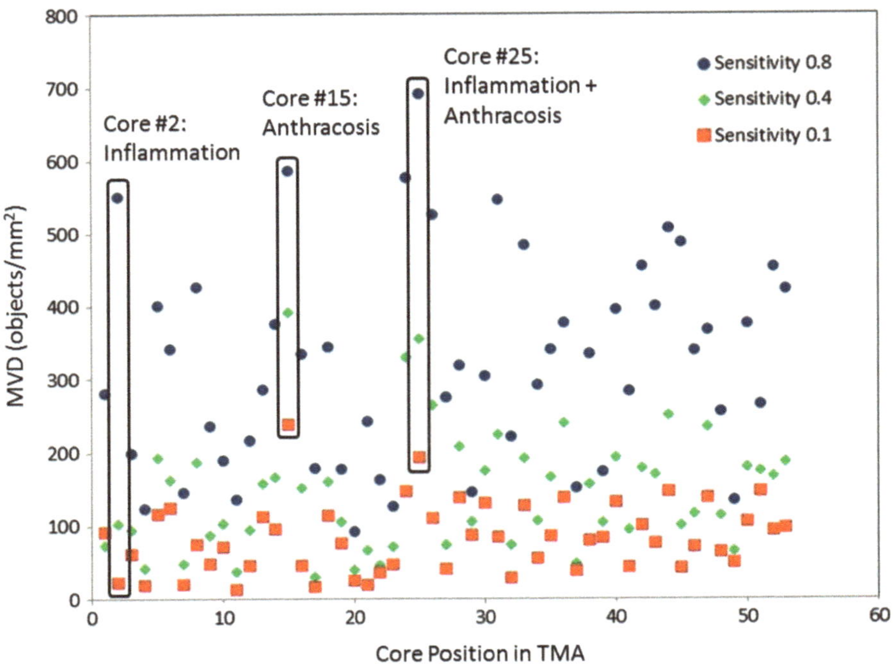

Fig. 3 A scatterplot of the object counts of all the samples in a CD31-stained non-small cell lung cancer (NSCLC) tissue microarray, analyzed over a range of different endothelial stain detection sensitivity levels, allows the easy identification of particular samples that show unusual trends suggesting different types of nonspecific staining

tissue section is only one sample from the three-dimensional tumor, a number of theoretical problems present themselves. Anything observed on a section should be considered a profile, rather than the actual object. Recording the number of vessel profiles per area is not a measurement with roots in reality. Thicker or thinner sections, under or overstraining, higher or lower cellularity in the sample, will all effect this endpoint. One violates all stereological considerations when trying to extrapolate this vessels per area measurement to volume; the best that the statistic can be used for is to compare the effect of one treatment with another, or before/after treatments, rather than as an absolute physical observation.

Experimentally, the difficulty with vessel densities is the ability to adequately number vessels with image analysis (Fig. 1). In tumors with limited vascularity, and only microcapillaries, it may be possibly, but as vascularity increases, it becomes difficult for the pathologist (and especially the computer) to determine which vessel profiles should be part of only one vessel. Many researchers resort to an area measurement to overcome this problem, the area of vessel profiles/area of tissue.

Despite the growing usage of stereology, there have been few publications utilizing stereology in the field of angiogenesis. Techniques like volume-length, as described in the chapter in this

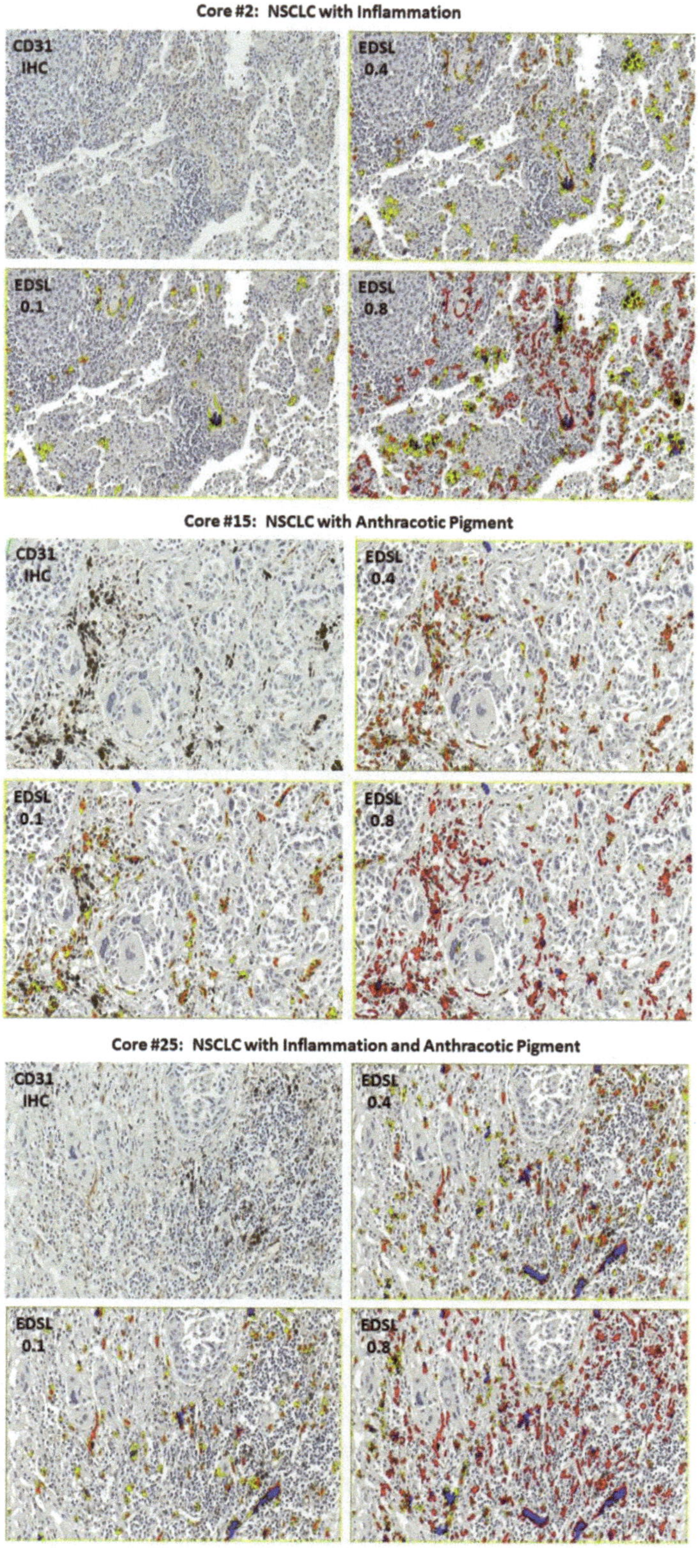

Fig. 4 Images of the outliers in the CD31-stained NSCLC tissue microarray shown in Fig. 3, demonstrating nonspecific detection of inflammation and anthracotic (carbon) pigment at various Endothelial Detection Sensitivity Level settings (EDSL; range 0–1) in the automated image analysis software

series on reticulin, would be useful in many measurements of angiogenesis.

One of the major advantages of using vessel proximity rather than a vessel density count is that errors associated with vessel segmentation are not included in the overall analysis. Previous work with an earlier algorithm [19] evaluated error rates in CD31 stained xenografts and reported an overall error rate of 18 %, with the errors associated with vessel segmentation a substantial contribution to this amount.

10 Practical Recommendations

Use caution when utilizing number estimates of vessel profiles in an endpoint. First conduct an error analysis between computer and pathologist scoring to determine the actual levels of error involved. Generally, area-based measurements will be less prone to error, or use an alternative approach such as proximity analysis.

Consider stereological principals when deciding on a given measurement strategy. The researcher must consider that at best cross sections of a volume of tissue are profiles, and various biases are introduced. Consider using stereology probes or other techniques when possible and ensure that random sampling has been conducted to every degree possible. Avoid reporting measurements like number of vessels per area of slide, and if they are used, limit them to relative comparisons between two treatments.

Utilize a random sampling scheme to avoid bias with pathologists choosing regions of interest. The human eye is naturally drawn toward areas of dilated vessels as well as necrotic areas, and these must be avoided if hot spot identification by human eye is desired. Users must be trained to look for areas with the most number of discrete vessels.

Hot spot approaches, whole tissue analysis, and proximity measurements are fundamentally different measurements. The best approach fits the biology, there may be cases where evaluating hot spots is most appropriate, other cases where statistics of vessels on the entire slide is best, and others where the number of cells (normal or tumor or immune) with access to vessels is most appropriate. One should careful consider which approach to take. Other cases may require only vascular interest in the tumor edges.

Ensure that the right kind of vessels is being utilized in the measurement. Arteries and large veins should not be part of measurements dealing with oxygenation. Software should be used that allows large vessels to be either thresholded out initially or deleted as necessary from the calculation during quality assurance.

Avoid overestimating the value of xenograft data or underestimating the difficulties of angiogenesis measurements in clinical samples. Patient derived or other forms of xenografts are frequently very limited models in angiogenesis. While they are far easier for running angiogenesis histopathology experiments, the information obtained may not be clinically translatable. It is worth the value working in clinical biopsy samples whenever possible. The algorithms will need to be more robust and a greater amount of pathologist supervision is required in clinical biopsies.

Avoid overreliance on one angiogenic factor in developing a tissue endpoint. What is clear from the anti-angiogenesis trial results is that tissues have many ways of stimulating and suppressing angiogenesis, and when one factor is knocked down (e.g., VEGF), inevitably other factors will take its place. The tumor does not rely on only one factor to create a favorable vascular network; neither should the researcher when attempting to monitor that network.

Decouple the evolution of technology and technique with the biological question. If whole slide imaging and analysis had been available to Dr. Judah Folkman's team four decades ago, would the need to measure hotspots still exist? If so, then this is an important biological endpoint to measure, but if not, we need to move on to other approaches. One has to appreciate the technology that was available at the time to previous researchers, but not have our imagination be bound to that technology when matching an endpoint with a biological question.

References

1. Brem S, Cotran R, Folkman J (1972) Tumor angiogenesis: a quantitative method for histologic grading. J Natl Cancer Inst 48(2):347–356
2. Weidner N et al (1991) Tumor angiogenesis and metastasis—correlation in invasive breast carcinoma. N Engl J Med 324(1):1–8
3. Chalkley H (1943) Method for the quantitative morphologic analysis of tissues. J Natl Cancer Inst 4(1):47–53
4. Fox SB et al (1994) Tumor angiogenesis in node-negative breast carcinomas—relationship with epidermal growth factor receptor, estrogen receptor, and survival. Breast Cancer Res Treat 29(1):109–116
5. Uzzan B et al (2004) Microvessel density as a prognostic factor in women with breast cancer a systematic review of the literature and meta-analysis. Cancer Res 64(9):2941–2955
6. Martin L et al (1997) Examining the technique of angiogenesis assessment in invasive breast cancer. Br J Cancer 76(8):1046
7. Van der Laak J et al (1998) An improved procedure to quantify tumor vascularity using true color image analysis: comparison with the manual hot-spot procedure in a human melanoma xenograft model. J Pathol 184:136–143
8. Belien J et al (1999) Fully automated microvessel counting and hot spot selection by image processing of whole tumour sections in invasive breast cancer. J Clin Pathol 52(3):184–192
9. Offersen B, Borre M, Overgaard J (2003) Quantification of angiogenesis as a prognostic marker in human carcinomas: a critical evaluation of histopathological methods for estimation of vascular density. Eur J Cancer 39(7):881–890
10. Sharma S, Sharma M, Sarkar C (2005) Morphology of angiogenesis in human cancer: a conceptual overview, histoprognostic perspective and significance of neoangiogenesis. Histopathology 46(5):481–489
11. Hansen S et al (1998) Angiogenesis in breast cancer: a comparative study of the observer variability of methods for determining microvessel density. Lab Invest 78(12):1563
12. Fox SB et al (1995) Quantitation and prognostic value of breast cancer angiogenesis: comparison of microvessel density, Chalkley count, and computer image analysis. J Pathol 177(3):275–283
13. Korkolopoulou P et al (2003) Clinicopathologic correlations of bone marrow angiogenesis in chronic myeloid leukemia: a morphometric study. Leukemia 17(1):89–97

14. Korkolopoulou P et al (2003) A morphometric study of bone marrow angiogenesis in hairy cell leukaemia with clinicopathological correlations. Br J Haematol 122(6):900–910
15. Safali M et al (2010) A distinct microvascular pattern accompanied by aggressive clinical course in breast carcinomas: a fact or a coincidence? Pathol Res Pract 206(2):93–97
16. van Niekerk CG et al (2009) Computerized whole slide quantification shows increased microvascular density in pT2 prostate cancer as compared to normal prostate tissue. Prostate 69(1):62–69
17. Bigler SA, Deering RE, Brawer MK (1993) Comparison of microscopic vascularity in benign and malignant prostate tissue. Hum Pathol 24(2):220–226
18. Hlatky L, Hahnfeldt P, Folkman J (2002) Clinical application of antiangiogenic therapy: microvessel density, what it does and doesn't tell us. J Natl Cancer Inst 94(12):883–893
19. Potts SJ et al (2009) Performance of a novel automated microvessel analysis algorithm across whole slide digital images. In: Toxicologic pathology. Sage Publications Inc, Thousand Oaks, CA
20. Salama ME, Lange H, Tripp SR, Kohan J, Landis ND, Krueger JS, Potts SJ (2014) AngioMap is a novel image analysis algorithm for assessment of plasma cell distribution within bone marrow vascular niche. Appl Immunohistochem Mol Morphol 22 (7):550–554
21. Potts SJ, Krueger JS, Landis ND, Eberhard DA, Young GD, Schmechel SC, Lange H (2012) Evaluating tumor heterogeneity in immunohistochemistry-stained breast cancer tissue. Lab Invest 92(9):1342–1357
22. Potts SJ et al (2013) Tissue pattern recognition error rates and tumor heterogeneity in gastric cancer. Appl Immunohistochem Mol Morphol 21(1):21–30
23. Chantrain CF, DeClerck YA, Groshen S, McNamara G (2003) Computerized quantification of tissue vascularization using high-resolution slide scanning of whole tumor sections. J Histochem Cytochem 51(2):151–158
24. Mohammed ZM, Orange C, McMillan DC, Mallon E, Doughty JC, Edwards J, Going JJ (2013) Comparison of visual and automated assessment of microvessel density and their impact on outcome in primary operable invasive ductal breast cancer. Hum Pathol 44 (8):1688–1695

Methods in Pharmacology and Toxicology (2015): 101–116
DOI 10.1007/7653_2014_29

Published online: 03 December 2014

Quantitative Histopathology and Alternative Approaches to Assessment of Fibrosis for Drug Development in Hepatitis C and Nonalcoholic Steatohepatitis

Steven J. Potts and Johanna K. DiStefano

Abstract

Fibrosis plays a role in a wide number of diseases and organs, including pulmonary fibrosis, liver cirrhosis, pancreatic cancer, cardiovascular and kidney diseases, macular degeneration, cancer metastasis, and chronic organ transplant rejection. Despite many efforts to develop effective and accurate noninvasive diagnostic and prognostic techniques, histopathology remains the gold standard. This chapter discusses the use of quantitative histology and alternative non-histological measurements, including serum protein and recently identified microRNA markers, as complementary approaches for the diagnosis and staging of liver fibrosis in pharmacological and clinical studies.

Keywords Tissue image analysis, Fibrosis, Cirrhosis, Quantitative assessment, Liver fibrosis

1 Introduction

Fibrosis can be considered an overreaction to normal tissue regeneration. Fibrosis is preceded by chronic inflammation in tissue, with a normal healing process involving components of the extracellular matrix (ECM) that assist in replacing dead or damaged cells. Under sustained aggravation, existing cells cease to be replaced, and instead, growth, angiogenic, and fibrogenic factors combine with proteolytic enzymes to produce connective tissue deposition that inhibits and destroys the normal tissue environment. Hepatic fibrosis develops as a response to chronic liver injury, which can be initiated from a wide range of diseases. ECM accumulates in the liver, hepatic lobules collapse, fibrous septae form, and nodules appear in hepatocytes. Hepatic fibrosis can be defined as the abnormal deposition of extracellular matrix (ECM) as a result of liver injury, and cirrhosis as a continuum of fibrogenesis, characterized by the widespread conversion of liver architecture into structurally fibrotic nodules. The hepatic stellate cell is the central cellular player in fibrogenesis, with other effector cells like resident fibroblasts, immune cells, fibrocytes, and potentially epithelial–mesenchymal transitory cells all playing a role.

In recent years, both hepatitis C (HCV) and nonalcoholic steatohepatitis (NASH) have received substantial attention from drug developers, with remarkable clinical and commercial success in hepatitis C programs. NASH is the progressive form of the relatively benign nonalcoholic fatty liver disease (NAFLD), and is increasing in prevalence concomitantly with obesity and type 2 diabetes in both developed and developing countries worldwide.

In histopathology, the viral infection and immune results and result fibrosis battleground is frequently relatively uniform in viral fibrosis samples. However, NAFLD fibrosis exhibits more heterogeneity in results, where fibrosis scores at different ends of a core needle biopsy can even present as different pathology scores [19]. Fibrosis tends to be highest in the deepest region of the sampled parenchyma, farthest from the capsule. With adequate sampling procedures, the impact of heterogeneity on diagnosis can be minimized in NAFLD [5].

One of the most critical obstacles to the successful launch of antifibrotic drugs is a lack of quantitative methodology for measuring clinical trial endpoints. Qualitative liver histology is still the gold standard, and common grading paradigms do not relate directly to the level of fibrosis. Another obstacle is the lack of measurement of early stage fibrosis, where patients could be identified at a time when they would be most likely to respond to new therapies. A final challenge in clinical trials is that the various stages of pathology grading cannot be appropriately treated as continuous variables in statistical analyses.

2 Histopathologic Evaluation of Liver Fibrosis

Liver biopsies are typically ordered when noninvasive liver tests show elevated levels for longer than several weeks, or to determine efficacy or toxicity of a given therapy. Historically, liver biopsy was used exclusively for diagnostic purposes, but recently, new treatment strategies for liver disease, as well as use of other pharmacological therapies with known liver toxicity have expanded the need for biopsy. Biopsy techniques include percutaneous, transvenous (tranjugular or transfemoral), and surgical/laporoscopic [44]. A plugged needle technique helps to obtain longer length samples with the percutaneous method. The most common needle is 16 G (1.6 mm), with a trough length of 1.6–1.8 cm. The biopsy must be large enough to view a representative amount of parenchyma and number of portal tracts; several studies have investigated the adequate number of portal tracts and proposed at least 11 [12]. Optimal biopsy length is a subject of substantial ongoing discussion with some recommendations of 3 cm length 16 G needle as desirable [43].

Biopsy complications are rare but very serious. Approximately 2–4 % of patients require hospitalization for an adverse event, with 96 % of adverse events occuring within the first 24 h of the procedure [39].

Several histopathology grading schemes for disease staging are shown in Table 1; none of these include a quantitative measure of the amount of fibrosis. Disease staging is important for patient management, but a measure of fibrosis is critical to both efficacy measuring in antifibrotic drug trials and to correlate new noninvasive biomarkers for liver fibrosis in clinical studies.

Table 1
Common classifications of liver fibrosis

METAVIR Five-stage classification, specifically designed for Hepatitis C patients: F0: no fibrosis F1: portal and periportal fibrosis with no septum F2: portal and periportal fibrosis with rare septum F3: portal and periportal fibrosis with many septum F4: cirrhosis In addition, necroinflammation activity (A) is graded as A0 (absent), A1 (mild), A2 (moderate), or A3 (severe)
KNODELL SCORE (histology activity index—HAI) The Knodell score helps define the extent of liver inflammation and damage. It is composed of four individually assigned numbers that make up a single score. The first component (periportal and/or bridging necrosis) is scored 0–10. The next two components (intralobular degeneration and portal inflammation) are scored 0–4. The combination of these three markers indicates the amount of inflammation in the liver: 0 = no inflammation 1–4 = minimal inflammation 5–8 = mild inflammation 9–12 = moderate inflammation 13–18 = marked inflammation The fourth component indicates the amount of scarring in the liver and is scored from 0 (no scarring) to 4 (extensive scarring or cirrhosis)
SCHEUER Grade: degree of inflammation, piecemeal or bridging necrosis Grade 0: no/minimal inflammation Grade 1: portal inflammation or lobular inflammation without necrosis Grade 2: mild periportal inflammation and piecemeal necrosis or focal hepatocellular necrosis Grade 3: moderate periportal inflammation and piecemeal necrosis or severe focal cell damage Grade 4: severe periportal inflammation and piecemeal necrosis or bridging necrosis Stage: degree of fibrosis Stage 0: no fibrosis Stage 1: enlarged fibrotic portal tracts Stage 2: periportal fibrosis or portal to portal septa, without architectural distortion Stage 3: bridging fibrosis with architectural distortion, no obvious cirrhosis Stage 4: cirrhosis (probable or definite)

(continued)

Table 1
(continued)

ISHAK
(A) Periportal or periseptal interface hepatitis (piecemeal necrosis)
0 Absent
1 Mild (focal, few portal areas)
2 Mild/moderate (focal, most portal areas)
3 Moderate (continuous around <50 % of tracts or septa)
4 Severe (continuous around >50 % of tracts or septa)
(B) Confluent necrosis
0 Absent
1 Focal confluent necrosis
2 Zone 3 necrosis in some areas
3 Zone 3 necrosis in most areas
4 Zone 3 necrosis + occasional portal-central (P-C) bridging
5 Zone 3 necrosis + multiple P-C bridging
6 Panacinar or multiacinar necrosis
(C) Focal (spotty) lytic necrosis, apoptosis and focal inflammation
0 Absent
1 One focus or less per 10× objective
2 Two to four foci per 10× objective
3 Five to ten foci per 10× objective
4 More than ten foci per 10× objective
(D) Portal inflammation
0 Absent
1 Mild, some or all portal areas
2 Moderate, some or all portal areas
3 Moderate/marked, all portal areas
4 Marked, all portal areas

3 Centralized Virtual Review for Antifibrosis Clinical Trials

When a manual pathology scoring system is used to support a pharmaceutical clinical trial, substantial preparation must be made that is intrinsically different from what a pathologist faces in everyday practice. A study of 254 chronic viral hepatitis liver biopsies were evaluated by a team of pathologists to examine ways to improve consensus scoring for clinical trials. Mean agreement was better for senior than junior pathologists, mean intra-observer agreement was better than inter-observer agreement (0.77 ± 0.18 vs. 0.58 ± 0.26, $P < .01$), and agreement with Metavir score greatly improved after consensus reading. Interestingly, the authors noted that "the level of experience (specialization, duration, and location of practice) had more influence on agreement than the characteristics of the specimen (length, fibrosis class number, miscellaneous factors)" [48].

Centralized consensus scoring can greatly help in clinical trials, and the use of a virtual digital pathology server facilitates logistics, although the glass slides should be available when requested. The ideal approach to minimizing observer variability could include the following, and reviews should be conducted as soon as possible:

1. A pretrial consensus meeting with pathologists experienced in the given disease area, fibrosis staging categories and borders agreed on.
2. Study assessed independently by two hepatopathologists, with consensus sessions devoted to disagreements.
3. Blinded review of biopsy pairs.

4 Quantitative Assessment of Fibrosis in Clinical Biopsies

There is a growing consensus of the value of morphological techniques for use in measuring liver fibrosis. While trichrome and reticulin stains are used in subjective evaluations, the degree of staining has not been found to correspond well with hepatic collagen content [41, 56]. Picrosirius red technique, is the recommended stain for image analysis in clinical biopsies [23, 26, 35]. The most common image-analysis-based technique uses Picrosirius stain to measure the ratio of the area occupied by collagen versus the entire area of liver tissue, which is referred to as the collagen proportionate area or CPA [3], after excluding structural collagen from blood vessel walls, large septal portal tracts, and other confounding areas. While it is difficult to compare numbers across studies, values for CPA have ranged from 1 to 7 % in normal and 12 to 36 % in cirrhotic liver, respectively [17]. Normal human liver is estimated to contain approximately 5.5 mg/g collagen (wet weight), and cirrhotic liver contains approximately 30 mg/g [46]. Others have estimated normal collagen concentration at 2–8 mg/g [4]. A large number of studies have compared CPA to the various clinical pathology grading schemes and other laboratory variables. Earlier publications tended to stress comparison of CPA to grading; however, fibrosis content and pathologist grading are not interchangeable [16]. CPA correlates with hepatic venous pressure gradient (HVPG) [6, 54], which is of useful prognostic value. CPA also predicts decompensation effectively in cirrhotic patients.

In one study, liver samples from 115 hepatitis C patients who had undergone transplantation were measured for CPA with sirius red, as well as HVPG, and graded with the Ishak system [6]. CPA correlated well with Ishak stage scores, and was a better histological correlate with HVPG than Ishak stage. In another study, biopsies from chronic hepatitis B patients were stained with picosirius red and the areas of collagen measured, and correlated with Ishak modified histologic scores. High correlations were reported between the two methods (Spearman's $r = .68–78$, $P < 0.001$) and morphometry had excellent discrimative power in differentiating mild and advanced fibrosis, and cirrhosis from non-cirrhosis [24].

Recently, 249 chronic hepatitis C patients with liver biopsies and serum fibrosis markers were evaluated with CPA, measured with both serius red and trichrome [23]. While both stains correlated well with Metavir stage, sirius red detected fibrous septa and spurs better than trichrome, had better detection of cirrhosis, and had superior correlation with serum markers of fibrosis.

A recent study looked empirically at the amount of tissue required to make accurate CPA measurements [21]. 10 mm × 10 mm regions from 120 cirrhotic tissue blocks were evaluated with serius red and CPA, and divided into 1 mm × 1 mm images from which virtual biopsies could be obtained. From these virtual images probabilities were calculated that the virtual biopsy CPA would be within 5 % of the reference CPA. Results showed that 22–28 mm^2 of analyzable tissue is required to be within 5 % of the reference CPA, and 12–15 mm^2 of analyzable tissue for 25 % of the reference CPA. These values were heavily dependent on the severity of cirrhosis, which illustrates empirically why sampling errors increase in lower severity disease.

There were also early studies in the use of modified stereological approaches that had early promise, prior to the availability of image analysis techniques and better approaches for acquiring microscopy data. These stereological techniques are worth revisiting. For example, core needle biopsies found stereology to be superior to image-analysis in quantifying fibrosis in alcoholic and nonalcoholic liver disease [65]. In a follow-up study, the same authors measured area fraction with a Weibel overlay grid with a 10× objective, counting 1,000 points per biopsy [64]. At each point, fat globules, unchanged parenchyma, portal tract areas, central veins, and fibrosis (Picro-Mallory stain), and stereological area fractions from each were computed. While the method was highly reproducible, the authors did not record the length of time required to perform each stereological analysis (usually 200 or 300 points are adequate for most stereological studies). Based on these combined studies, this method may be valuable for the measurement of both fibrosis and steatosis in the evaluation of new treatment strategies for patients with chronic hepatitis C or NASH.

5 Further Developments in Histological Quantitation Methods

The most common fibrosis measurement is simply an area ratio, and does not include architectural changes such as nodularity, fibrous portal linking, or portal-central fibrous bridging: features that are included in stage scoring systems. Further work is required in image analysis to elucidate techniques to incorporate these features. Stereological techniques should also be more widely considered in clinical biopsy samples, even if the sampling technique is limited to what can be obtained from patient-donors. Laboratories supporting clinical trials should consider both whole slide image analysis techniques, as well as stereological endpoints.

6 Alternative Approaches to Histologic Assessment of Liver Fibrosis

While histological examination of biopsied hepatic tissue remains the reference standard for detection and staging of liver fibrosis [5, 57], there are several well-recognized drawbacks to the technique, including risk of hemorrhage and infections [39], patient discomfort and complications, particularly among the growing population of obese patients requiring biopsy [22, 53], potential sampling error due to heterogeneous distribution of lesions [25, 42], variability in histopathologic interpretation [32, 63], and financial cost [13, 50] that have led to a growing interest in alternative noninvasive strategies for the evaluation of hepatic fibrosis [8]. For example, noninvasive liver-elasticity-based imaging techniques, such as transient elastography, have shown value in the diagnosis and staging of hepatic fibrosis; however, this method has certain limitations in obese patients, in whom accuracy is significantly reduced [7]. Excellent reviews of the advantages and disadvantages of various imaging techniques, including transient elastography, can be found elsewhere [2, 9] and they are not discussed further here.

The optimal noninvasive assay should be simple to perform, yield reproducible results, be more cost effective than liver biopsy, and accurately predict the full range of hepatic fibrosis. Although many proteins have been explored as potential noninvasive markers of liver disease, only a few have been able to distinguish among the different stages of fibrosis at a level comparable to that achieved with liver biopsy. In this section, we review those proteins showing the most promise as direct biomarkers of liver fibrosis, focusing on components of the extracellular matrix and hepatocyte apoptosis. In addition, we discuss the emerging potential of noncoding RNA molecules, namely microRNAs (miRNAs), to serve as markers of severity and progression of fibrosis in chronic liver disease.

6.1 Serum Biomarkers of Liver Fibrogenesis

The development of hepatic fibrosis results from complex interactions among factors that regulate synthesis and degradation of the extracellular matrix. Some of these factors, including hyaluronic acid (HA) and type IV collagen 7S domain, are present in serum at low levels and have been investigated as potential predictors of liver fibrosis. With the development of liver fibrosis, increased production of collagen, in combination with reduced functioning of sinusoidal endothelial cells, leads to increased serum HA levels. A number of studies have investigated the potential use of serum HA levels in the diagnosis and staging of NAFLD-related liver fibrosis. Suzuki et al. [58] examined serum HA in 79 NAFLD patients and found that levels of the marker were significantly different relative to fibrosis stage ($P < 0.0001$), showing a positive correlation with degree of fibrosis after adjusting for age and serum albumin. The calculated area under the receiver operator characteristic curve (i.e., AUROC) was 0.87, 0.89, and 0.92 for $\geq$ moderate fibrosis, $\geq$ severe fibrosis, and cirrhosis, respectively, and the corresponding positive (PPV) and negative predictive values (NPV) were 51 and 96 %. Similar associations of serum HA and fibrosis have been reported, although with varying levels of sensitivity and predictive value [30, 49]. In an investigation of type IV collagen 7S domain in 112 NAFLD patients with varying levels of fibrosis, a cutoff point of 5.0 ng/ml provided a PPV and NPV of 68 % and 84 %, respectively, for the presence of advanced fibrosis [49]. An independent study also found an association (AUROC = 0.77) between this marker and advanced fibrosis, prior to progression to cirrhosis [62]. However, despite the apparent promise of serum levels of HA and type IV collagen 7S domain, the studies published to date have reported different proposed cutoff values and diagnostic accuracies, which may be a result of the relatively limited sample sizes comprising these studies, as well as differences in cohort ethnicities, experimental design, and methods of measurement. In the case of HA, it is worth noting that levels are often increased in systemic inflammation, which may confound measures associated with fibrosis.

Serum levels of extracellular matrix function, including HA and type IV collagen, have not been useful for detecting mild stages of fibrosis, with AUROC for any degree of fibrosis ranging between ~0.60 and 0.70 [20, 30, 49, 51]. Further, while a combination of different parameters involved in fibrogenesis, including HA and type IV collagen 7S domain, showed good ability (i.e., >90 %) to detect fibrosis, the power of these combined markers to discriminate between early fibrosis stages was limited [47, 52]. As a result, the search for markers capable of discriminating effectively among the different stages of fibrosis remains a key area of focus in liver research.

In addition to single measures of ECM components, several multicomponent panels that include various clinical and laboratory assessments, have been developed for use in the diagnosis and staging of liver fibrosis. We highlight some of the most common commercial panels in Table 2. However, even when used in combination, current noninvasive, multiplexed serum scores only give reliable information on liver fibrosis in approximately 30 % of chronic hepatitis C patients [34]. In addition, these tests perform poorly at early stages of liver fibrosis, when intervention would be expected to have the greatest clinical impact. While the two largest US commercial reference laboratories actively market multi-analyte tests for viral hepatitis and NAFLD, the insurance guidance from a major carrier is instructive of the challenges in this area:

Table 2
Serum marker systems of liver fibrosis

Name	Analytes	Strengths	Weaknesses
AST/ALT	Aspartate aminotransferase, alanine aminotransferase	Ratio of >1 indicates cirrhosis, useful for diagnosis [23]	Unreliable for fibrosis staging, due to alcoholic and other acute and chronic fatty infiltrating liver diseases
AST-to-platelet ratio index (APRI)	AST, platelet count	Determines risk for advanced liver disease [24]	Not as useful for staging, AST and platelet can vary for other reasons than platelet stage
Fibrotest	Alpha-2-macroglobulin, apolipoprotein A1, haptoglobin, γ-glutamyl transpeptidase, bilirubin	Score is correlated with the degree of liver damage in a variety of liver diseases [25]	Not widely available, high cost, confounding with steatosis
Fibrometer	Number of platelets, prothrombin time, AST, α2-macroglobulin, hyaluronate, urea and age	[26]	

"Aetna considers serum marker tests (e.g., FibroMAX, FibroSpect, FibroTest-ActiTest, HCV-FibroSure, HepaScore, and NASH FibroSure) experimental and investigational for detecting or monitoring hepatic fibrosis in persons with hepatitis C or other chronic liver diseases (e.g., nonalcoholic fatty liver disease) because their effectiveness for these indications has not been established."[1]

Because hepatocyte apoptosis is a key factor in the pathogenesis of NAFLD and also plays a significant role in the development of fibrosis, markers in this pathway have been heavily explored. In specific, apoptosis involves the activation of the caspase family of proteases, which function to cleave specific intracellular proteins, including the well-studied cytokeratin 18 (CK18), which is the most common intermediate filament protein in hepatocytes. Specific studies of CK18 in liver fibrosis were based on the idea that serum levels of caspase-cleaved CK18 fragments, which reflect hepatocyte apoptosis, could be used to quantify cell death. Early studies showed a relatively high diagnostic accuracy in predicting NASH and/or differentiating simple steatosis from nonalcoholic steatohepatitis [14, 60, 61], including the ability to discriminate between advanced and early stage fibrosis with a sensitivity of ~70 % and a specificity of ~77 % [61]. In HCV patients, concentration of CK18 fragments increased with presence of liver fibrosis, even in

[1] Noninvasive Tests for Hepatic Fibrosis, Clinical Policy Bulletin 0690, Aetna. http://www.aetna.com/cpb/medical/data/600_699/0690.html.

the absence of high levels of aminotransferase [1]. In patients with NASH, an initial study showed that CK18 levels increased in a stepwise manner with fibrosis stage [59]. In a cohort of 127 obese patients, levels of CK18 measured using the M30 assay, which detects a specific epitope of CK18, were higher in NAFLD patients with advanced fibrosis [29]. In a separate study, Joka et al. [27] assessed serum levels of caspase-cleaved CK18 fragments (M30) and total CK18 (M65 and M65ED) in 121 patients with chronic liver diseases. The major findings showed that all three markers discriminated among patients across all levels of fibrosis compared with controls ($P < 0.01$). Measurement of cleaved CK18 (M30 assay) could significantly discriminate between low (F0–1) and high (F5) fibrosis stages and also between moderate (F2–4) and high fibrosis stages, but not between low and moderate stages, while the M65 and M65ED assays could discriminate between low and moderate and moderate and high fibrosis stages. Total CK18 levels at a 353.0 U/L cutoff correctly predicted fibrosis stage 2 (F2) with a sensitivity of ~71 % and AUROC of ~0.73. Together, these findings showed that total CK18 (measured by M65 and M65ED) provided better diagnostic accuracy for the detection of stage F2 fibrosis compared with cleaved CK18, whereas all three markers showed similar sensitivity and specificity to detect advanced fibrosis (F5). Similar results were reported in a cohort of 147 with biopsy-proven NAFLD and 72 healthy controls, in which M65ED, at a cutoff of 236 U/L had 88 % NPV and 59 % PPV to exclude and predict progression of fibrosis [55]. Combined, these studies suggest that measurement of markers of cell death may have potential as sensitive, discriminating, noninvasive means to assess and stage liver fibrosis, but additional studies are required for validation in larger study samples and in fibrosis originating from other forms of chronic liver disease.

6.2 Circulating miRNAs as Biomarkers of Liver Fibrosis

Recently, a class of noncoding RNAs, microRNAs (miRNAs), has emerged as a potential new kind of biomarker for liver fibrosis. MiRNAs are endogenous, single-stranded RNAs (21–25 nucleotides in length) that regulate gene expression either posttranscriptionally, by blocking translation or promoting cleavage of specific target mRNAs, or transcriptionally, through targeting of the promoter region [31]. In humans, miRNAs silence the expression of target genes predominantly at the posttranscriptional level by imperfectly base-pairing to the 3′ untranslated region (3′UTR) of target mRNAs, leading to translational inhibition and/or mRNA deadenylation and decay [31]. In addition to expression in tissue and free circulation in the blood, miRNAs can also exist in a stable form protected from RNase-dependent degradation in many biological fluids, including serum, plasma, cerebrospinal fluid, saliva, and urine [18].

An altered miRNA expression profile has been observed during liver fibrogenesis arising from different etiologies [11, 33, 37, 45]. In normal liver, miR-122 is the most abundant miRNA, accounting for >70 % of the total miRNA content in hepatocytes [28], but during the development of fibrosis in chronic liver diseases such as NAFLD or chronic HCV infection, miR-122 levels decrease sharply. Interestingly, levels of miR-122 in serum and liver are significantly correlated ($R = 0.461$; $P = 0.005$) in patients with NAFLD [36], suggesting that miR-122 released from hepatic cells enters the bloodstream. In this report, serum levels of miR-122 were found to be lower in individuals with mild steatosis, compared to those with severe steatosis, but higher in patients with mild fibrosis compared to those with severe fibrosis. This result is in agreement with those of previous studies, reporting decreased levels of hepatic miR-122 at advanced stages of fibrosis in patients with liver disease [10]. The reason for the discrepancy in miR-122 levels in fibrosis stage may represent the increasing loss of hepatocytes concomitant with escalating liver injury. Because hepatocytes are the primary source of miR-122 and because worsening of liver fibrosis results in the replacement of hepatocytes with extracellular matrix, hepatic miR-122 levels may be expected to decrease with severe fibrosis.

Cermelli et al. [10] investigated serum levels of four miRNAs commonly dysregulated in liver fibrosis: miR-122, miR-34a, miR-16, and miR-21. In a study sample comprising 34 individuals with NAFLD and 19 healthy controls, serum levels of miR-122, miR-34a, and miR-16 were significantly higher in NAFLD patients. Levels of miR-21 showed no difference between the two groups. Levels of miR-122 and miR-34a were positively correlated with disease severity from simple steatosis to steatohepatitis, supporting the potential value of these two miRNAs to serve as noninvasive biomarkers for progressive NAFLD. In a study of 43 HCV-positive liver transplant recipients, a set of miRNAs, including miR-155, miR-23b, and miR-30c, miR-27 and miR-34a, miR-222 was capable of identifying early post-liver transplant HCV patients at high risk of developing severe HCV leading to fibrosis and cirrhosis [15].

Two studies have recently investigated the potential of circulating miRNAs to determine disease parameters in patients with chronic liver disease. In the first study, RNA was extracted from serum-derived microvesicles obtained from 64 chronic HCV patients and 24 controls with normal liver function [38]. Patient samples were stratified according to stage of fibrosis (F0–F3). Expression levels of miR-483-5p and miR-671-5p were significantly upregulated with increasing stage of fibrosis, while levels of 14 miRNAs, let-7a, miR-106b, miR-1274a, miR-130b, miR-140-3p, miR-151-3p, miR-181a, miR-19b, miR-21, miR-24, miR-374, miR-548, miR-93, and miR-941 were downregulated with increasing stage of fibrosis.

Using the Leave One Out Cross-Validation method, the authors were able to distinguish stage F0 from other stages with an accuracy of 87.5 % and an odds ratio of 14.25. The classification of the remaining stages was 65.6, 70.3, and 73.4 with odds ratios of 3.16, 6.39, and 5.80 for F1, F2, and F3, respectively. Interestingly, the miRNA signature observed with chronic HCV patients was distinguishable from those of either HBV or NASH patients, suggesting that specific kinds of chronic liver diseases have their own unique profiles. These results showed that dysregulated miRNA expression was closely related to progression of HCV and provided preliminary evidence that these miRNAs are potential biomarkers for grading and staging fibrosis in chronic liver diseases. Additional studies, in larger and more ethnically diverse samples, will be necessary to validate and confirm these findings.

In the second study, a panel of 84 circulating miRNAs was screened in 47 NASH patients, 30 individuals with simple steatosis, and 19 healthy controls [40]. The authors found that levels of miR-122, miR-192, and miR-375 were upregulated in patients with NASH compared to those with simple steatosis, and associated with histological disease severity; however, only serum miR-122 levels were significantly associated with the presence of advanced fibrosis in comparisons of F0–1 and F2–3. The accuracy of miR-122, miR-192, and miR-375 to discriminate advanced from mild disease was fair (AUROC = ~0.7). In a cross-comparison analysis, miR-122 was able to predict NASH and fibrosis with greater sensitivity than CK18, and both serum miR-122 and miR-192 levels were positively correlated with serum CK18 concentration ($P < 0.03$).

Although our knowledge of the potential value of miRNAs to serve as biomarkers for liver fibrosis is still in its infancy, current findings, combined with the rapidly expanding field of miRNA research, are expected to lead to the identification of novel, noninvasive biomarkers for the disease. Notably, identification and validation of circulating miRNA signatures may facilitate the development of improved methods for diagnosis and clinical monitoring of disease progression.

7 Conclusion

In this chapter, approaches in quantitative histology and alternatives to biopsy have been reviewed. The rapid progress in drug development in both NAFLD and viral fibrosis will require substantial investment and innovation in improving diagnostic and prognostic quantitation. As in other areas of anatomic pathology, the morphological measurements will become more digital and quantitative, while expanding to integrate analytes beyond the tissue biopsy.

References

1. Bantel H, Lugering A, Heidemann J, Volkmann X, Poremba C, Strassburg CP, Manns MP, Schulze-Osthoff K (2004) Detection of apoptotic caspase activation in sera from patients with chronic HCV infection is associated with fibrotic liver injury. Hepatology 40(5):1078–1087. doi:10.1002/hep.20411
2. Barsic N, Lerotic I, Smircic-Duvnjak L, Tomasic V, Duvnjak M (2012) Overview and developments in noninvasive diagnosis of nonalcoholic fatty liver disease. World J Gastroenterol 18(30):3945–3954. doi:10.3748/wjg.v18.i30.3945
3. Bedossa P, Dargère D, Paradis V (2003) Sampling variability of liver fibrosis in chronic hepatitis C. Hepatology 38(36):1449–1457
4. Biagini G, Ballardini G (1989) Liver fibrosis and extracellular matrix. J Hepatol 8 (1):115–124
5. Brunt EM (2011) Non-alcoholic fatty liver disease: what's new under the microscope? Gut 60(8):1152–1158. doi:10.1136/gut.2010.218214
6. Calvaruso V, Burroughs AK, Standish R et al (2009) Computer-assisted image analysis of liver collagen: relationship to Ishak scoring and hepatic venous pressure gradient. Hepatology 49:1236–1244
7. Castera L, Foucher J, Bernard PH, Carvalho F, Allaix D, Merrouche W, Couzigou P, de Ledinghen V (2010) Pitfalls of liver stiffness measurement: a 5-year prospective study of 13,369 examinations. Hepatology 51 (3):828–835. doi:10.1002/hep.23425
8. Castera L, Pinzani M (2010) Non-invasive assessment of liver fibrosis: are we ready? Lancet 375(9724):1419–1420. doi:10.1016/S0140-6736(09)62195-4
9. Castera L, Vilgrain V, Angulo P (2013) Noninvasive evaluation of NAFLD. Nat Rev Gastroenterol Hepatol 10(11):666–675. doi:10.1038/nrgastro.2013.175
10. Cermelli S, Ruggieri A, Marrero JA, Ioannou GN, Beretta L (2011) Circulating microRNAs in patients with chronic hepatitis C and nonalcoholic fatty liver disease. PLoS One 6(8): e23937. doi:10.1371/journal.pone.0023937
11. Cheung O, Puri P, Eicken C, Contos MJ, Mirshahi F, Maher JW, Kellum JM, Min H, Luketic VA, Sanyal AJ (2008) Nonalcoholic steatohepatitis is associated with altered hepatic MicroRNA expression. Hepatology 48 (6):1810–1820. doi:10.1002/hep.22569
12. Crawford AR, Lin XZ, Crawford JM (1998) The normal adult human liver biopsy: a quantitative reference standard. Hepatology 28:323–331
13. Crockett SD, Kaltenbach T, Keeffe EB (2006) Do we still need a liver biopsy? Are the serum fibrosis tests ready for prime time? Clin Liver Dis 10(3):513–534. doi:10.1016/j.cld.2006.08.009, viii
14. Feldstein AE, Wieckowska A, Lopez AR, Liu YC, Zein NN, McCullough AJ (2009) Cytokeratin-18 fragment levels as noninvasive biomarkers for nonalcoholic steatohepatitis: a multicenter validation study. Hepatology 50 (4):1072–1078. doi:10.1002/hep.23050
15. Gehrau RC, Mas VR, Villamil FG, Dumur CI, Mehta NK, Suh JL, Maluf DG (2013) MicroRNA signature at the time of clinical HCV recurrence associates with aggressive fibrosis progression post-liver transplantation. Am J Transplant 13(3):729–737. doi:10.1111/ajt.12047
16. Germani G, Burroughs AK, Dhillon AP (2010) The relationship between liver disease stage and liver fibrosis: a tangled web. Histopathology 57(56):773–784
17. Germani G, Hytiroglou P, Fotiadu A et al (2011) Assessment of fibrosis and cirrhosis in liver biopsies: an update. Semin Liver Dis 31 (31):82–90
18. Gilad S, Meiri E, Yogev Y, Benjamin S, Lebanony D, Yerushalmi N, Benjamin H, Kushnir M, Cholakh H, Melamed N, Bentwich Z, Hod M, Goren Y, Chajut A (2008) Serum microRNAs are promising novel biomarkers. PLoS One 3 (9):e3148. doi:10.1371/journal.pone.0003148
19. Goldstein NS, Hastah F, Galan MV, Gordon SC (2005) Fibrosis heterogeneity in nonalcoholic steatohepatitis and hepatitis C virus needle core biopsy specimens. Am J Clin Pathol 123(3):382–387. doi:10.1309/ey72f1en9xcb1kxx
20. Guha IN, Parkes J, Roderick P, Chattopadhyay D, Cross R, Harris S, Kaye P, Burt AD, Ryder SD, Aithal GP, Day CP, Rosenberg WM (2008) Noninvasive markers of fibrosis in nonalcoholic fatty liver disease: validating the European liver fibrosis panel and exploring simple markers. Hepatology 47(2):455–460. doi:10.1002/hep.21984
21. Hall AR, Tsochatzis E, Morris R, Burroughs AK, Dhillon AP (2013) Sample size requirement for digital image analysis of collagen. Histopathology 62:421–430
22. Harwood J, Bishop P, Liu H, Nowicki M (2010) Safety of blind percutaneous liver

biopsy in obese children: a retrospective analysis. J Clin Gastroenterol 44(10):e253–e255. doi:10.1097/MCG.0b013e3181cf8358

23. de Huang Y, Boer WB, Adams LA et al (2013) Image analysis of liver collagen using sirius red is more accurate and correlates better with serum fibrosis markers than trichrome. Liver Int 33:1249–1256
24. Hui AY, Liew CT, Go MY et al (2004) Quantitative assessment of fibrosis in liver biopsies from patients with chronic Hepatitis B. Liver Int 24(26):611–618
25. Janiec DJ, Jacobson ER, Freeth A, Spaulding L, Blaszyk H (2005) Histologic variation of grade and stage of non-alcoholic fatty liver disease in liver biopsies. Obes Surg 15 (4):497–501. doi:10.1381/0960892053723268
26. Jimenez WPA, Caballeréa J et al (1985) Measurement of fibrosis in needle liver biopsies: evaluation of a colorimetric method. Hepatology 5(5):815–818
27. Joka D, Wahl K, Moeller S, Schlue J, Vaske B, Bahr MJ, Manns MP, Schulze-Osthoff K, Bantel H (2012) Prospective biopsy-controlled evaluation of cell death biomarkers for prediction of liver fibrosis and nonalcoholic steatohepatitis. Hepatology 55(2):455–464. doi:10.1002/hep.24734
28. Jopling C (2012) Liver-specific microRNA-122: biogenesis and function. RNA Biol 9 (2):137–142. doi:10.4161/rna.18827
29. Kalsch J, Bechmann LP, Kalsch H, Schlattjan M, Erhard J, Gerken G, Canbay A (2011) Evaluation of biomarkers of NAFLD in a cohort of morbidly obese patients. J Nutr Metab 2011:369168. doi:10.1155/2011/369168
30. Kaneda H, Hashimoto E, Yatsuji S, Tokushige K, Shiratori K (2006) Hyaluronic acid levels can predict severe fibrosis and platelet counts can predict cirrhosis in patients with nonalcoholic fatty liver disease. J Gastroenterol Hepatol 21(9):1459–1465. doi:10.1111/j.1440-1746.2006.04447.x
31. Kim DH, Saetrom P, Snove O Jr, Rossi JJ (2008) MicroRNA-directed transcriptional gene silencing in mammalian cells. Proc Natl Acad Sci U S A 105(42):16230–16235
32. Kleiner DE, Brunt EM, Van Natta M, Behling C, Contos MJ, Cummings OW, Ferrell LD, Liu YC, Torbenson MS, Unalp-Arida A, Yeh M, McCullough AJ, Sanyal AJ (2005) Design and validation of a histological scoring system for nonalcoholic fatty liver disease. Hepatology 41(6):1313–1321. doi:10.1002/hep.20701
33. Lakner AM, Steuerwald NM, Walling TL, Ghosh S, Li T, McKillop IH, Russo MW, Bonkovsky HL, Schrum LW (2012) Inhibitory effects of microRNA 19b in hepatic stellate cell-mediated fibrogenesis. Hepatology 56 (1):300–310. doi:10.1002/hep.25613
34. Leroy V, Hilleret MN, Sturm N, Trocme C, Renversez JC, Faure P, Morel F, Zarski JP (2007) Prospective comparison of six non-invasive scores for the diagnosis of liver fibrosis in chronic hepatitis C. J Hepatol 46 (45):775–782
35. López-De León A, Rojkind M (1985) A simple micromethod for collagen and total protein determination in formalin-fixed paraffin-embedded sections. J Histochem Cytochem 33(38):737–743
36. Miyaaki H, Ichikawa T, Kamo Y, Taura N, Honda T, Shibata H, Milazzo M, Fornari F, Gramantieri L, Bolondi L, Nakao K (2013) Significance of serum and hepatic microRNA-122 levels in patients with non-alcoholic fatty liver disease. Liver Int. doi:10.1111/liv.12429
37. Morita K, Taketomi A, Shirabe K, Umeda K, Kayashima H, Ninomiya M, Uchiyama H, Soejima Y, Maehara Y (2011) Clinical significance and potential of hepatic microRNA-122 expression in hepatitis C. Liver Int 31 (4):474–484. doi:10.1111/j.1478-3231.2010.02433.x
38. Murakami Y, Toyoda H, Tanahashi T, Tanaka J, Kumada T, Yoshioka Y, Kosaka N, Ochiya T, Taguchi YH (2012) Comprehensive miRNA expression analysis in peripheral blood can diagnose liver disease. PLoS One 7(10): e48366. doi:10.1371/journal.pone.0048366
39. Piccinino F, Sagnelli E, Pasquale G, Giusti G (1986) Complications following percutaneous liver biopsy. A multicentre retrospective study on 68,276 biopsies. J Hepatol 2(2):165–173
40. Pirola CJ, Fernandez Gianotti T, Castano GO, Mallardi P, San Martino J, Mora Gonzalez Lopez Ledesma M, Flichman D, Mirshahi F, Sanyal AJ, Sookoian S (2014) Circulating microRNA signature in non-alcoholic fatty liver disease: from serum non-coding RNAs to liver histology and disease pathogenesis. Gut. doi:10.1136/gutjnl-2014-306996
41. Puchtler H, Meloan S, Waldrop FS (1988) Are picro-dye reactions for collagens quantitative? Chemical and histochemical considerations. Histochemistry 88(83–86):243–256
42. Ratziu V, Charlotte F, Heurtier A, Gombert S, Giral P, Bruckert E, Grimaldi A, Capron F, Poynard T (2005) Sampling variability of liver biopsy in nonalcoholic fatty liver disease. Gastroenterology 128(7):1898–1906

43. Regev A, Berho M, Jeffers LJ et al (2002) Sampling error and intraobserver variation in liver biopsy in patients with chronic HCV infection. Am J Gastroenterol 97:2614–2618
44. Rockey DC, Caldwell S, Goodman ZD (2009) Liver biopsy. Hepatology 49(43):1017–1044
45. Roderburg C, Urban GW, Bettermann K, Vucur M, Zimmermann H, Schmidt S, Janssen J, Koppe C, Knolle P, Castoldi M, Tacke F, Trautwein C, Luedde T (2011) Micro-RNA profiling reveals a role for miR-29 in human and murine liver fibrosis. Hepatology 53 (1):209–218. doi:10.1002/hep.23922
46. Rojkind M, Ponce-Noyola P (1982) The extracellular matrix of the liver. Coll Relat Res 2:151–175
47. Rosenberg WM, Voelker M, Thiel R, Becka M, Burt A, Schuppan D, Hubscher S, Roskams T, Pinzani M, Arthur MJ (2004) Serum markers detect the presence of liver fibrosis: a cohort study. Gastroenterology 127(6):1704–1713
48. Rousselet MC, Michalak S, Dupre F et al (2005) Sources of variability in histological scoring of chronic viral hepatitis. Hepatology 41(42):257–264
49. Sakugawa H, Nakayoshi T, Kobashigawa K, Yamashiro T, Maeshiro T, Miyagi S, Shiroma J, Toyama A, Nakayoshi T, Kinjo F, Saito A (2005) Clinical usefulness of biochemical markers of liver fibrosis in patients with nonalcoholic fatty liver disease. World J Gastroenterol 11(2):255–259
50. Sanai FM, Keeffe EB (2010) Liver biopsy for histological assessment: the case against. Saudi J Gastroenterol 16(2):124–132. doi:10.4103/1319-3767.61244
51. Santos VN, Leite-Mor MM, Kondo M, Martins JR, Nader H, Lanzoni VP, Parise ER (2005) Serum laminin, type IV collagen and hyaluronan as fibrosis markers in non-alcoholic fatty liver disease. Brazilian journal of medical and biological research=Revista brasileira de pesquisas medicas e biologicas/Sociedade Brasileira de Biofisica [et al] 38(5):747–753. doi:/S0100-879X2005000500012
52. Sebastiani G, Alberti A (2006) Non invasive fibrosis biomarkers reduce but not substitute the need for liver biopsy. World J Gastroenterol 12(23):3682–3694
53. Seeff LB, Everson GT, Morgan TR, Curto TM, Lee WM, Ghany MG, Shiffman ML, Fontana RJ, Di Bisceglie AM, Bonkovsky HL, Dienstag JL (2010) Complication rate of percutaneous liver biopsies among persons with advanced chronic liver disease in the HALT-C trial. Clin Gastroenterol Hepatol 8(10):877–883. doi:10.1016/j.cgh.2010.03.025
54. Sethasine S, Jain D, Groszmann RJ, Garcia-Tsao G (2012) Quantitative histological-hemodynamic correlations in cirrhosis. Hepatology 55:1146–1153
55. Shen J, Chan HL, Wong GL, Choi PC, Chan AW, Chan HY, Chim AM, Yeung DK, Chan FK, Woo J, Yu J, Chu WC, Wong VW (2012) Non-invasive diagnosis of non-alcoholic steatohepatitis by combined serum biomarkers. J Hepatol 56(6):1363–1370. doi:10.1016/j.jhep.2011.12.025
56. Standish RA, Cholongitas E, Dhillon A, Burroughs AK, Dhillon AP (2006) An appraisal of the histopathological assessment of liver fibrosis. Gut 55(54):569–578
57. Straub BK, Schirmacher P (2010) Pathology and biopsy assessment of non-alcoholic fatty liver disease. Dig Dis 28(1):197–202. doi:10.1159/000282086
58. Suzuki A, Angulo P, Lymp J, Li D, Satomura S, Lindor K (2005) Hyaluronic acid, an accurate serum marker for severe hepatic fibrosis in patients with non-alcoholic fatty liver disease. Liver Int 25(4):779–786. doi:10.1111/j.1478-3231.2005.01064.x
59. Tarantino G, Conca P, Coppola A, Vecchione R, Di Minno G (2007) Serum concentrations of the tissue polypeptide specific antigen in patients suffering from non-alcoholic steatohepatitis. Eur J Clin Invest 37(1):48–53. doi:10.1111/j.1365-2362.2007.01745.x
60. Wieckowska A, Zein NN, Yerian LM, Lopez AR, McCullough AJ, Feldstein AE (2006) In vivo assessment of liver cell apoptosis as a novel biomarker of disease severity in nonalcoholic fatty liver disease. Hepatology 44(1):27–33. doi:10.1002/hep.21223
61. Yilmaz Y, Dolar E, Ulukaya E, Akgoz S, Keskin M, Kiyici M, Aker S, Yilmaztepe A, Gurel S, Gulten M, Nak SG (2007) Soluble forms of extracellular cytokeratin 18 may differentiate simple steatosis from nonalcoholic steatohepatitis. World J Gastroenterol 13(6):837–844
62. Yoneda M, Mawatari H, Fujita K, Yonemitsu K, Kato S, Takahashi H, Kirikoshi H, Inamori M, Nozaki Y, Abe Y, Kubota K, Saito S, Iwasaki T, Terauchi Y, Togo S, Maeyama S, Nakajima A (2007) Type IV collagen 7s domain is an independent clinical marker of the severity of fibrosis in patients with nonalcoholic steatohepatitis before the cirrhotic stage. J Gastroenterol 42 (5):375–381. doi:10.1007/s00535-007-2014-3

63. Younossi ZM, Gramlich T, Liu YC, Matteoni C, Petrelli M, Goldblum J, Rybicki L, McCullough AJ (1998) Nonalcoholic fatty liver disease: assessment of variability in pathologic interpretations. Mod Pathol 11(6):560–565
64. Zaitoun AM, AI Mardini H, Awad S et al (2001) Quantitative assessment of fibrosis and steatosis in liver biopsies from patients with chronic hepatitus C. J Clin Pathol 54:461–465
65. Zaitoun AM, AI Mardini H, Record CO (1999) Quantitative assessment of fibrosis in liver biopsies from patients with alcoholic and non alcoholic liver disease. J Cell Pathol 4:11–19

Methods in Pharmacology and Toxicology (2015): 117–126
DOI 10.1007/7653_2014_36

Published online: 11 March 2015

Stereology and Computer-Based Image Analysis Quantifies Heterogeneity and Improves Reproducibility for Grading Reticulin in Myeloproliferative Neoplasms

Mohamed E. Salama, Erik Hagendorn, Sherrie L. Perkins, Jeff L. Kutok, A. Etman, Josef T. Prchal, and Steven J. Potts

Abstract

Evaluation of bone marrow fibrosis in myeloproliferative neoplasms (MPN) is subject to interobserver inconsistency, and cutpoints for determination of therapy are dependent on manual pathologist grading of reticulin. To provide a more consistent approach to cutpoint determination, we developed a stereology-based method of calculating length fiber density (the length of the fiber network divided by the volume of bone marrow hematopoietic tissue).

Forty-eight thin needle core bone marrow biopsy samples from patients with MPN were obtained from the University of Utah archives after institutional review board approval and stained for reticulin and then scanned using whole slide imaging. To determine interpathologist concordance, the blinded cases were scored according to the European consensus system by four pathologists. Two pathologists scored twice with a 1-month washout period to determine intrapathologist concordance.

Using systematic uniform random sampling and line counting, two techniques adapted from the stereology field, the length density of the reticulin network was measured as well as a measure of heterogeneity across the bone marrow sample. Image analysis was used to measure the area of the reticulin fiber as a two-dimensional profile. To build a model for predicting scoring by stereology, 16 samples were chosen randomly and a linear regression relationship determined. This score was then rounded to 0, 1, 2, or 3, and concordance rates between pathologists and between pathologist and the computer stereology score determined.

Results: In the 48 patient cohort, stereology assessment was well correlated with the average of manual pathologist scoring (linear regression, $R^2 = 0.7038$). While the European consensus scoring system ranges from 0 to 3, the cutpoint between 0/1 and 2/3 is generally used for guiding therapeutic decision as myelofibrosis. In 7 of the 48 cases (15 %), pathologists had differing scores across this cutpoint, meaning that the ultimate therapeutic outcome would have been impacted. The stereology score differed with the average of the pathologist scores in only 5 of the 48 cases across this cutpoint (10 %). Precision rates for the stereology analysis were 92 %, compared with interpathologist concordance ranging from 54.2 to 65.2 % and intrapathologist at 60.9 %. Computer-based stereology proved to be more reproducible at predicting therapeutic cutpoint than manual scoring. The new technique can be run using standard histochemistry and provides both a nonbiased systematic measure of reticulin and a new measure of reticulin heterogeneity.

Key words Myelofibrosis, Bone marrow fibrosis, Reticulin scoring, Digital imaging, Myeloproliferative neoplasms

1 Background and Introduction

The three major ph(-) myeloproliferative disorders (MPDs) are essential thrombocythemia (ET), polycythemia vera (PV), and primary myelofibrosis (PMF). They share numerous phenotypic similarities. The bone marrow (BM) in MPNs shows megakaryocyte proliferation with overlapping histo-topography and varying degrees of reticulin fibrosis. In 2008 WHO criteria recognized bone marrow histologic features as one of the criteria for MPD classification with BM fibrosis as a major element. The BM stroma is typically comprised, in part, of a structural framework of connective tissue fibers on which hematopoiesis occurs. Among these fibers are reticulin and collagen. Reticulin is a normal BM component and is detectable by silver staining methods in 73–81 % of healthy subjects. However, increased reticulin deposition in the bone marrow (reticulin fibrosis) is associated with a number of benign and malignant conditions. In contrast to reticulin, any amount of collagen deposition in the bone marrow (collagen fibrosis), detected by trichrome staining, is pathologic. Collagen fibrosis is most often associated with MPD or solid tumors metastatic to the bone marrow and is characteristically irreversible.

Several different grading and reporting systems have been devised to quantify reticulin and collagen deposition in the bone marrow. The most frequently used grading systems to assess myelofibrosis (MF) are based on reticulin grade scored using Bauermeister system from 1971 (0–4 scale) and the recently revised European consensus system (0–3 scale). The currently accepted evaluation and scoring methods of BM fibrosis are primarily dependent on manual pathologist grading of reticulin and are subject to interobserver inconsistency. Reports that addressed these inconsistencies are scarce and attempts to study heterogeneity in reticulin fibrosis are lacking. This is further confounded by the variability of pre-analytical processing, staining inconsistency, and subjective assessment from lack of a positive staining internal standard and inadequate guidance to disregard lymphoid nodules, vessels, as well as fibers framing adipocytes. Thus, an objective scoring method remains an unmet need. In our search for a more consistent approach, we investigated the utility of stereology-based approach for calculating length fiber density (L_V) in BM samples with MF.

L_V is defined as the linear length of an object in a volume of space. A combination of vertical uniform random slices and sine-weighted lines (cycloids, also known as a Merz grid) permits unbiased estimation of total L_V from projected light microscope images of linear structures [1]. Several studies have used this approach to estimate length of biological objects, including dendritic processes from single neurons [2], capillaries in cerebral cortex [3], and immunostained epidermal nerve fibers in skin punch biopsies [4]. In strict stereological sampling, the technique would require taking a sample like a bone marrow as a cylinder and taking long vertical sections through it, randomly rotating and taking another long vertical section, so that the sections have the same lengthwise dimension, but the other two dimensions are random [5, 6]. To modify this technique and make it applicable to BM fibrosis evaluation, where only a single section is often available, we utilized only a single section to calculate the length density.

The objective of this study is to develop a stereology-based method to calculate length fiber density (the length of the fiber network divided by the volume of bone marrow hematopoietic tissue) and compare the performance of the computer-based stereology to subjective pathologist grading. In addition we aim at using this tool to investigate the heterogeneity in reticulin fibrosis and its possible contribution to traditional subjective grading systems.

2 Materials and Methods

2.1 Samples Selection and Study Design

Archived histologic slides from cases with documented diagnosis of MPN covering the spectrum of fibrosis grading were selected from ARUP and the Department of Pathology. Samples with adequate core biopsy materials including H&E, reticulin, and trichrome stained slides were included in the evaluation. Peripheral blood, bone marrow aspirate, and core biopsy slides were reviewed for confirmation of the reported MPN diagnosis and to assess the technical adequacy of the reticulin stained BM core biopsy slides. Cases without adequate biopsy materials necessary for performing the study were excluded. Following case selection, reticulin stained BM core biopsy slides were morphologically scored using the revised European consensus system and subsequently digitally scanned using the Aperio XT Scan-scope (Aperio Technologies, Inc., Vista, CA, USA).

2.2 Morphologic Characteristics of Bone Marrow Fibrosis Score

Each reticulin stained slide was scored for fibrosis using the revised European consensus system with 0–3 scale. Morphologic grading was performed in an independent fashion on each case by four experienced hematopathologists across two sites for interpathologist concordance. For intrapathologist concordance, two hematopathologists reviewed the same set of cases twice with a 1-month washout period. The pathologists were blinded for the clinical characterization of the patients from whom the samples had been taken.

2.3 Use of Stereology to Estimate Fiber Length Density

The length density (L_v) of a feature is its length per unit volume of a given reference space as shown in the equation:

$$L_v(\Upsilon, \text{ref}) = \frac{\text{Length of feature } \Upsilon \text{ in reference space}}{\text{Volume of reference space}}$$

The dimensions of length density are L^1/L^3, which simplifies to L^{-2} [7]. Length density is defined as the linear length of an object in a volume of space. A combination of vertical uniform random slices and sine-weighted lines (cycloids, also known as a Merz grid) permits unbiased estimation of total L_V from projected light microscope images of linear structures [1]. Several studies have used this approach to estimate length of biological objects [2–4]. In strict stereological sampling, the technique would require taking a sample like a bone marrow as a cylinder and taking long vertical sections through it, randomly rotating and taking another long vertical section, so that the sections have the same lengthwise dimension, but the other two dimensions are random. To modify this technique into a clinical environment, where only a single section is often available, we utilized only a single section to calculate the length density. It should be noted that this is not strictly a pure stereological method, as an assumption is made that this single section is a representative random sample of the entire bone marrow.

A two-step approach was implemented to explore quantitative evaluation of reticulin fibers. The entire area of bone marrow, excluding bone, large interstitial spaces, and histology and whole slide scanning artifact areas, was drawn as an electronic region of interest (ImageScope, Aperio, Vista, CA). A program for systematic random sampling called GridMap (Flagship Biosciences, Westminster, CO) allows input of number of areas, percent coverage, and other parameters for generating square sampling areas within a larger region of interest or whole slide image. Ten regions of interest were selected randomly on each slide using a systematic uniform random sampling algorithm (Flagship Biosciences). Areas of bone were avoided (Figs. 1 and 2).

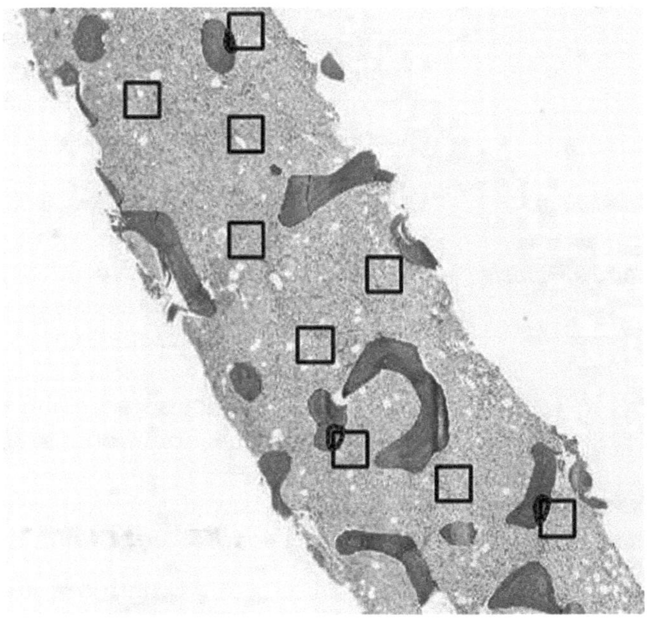

Fig. 1 Systematic uniform random sampling was utilized to identify regions of interest in a given sample. Ten areas were selected randomly, with bone sections eliminated

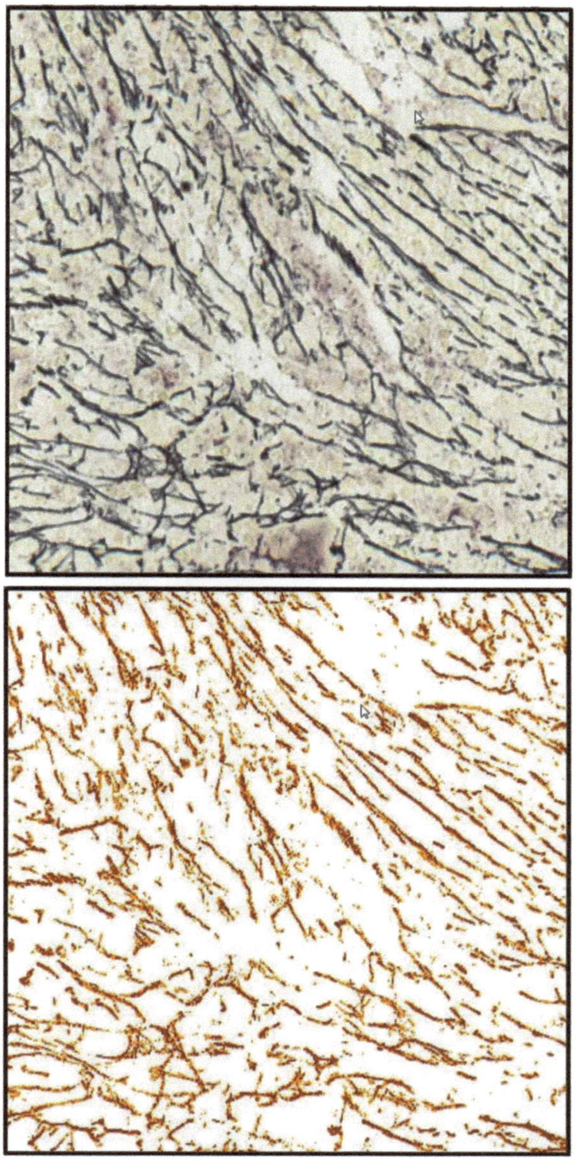

Fig. 2 Area profile of reticulin fibers. Reticulin staining above a given threshold (*red* and *orange* markup) was considered positive for reticulin fiber

A length density stereology estimator known as a Merz guide was used to evaluate length density, counting the number of fibers that crossed the semicircular lines [8]. Ten counts were performed in randomly selected areas (Fig. 3), and length density calculated. The average of the ten counts for length density and the average of the cross-sectional area estimate along with the standard deviations and coefficient of variations for each measurement were recorded. In addition, for each slide, the pathologist scored heterogeneity as a

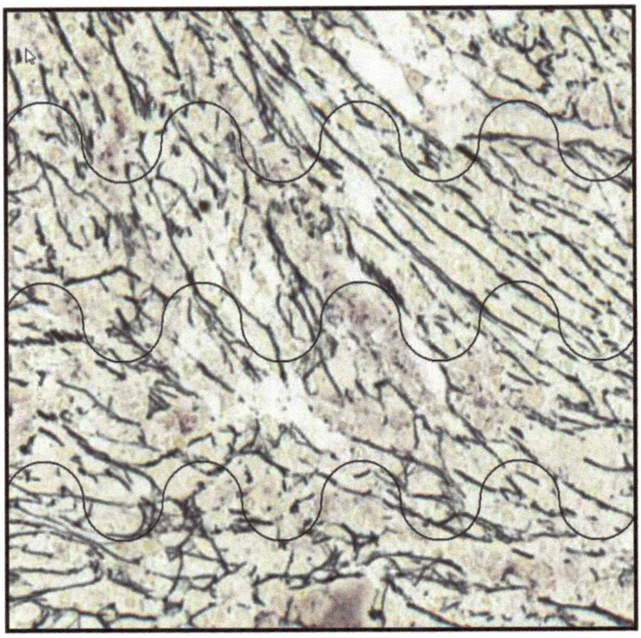

Fig. 3 Length density measurement using sine-weighted lines known as a Merz grid evaluated at 400× magnification. The number of intersections on the central line of reticulin was counted by an observer. The first line that had no artifacts (e.g., bone, fat, scanning, or histology defects) was utilized in this order—*middle*, *top*, *bottom*, repositioned grid one field of view to the left (*middle*, *top*, then *bottom*). The *middle line* interests reticulin 26 times in the example below

number from 1 to 10, with 1 being highly homogeneous and 10 being extremely heterogeneous in reticulin fibers present.

3 Results

3.1 Morphologic Scores Concordance Rates

The interpathologist and intrapathologist concordance rates are shown in Fig. 4. Interestingly, both intra- and interpathologist concordance were comparable, with interpathologist concordance ranging from 54.2 to 65.2 % and intrapathologist at 60.9 %. The pathologists were almost always within one grade scoring of each other, with only one instance out of 186 total reads where a pathologist differed from another pathologist or from himself by more than one grade (e.g., 1 versus 3 or 0 versus 2).

3.2 Concordance of Stereology-Based Analysis and Subjective Microscope Scoring

Cases were stratified according to computer-generated stereology score into four groups corresponding to 0–3 score groups generated by subjective morphology scores. Computer-generated stereology scoring classification and pathologist subjective classification showed concordance in 56 % of cases when individual 0–3 classes are compared. However, the concordance reached

INTERPATHOLOGIST CONCORDANCE

P1 versus P2 Percent agreement 65.2%

	0	**1**	**2**	**3**
0	2	3	0	0
1	2	4	1	0
2	0	0	10	4
3	0	0	6	14

P1 versus P4 Percent agreement 54.2%

	0	**1**	**2**	**3**
0	3	4	0	0
1	2	5	0	0
2	0	3	5	6
3	0	0	7	13

P2 versus P4 Percent agreement 63.0%

	0	**1**	**2**	**3**
0	4	0	0	0
1	1	6	0	0
2	0	3	7	7
3	0	1	5	12

INTRAPATHOLOGIST CONCORDANCE

P1 versus P1 after washout period. Percent agreement 60.9%

	0	**1**	**2**	**3**
0	1	4	0	0
1	2	4	1	0
2	0	1	8	5
3	0	0	5	15

Fig. 4 Interpathologist (three pathologists) and intrapathologist (one pathologist after 2-month washout period) concordance with manual reticulin scoring of 47 slides

89 % when 0/1 and 2/3 were grouped together, with a Positive Predictive Value of 96 % and a Negative Predictive Value of 75 % (Table 1).

3.3 Practical Applicability of the Stereology-Based Method

To address the practical applicability of the stereology-based method, we evaluated the concordance rates of stereology-based analysis between different laboratories. The correlation between observers in two different laboratories conducting independent length density scoring (Fig. 5) showed high concordance ($r^2 = 0.85$). The length density measurements averaged approximately 21 for the two labs and had a median standard deviation of 1.8 and median coefficient of variation of 8.9 % between the two observers.

Table 1
Interpathologist concordance and pathologist to computer concordance

Pathologist to Computer Concordance

	0	1	2	3
0	3	7	0	0
1	1	15	4	0
2	0	9	10	24
3	0	1	10	44
	Percent Agreement =		56%	

	0	1	2	3
0	3	7	0	0
1	1	15	4	0
2	0	9	10	24
3	0	1	10	44
	Percent Agreement =		63%	

	0	1	2	3
0	3	7	0	0
1	1	15	4	0
2	0	9	10	24
3	0	1	10	44
	Percent Agreement =		89%	

Rows are stereological scores, binned against a regression equation, and columns are the combined adjudicated score of four pathologists. Top shows concordant reads in green for a four-class scoring system, middle for 0/1 grouped (three class), and bottom for 0/1 and 2/3 grouped together (two class). With 2/3 considered positive, the PPV is 96 % and the NPV is 75 % for the two-class scoring system

4 Discussion

Reticulin staining is routinely used to assess the presence or absence of marrow fibrosis in bone marrow core biopsies of MPN patients. The most frequently used grading systems to assess myelofibrosis are based on the Bauermeister system (0–4 scale) [9] or the recently revised European consensus system (0–3 scale) [10, 11]. Both systems are problematic, semiquantitative, and suffer major limitations related to subjectivity [12]. In addition, the condensation of fibers is often not homogeneous. This inhomogeneity can further contribute to the complexity of the evaluation. For this reason, a new and optimal system for routinely and reproducibly determining reticulin fiber content in the bone marrow is required. We assessed the utility of a new method adopted from the field of stereology to assess

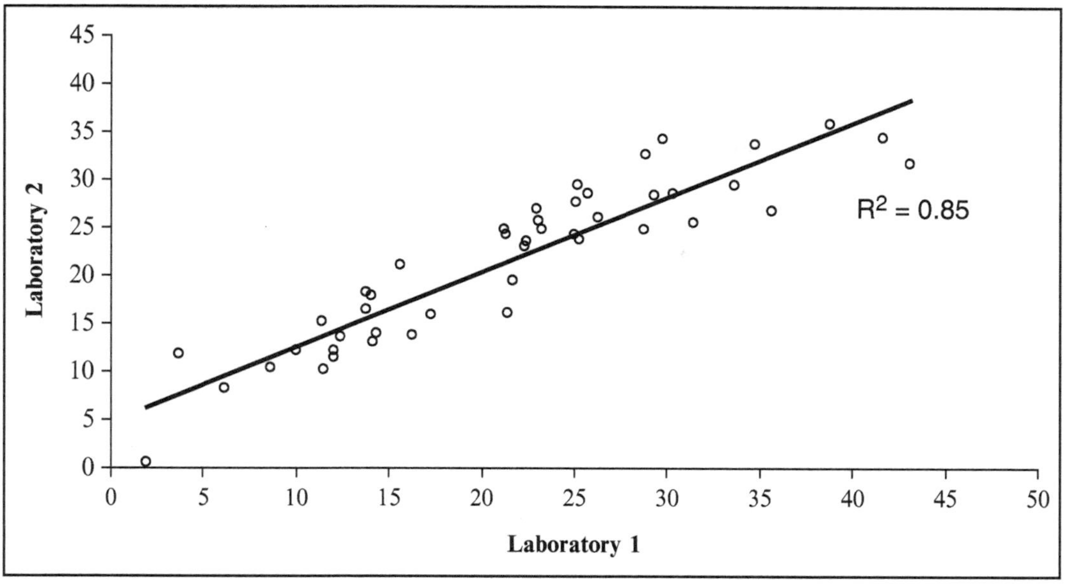

	Lab 1	Lab 2
Min	0.6	1.9
Max	36.7	43.1
Average	21.4	21.5
Median St. Dev	1.8	
Median CV	8.9%	

Fig. 5 Interlaboratory concordance for length density stereology measurement

marrow fibrosis. Our results show equivocal interobserver variability of the European consensus classification with concordance ranging from 54.2 to 65.2 % and an intrapathologist concordance at 60.9 %. These findings underscore the fact that semiquantitative estimation of the average fiber density in the bone marrow is not optimal. The grading of fibrosis is important not only to diagnose MPN but also to guide treatment decisions and stratify patients in clinical trials. In comparison, the stereology-based analysis showed high interlaboratory concordance of ($r^2 = 0.85$, CV = 8.9 %). While the concordance of the stereology-based analysis and the subjective European consensus scoring was at 56 % when individual 0–3 classes are compared, this correlation was significantly higher at 89 % when two categories only (0/1 versus 2/3) were compared (PPV of 96 % and NPV of 75 % for the two-class scoring system). This is a clinically important cutoff that guides treatment decisions as well as classification of a given case as myelofibrosis or not.

This novel stereology-based method is fast and can be easily implemented in the clinical laboratory with high level of reproducibility. It provides much needed objectivity to this type of analysis. A final advantage of this approach is that it can provide a measure of the heterogeneity and can be applied over the total range of fiber contents in the bone marrow. In conclusion, computer-based stereology proved to be more reproducible at predicting therapeutic cutpoint than manual scoring. The new technique can be run using standard histochemistry and provides both a nonbiased systematic measure of reticulin and a new measure of reticulin heterogeneity.

References

1. Gokhale AM (1990) Unbiased estimation of curve length in 3-D using vertical slices. J Microsc 159(2):133–141
2. Cruz-Orive LM, Howard CV (1991) Estimating the length of a bounded curve in three dimensions using total vertical projections. J Microsc 163(1):101–113
3. McMillan PJ, Archambeau JO, Gokhale AM (1994) Morphometric and stereological analysis of cerebral cortical microvessels using optical sections and thin slices. Acta Stereol 13:33–38
4. Stocks E, McArthur J, Griffen J, Mouton P (1996) An unbiased method for estimation of total epidermal nerve fibre length. J Neurocytol 25(1):637–644
5. Vesterby A, Kragstrup J, Gundersen HJG, Melsen F (1987) Unbiased stereologic estimation of surface density in bone using vertical sections. Bone 8:13–17
6. Buesche G, Georgii A, Duensing A, Schmeil A, Schlue J, Kreipe HH (2003) Evaluating the volume ratio of bone marrow affected by fibrosis: a parameter crucial for the prognostic significance of marrow fibrosis in chronic myeloid leukemia. Hum Pathol 34(4):391–401
7. Howard CV, Reed MG (2005) Unbiased stereology. Garland Science, New York
8. Calhoun ME, Mouton PR (2000) Length measurement: new developments in neurostereology and 3D imagery. J Chem Neuroanat 20(1):61–69
9. Bauermeister BE (1971) Quantification of bone marrow reticulin: a normal range. Am J Clin Pathol 56:24–31
10. Thiele J, Kvasnicka HM, Facchetti F et al (2005) European consensus on grading bone marrow fibrosis and assessment of cellularity. Haematologica 90:1128–1132
11. Thiele J, Kvasnicka HM (2007) Myelofibrosis—what's in a name? Consensus on definition and EUMNET grading. Pathobiology 74:89–96
12. Teman C et al (2010) Quantification of fibrosis and osteosclerosis in myeloproliferative neoplasms: a computer-assisted image study. Leuk Res 34(7):871–876

Methods in Pharmacology and Toxicology (2015): 127–139
DOI 10.1007/7653_2014_15

Published online: 26 September 2014

Image Analysis Tools for Quantification of Spinal Motor Neuron Subtype Identities

Mirza Peljto and Hynek Wichterle

Abstract

Discovery of neuronal subtype-specific markers allows for precise identification of molecularly and functionally unique nerve cells in the central nervous system (CNS). High degree of neuronal diversity generally implies that single markers can rarely be used to define individual motor neuron subtypes. As we improve our knowledge of molecular heterogeneity of nerve cells, neuronal quantification increasingly depends on intersectional expression analysis of multiple markers. In the case of spinal cord motor neurons, dozens of motor neuron subtypes can be defined by unique patterns of expression of developmentally regulated transcription factors thus offering a unique opportunity to utilize image analysis approaches for identification and analysis of motor neuron subtype identities. Here, we describe a detailed approach for quantification and analysis of motor neuron subtype identities differentiated from embryonic stem (ES) cells using immunohistochemistry, immunocytochemistry, and Flagship Biosciences image analysis approaches. The discussed approaches simplify and accelerate neuronal subtype quantification and as such should benefit both basic and translational neuroscience research.

Key words ES cells, Embryonic stem cells, Spinal motor neurons, Image analysis, CellMap, Neuronal subtypes, Motor neuron subtypes, Transcription factor quantification

1 Introduction

Thousands of distinct subtypes of neurons are found in the CNS, jointly contributing to the diverse and complex functions of the CNS. Over the last 20 years, neuroscience community has invested an extensive amount of effort to the identification and characterization of molecular markers specific to individual neuronal subtypes. Spinal motor neuron subtypes are amongst the best characterized cells in the CNS [1, 2]. The majority of motor neuron subtype-specific markers are based on both combinatorial co-expression and mutually exclusive expression of transcription factors. For example, FoxP1 transcription factor, when expressed together with generic motor neuron markers Hb9 or Isl1 and brachial spinal cord marker Hoxc6, defines forelimb innervating motor neurons of the lateral motor column (LMC) [1] (for a list of motor neuron subtype-specific markers).

Embryonic stem (ES) cells have been shown to give rise to distinct subtypes of spinal cord motor neurons through directed differentiation [3–5]. ES cell-derived motor neuron (ES motor neuron) subtypes appear to follow similar developmental pathways as their in vivo counterparts, and express distinct sets of subtype-specific molecular markers found in the embryonic spinal cord during development. Moreover, functional analysis of individual motor neuron subtypes reveals that the ES motor neurons exhibit functional characteristics indistinguishable from their in vivo embryonic counterparts [3]. Thus, ES motor neurons provide an excellent cellular substrate for analysis of motor neuron development and motor neuron subtype identities. Importantly, ES cell-derived motor neurons can be used as platform for drug screening [6] and continue to display a strong potential in the context of cell replacement therapy for CNS diseases and injuries. In contrast to the developing spinal cord where functionally distinct motor neurons occupy specific territories, in vitro generated motor neurons are intermixed and therefore their classification cannot be aided by positional information and relies solely on marker co-expression analysis. Lack of a complete set of unique molecular markers pose a challenge in identification and determination of individual subtypes based on molecular marker expression. As a result, and due in part to limitations in quantitative analysis and characterization of neuronal subtypes, full diversity of in vitro generated motor neurons is not fully understood.

Image analysis (IA) approaches offer a robust tool for unbiased identification, quantification, and molecular analysis of definable neuronal subtypes. Flagship Biosciences IA tools can deliver quantitative assessments of individual biomarkers as well as robust high-throughput analysis and quantifications of cellular identities. User-defined endpoints can be catered to individual analysis needs including cell-by-cell based analysis of critical biological endpoints. Importantly, in the context of CNS, IA tools can be used reliably for quantitative analysis of neuronal subtypes through analysis of co-expression of markers specific to individual distinct neuronal subtypes.

Here, we describe in detail an approach that combines ES cell differentiation to distinct spinal cord motor neuron subtypes with robust high-throughput image analysis tools to directly demonstrate the effectiveness of image analysis approaches for dissecting and quantification of individual neuronal subtype diversity. Our focus is on: (1) pre-analytical processing of ES motor neurons; (2) immunohistochemistry (IHC) and immunocytochemistry (ICC); (3) imaging of embryoid bodies (EBs) and single cell ES motor neurons; and (4) CellMap™ algorithm-driven image analysis of ES motor neuron subtype identities.

2 Materials

Suspension culture dishes (35 mm, 430588; 60 mm, 430589; 10 cm, 430591; Corning); wide orifice pipette tips (21-197-2A, Fisher Scientific); 200 μl pipette tip (02-707-430, Fisher Scientific); 1.5 ml Eppendorf tube (0030 125.150, Eppendorf); 16 % PFA solution (15700, Electron Microscopy Sciences), sucrose (S0389, Sigma); OCT (62550-12, Electron Microscopy Diatome/Fisher Scientific); tabletop mini-centrifuge; embedding molds (70182, Electron Microscopy Sciences); 4-well Nunc plates (12-565-72; Fisher Scientific); 15 mm round glass coverslips (633031, Carolina Biological); 4 % PFA/10 % sucrose solution in PBS; parafilm; cryostat; slides; Superfrost Plus slides (12-550-15, Fisher Scientific); ImmunoPen (402176, Calbiochem/EMD); PBS (21-030-CV, Cellgro); Horse Serum; Triton X100 (T8787, Sigma); Aqua-PolyMount solution (18606, Polysciences Inc.), confocal microscope (LSM510, Zeiss); rabbit anti-FoxP1, mouse anti-Hoxc8, and mouse anti-Lhx3 antibodies were kindly provided by Susan Brenner-Morton and Dr. Tom Jessell.

3 Methods

3.1 Pre-analytical Processing of ES Motor Neurons

Embryoid Bodies (EBs)

1. After differentiation to ES motor neurons, swirl the dish containing EBs in a circular motion on a table top in order to collect EBs in the center of the dish.
2. Use wide orifice pipette tips mounted on a 200 μl pipette and gently aspirate EBs into the wide orifice pipette tip and transfer to a clean, labeled 1.5 ml Eppendorf tube.
3. Allow EBs to settle by gravity (generally takes about 3–5 min).
4. Gently aspirate the medium above EB pellet making sure not to remove the EBs.
5. Add ice-cold paraformaldehyde (4 % in PBS) to EBs and place on ice. Invert the Eppendorf tube containing EBs every 3–5 min and place back on ice. Fix EBs for 30 min.
6. Aspirate fixative and immediately replace it with 1× ice-cold PBS.
7. Wash fixed EBs with ice-cold PBS 3 times with 10 min incubation times between washes. EBs should be mixed every 5 min by gentle inversion of the Eppendorf tube (*see* **Note 1**).
8. Prepare 30 % sucrose in PBS (*see* **Note 2**). Aspirate PBS to 500 μl mark from Eppendorf tubes containing EBs. Gently tilt the tube to an angle (about 45°) and add 500 μl of 30 % sucrose against the side of the Eppendorf tube (*see* **Note 3**).

EBs will settle at the interface of sucrose (bottom) and PBS (top) solutions.

9. Allow EBs to settle by gravity and equilibrate with the sucrose solution (~30 min).
10. Aspirate the solutions above EBs ensuring to aspirate as much of the solution as possible without aspirating EBs.
11. Gently add OCT to EBs in the Eppendorf tube ensuring not to introduce bubbles during addition.
12. Take a clean 20 μl or 200 μl pipette tip and gently swirl EBs around in OCT.
13. Centrifuge Eppendorf tubes with EBs at 3,200 rpm for 2 min keeping track of orientation of individual tubes within the mini-centrifuge.
14. Turn the tube 180° and centrifuge again at 3,200 rpm for 2 min.
15. Label a freezing embedding mold and fill it with OCT.
16. Prepare a dry ice bucket by crushing dry ice into smaller pieces.
17. Mount a 200 μl wide orifice pipette tip onto a 200 μl pipette and gently aspirate some OCT medium into the pipette tip in order to coat it.
18. Aspirate 20 μl of OCT into the pipette tip in order to coat it and then gently aspirate EBs from the bottom of the Eppendorf tube.
19. Insert EBs into the embedding mold filled with OCT and immediately place in dry ice (*see* **Note 4**).
20. Frozen blocks can be kept at −80 °C for extended periods of time and can be re-stored after cutting without compromise to histology or tissue integrity of EBs.
21. Frozen blocks containing EBs are temperature equilibrated in the cryostat for 30–45 min prior to cutting.
22. 10–20 μm sections of EBs are collected onto Superfrost Plus slides and allowed to dry at room temperature for 30 min–1 h. Sections can be stored at −80 °C in a slide box.

Single Cells

1. Single cells are generally cultured on round coverslips contained within Nunc 4 well plates and coated with laminin or fibronectin and processed 48 h after culture [7].
2. To fix single cells, remove 300 μl of 500 μl of medium and replace it with room temperature 4 % PFA/10 % sucrose solution in PBS by gently adding the fixative to the wells.

3. After 5 min of fixation, again replace 300 μl of 500 μl of solution with room temperature 4 % PFA/10 % sucrose solution in PBS.
4. Continue fixation for another 10 min.
5. Wash the fixative with room temperature 1× PBS three times with at least 15 min intervals between individual washes.
6. Proceed with ICC immediately after processing (*see* **Note 5**).

3.2 Immunohistochemistry/ Immunocytochemistry

Immunohistochemistry (IHC) on embryoid bodies and immunocytochemistry (ICC) are performed in an identical fashion except that IHC on EBs is performed on tissue section slides and ICC on single dissociated ES motor neurons is performed on coverslips in tissue culture 4 well plates (*see* **Note 6**). The protocol here is for transcription factor IHC/ICC.

1. After defrosting slides and equilibrating them to room temperature by allowing them to sit at room temperature for ~15 min, incubate EBs or single cells in 10 % HS + 0.01 % Triton X100.
2. Wash with 1× PBS three times.
3. Dilute primary antibody according to manufacturer's specifications in 2 % horse serum (HS) with 0.01 % Triton X-100.
4. Incubate samples with primary antibody overnight at 4 °C.
5. Wash 3× with 1× PBS.
6. Dilute secondary antibody according to manufacturer's specifications in 2 % horse serum (HS) with 0.01 % triton X-100.
7. Incubate for 90 min at 4°c and wash 3× with 1× PBS.
8. Mount samples (*see* **Note 7**).

3.3 Imaging of EBs and Single Cells ES Motor Neurons

After ICC and IHC, ES motor neurons and EBs can be imaged using confocal microscopy or automated microscopy approaches. For the purpose of our methods paper, single cell ES cell derived motor neurons as well as EBs were imaged using Zeiss LSM 510 confocal microscope for Cy3, GFP, and Cy5 signals. Hb9-GFP marker is motor neuron specific and is green fluorescent protein (GFP) that is expressed from the motor neuron-specific Hb9 gene promoter. Cy3 and Cy5 signals correspond to individual motor neuron subtype-specific transcription factor signals.

3.4 CellMap™ Image Analysis of ES Cell-Derived Spinal Cord Motor Neuron Subtypes

Here we describe an image analysis protocol for identification and quantification of individual ES cell-derived motor neuron subtypes in the context of both dissociated single cells and three-dimensional aggregates named embryoid bodies (EBs). On an Aperio Image Scope platform for Flagship Biosciences CellMap™ algorithm was used to perform image analysis on ES cell-derived motor neurons in EBs and dissociated single cells using the following protocol:

Table 1
Hue × Saturation settings for individual fluorescent signals and combinations of individual signals

Stain	FITC (GFP)	Cy3	Cy5	FITC/Cy5	FITC/Cy3	Cy3/Cy5	FITC/Cy3/Cy5
HUE [DEG] Avrg.	116	5	238	140	64	267	150
HUE [DEG] Std. Dev.	5	4	2	24	7	22	100
Saturation [%] Avrg.	81	83	89	60	95	55	20
Saturation [%] Avrg.	20	20	11	23	7	21	20

1. Each image is to be processed using a fluorescent-image inverting algorithm "InvertFlrgb_0.1." This is done due to the black background of fluorescent images which is registered and picked up by the algorithm-based image analysis irrespective of the color settings (Table 1). Inverted images as well as the original images can be hosted and stored either on the server or can be stored and analyzed locally on the computer.
2. For images containing embryoid bodies, it is essential to create a region of analysis. Manual annotations are performed on each image using a Pen Tool (F2) in order to identify a region of analysis for each sample (*see* **Note 8**) (Fig. 2b). Annotations define the region of analysis that is to be analyzed by the algorithm. As demonstrated in Fig. 2b, a green line is apparent and outlines individual embryoid body thus rendering it as region of analysis. (This only has to be done for embryoid bodies and not for single cell containing images.)
3. For both EB images as well as single cell images (Figs. 1 and 2), proceed to identify color settings for individual transcription factor signals (GFP, Cy3, Cy5) as well as their co-expression-based combinations (GFP/Cy3, GFP/Cy5, GFP/Cy3/Cy5). This is done by selecting individual nuclei containing single or combinations of signals. To select individual nuclei, create a new annotation layer by opening the "Annotations" window and clicking on "New Layer" button. Use the Pen Tool (F2) to outline nuclear regions on the image containing individual signal or combination of signals you are interested in analyzing (*see* **Note 9**).
4. To further provide settings for different signals and their combinations, open CellMap™ 0.6 version. This is done with an already open image in Image Scope, selecting View, and then selecting Analysis (Ctrl + G). Once the analysis browser is open, choose Select Algorithm and select CellMap™ 0.6 version. Go to CellMap™ 0.6 parameters and select "Staining" in the Mode selection. Once step 5 is complete and individual

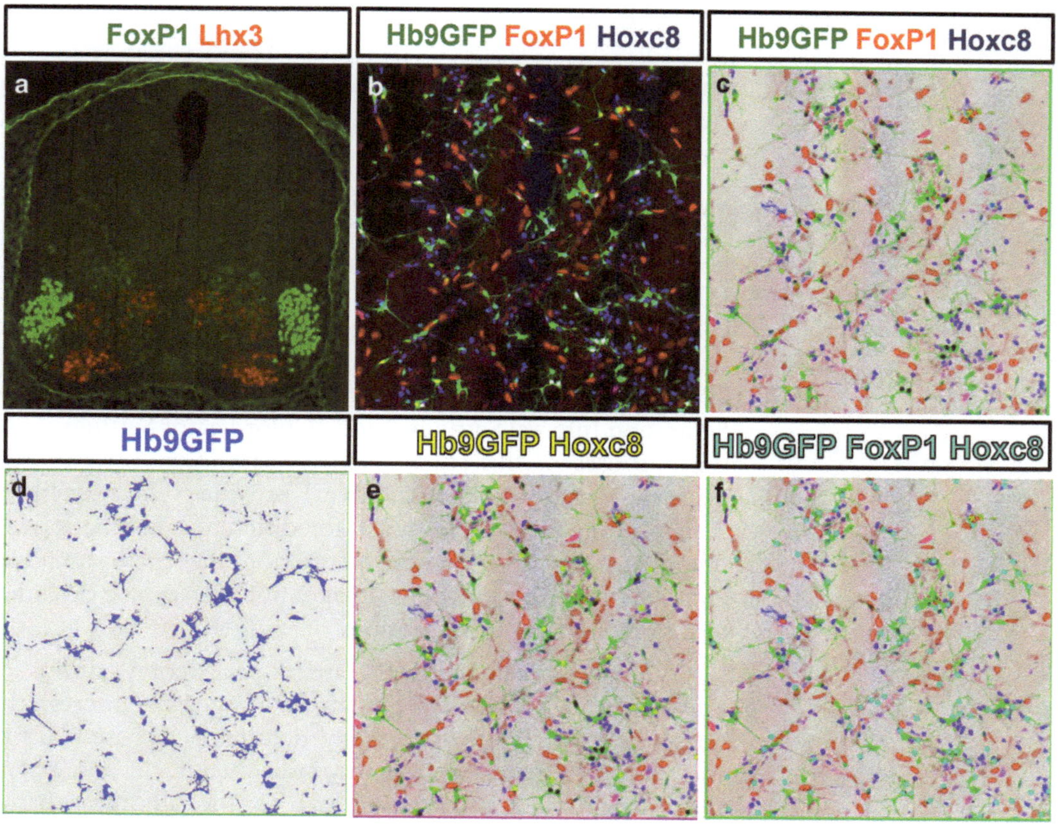

Fig. 1 CellMap™ image analysis and quantification of ES motor neuron subtypes upon dissociation into single cell cultures. (**a**) FoxP1 and Lhx3 expression in the mouse embryonic brachial level spinal cord at embryonic day 13.5 (E13.5); (**b**) An image of FoxP1 and Hoxc8 expression in dissociated brachial ES motor neurons; (**c**) An inverted image from panel (**b**); (**d**) separation of GFP channel and markup image of identified spinal cord motor neurons that express motor neuron-specific marker Hb9::GFP (GFP driven from the Hb9, motor neuron-specific promoter); (**e**) Image analysis generated markup image displaying Hoxc8 expressing ES motor neurons as *yellow dots*; (**f**) Image analysis generated markup image of FoxP1 and Hoxc8 expressing ES motor neurons (cyan)

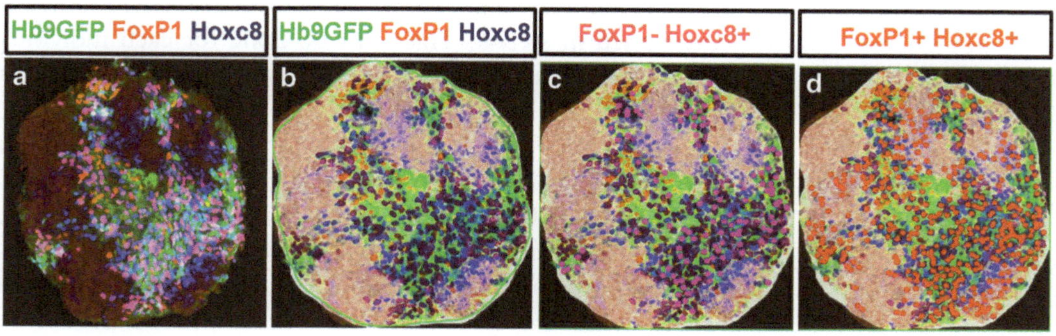

Fig. 2 CellMap™ image analysis and quantification of ES motor neuron subtypes upon dissociation into single cell cultures. (**a**) An example of an embryoid body differentiated into brachial level ES motor neurons and a representative IHC section of an embryoid body demonstrating spinal cord motor neurons (Hb9-GFP+) and co-expression of limb innervating motor neuron marker FoxP1 with caudal brachial marker Hoxc8. (**b**) Image inversion of the image in panel (**a**) displaying the region of analysis as a *green line* demarcating the boundary of the EB. (**c**) Image analysis of ES motor neurons that express Hoxc8 but do not express FoxP1 marked by *pink dots*. (**d**) Image analysis and quantification of caudal brachial LMC ES motor neurons that co-express Hoxc8 and FoxP1 marked by *red dots*

nuclei annotated, ensure that the Region of Analysis on the CellMap™ browser is selected to be "Selected Annotation Layer." Click Run.

5. In order to specify individual color settings, go to the "Layer Attributes" sub-window of the "Annotations" window and note the values for these four parameters: HUE [DEG] Avrg., HUE [DEG] Std. Dev., SATURATION [Percentage] Avrg., and SATURATION [Percentage] Std. Dev. There are indeed several distinct methods to specify color settings in CellMap™. These include "Hue × Saturation Interval Settings," "Hue × Saturation Interval Settings," and "Color Vector Settings." For the purposes of this study, "Hue × Saturation Interval Settings" appear to work best. Table 1 provides values for "Hue × Saturation Interval Settings" for different combinations of fluorescent signals. It is important to note that these should be identified based on individual images but may also be extended to different studies with caution.

   While the goals of steps 1–5 are to specify color settings for each individual signal or combinations of signals of interest, the goals of steps 6–13 are to correctly identify nuclei based on size and shape parameters, to quantify individual nuclei and generate markup images corresponding to the algorithm-based quantification. Note that step 7 can be adjusted for one of the signals or signal combinations as biometric parameters for nuclear detection should be identical irrespective of the transcription factor expression.

6. Go back to the CellMap™ algorithm parameters interface and select "Nucleus Analysis" settings for Mode. Select value of 1 for Nucleus Marker Stain [#] and value of 0 for Nucleus Quantification Stain [#]. For Stain 1, select "Hue × Saturation Interval Settings" and enter the individual signal values as determined in step 5. This is to be done for each individual signal or combinations of signals to be quantified in a step-wise fashion.

7. It is essential to provide the algorithm with specific biometric parameters in order to identify individual nuclei correctly irrespective of signals or signal combinations. This step is essential to ensure that nuclear identification is reliable and that individual nuclei are correctly identified. Adjust settings for Nucleus Detection Sensitivity, Nucleus Detection Diameter Maximum [μm], and Cell Neighborhood Radius [μm] in order to reliably identify and quantify individual nuclei. To do so, generate a small annotation layer by opening the "Annotations" window and selecting "New Layer" and drawing a small text box using a "Rectangle Tool" (F5) for testing algorithm performance on a small sample of an EB or single cells until the parameters are

idealized (*see* Note 10 on using Cell Selection Criteria). Once the parameters are set and lead to reliable and precise identification of nuclei, one can now run the algorithm on an entire image. At this point, it is useful to save individual algorithms. These can be saved locally or on the server. To save the custom-made algorithm locally, select "Export Macro" on the top of the algorithm analysis browser and identify the location to save the algorithm and name the algorithm appropriately. For a project similar to this, algorithms containing different color settings but identical biometric settings can be saved. Note that steps 1–7 are sufficient to obtain quantitative analysis of individual nuclei. Steps 8–12 are generally used in order to create markup images providing the user with the visual representation of analysis.

8. Data files are created in order to generate markup images that reflect algorithm-based analysis, nuclear identification, and quantification (Fig. 1c, e, f; Fig. 2b–d). Algorithm analysis can be saved as data files. Once the algorithm is sufficiently built to recognize the individual nuclei and to identify them, and to correctly quantify individual signals based on the color settings one can then proceed to generate data files. To generate data files, create a new folder on your Desktop and name it "Data Folder." In the CellMap™ 0.6 browser, scroll down to the bottom of CellMap™ 0.6 browser and parameter settings and select "Save CSV data files" under the "Save Data File" option. Under the "Save Data File Folder Name" enter the folder location that you created. In the case of a folder named "Data Folder" generated on your desktop you would enter C:\Users\YourName\Desktop\Data Folder (*see* **Note 11**).
9. Select "Run" and choose "Entire Image" for "Region of Analysis." After the run is completed, one can already obtain the number of nuclei with the specific staining characteristics in the annotations window under layer attributes "Nuclei: Nb Cells with Nucleus." This number can be recorded and transcribed to an excel data file. This step can be performed without step 8, namely saving the result as data files.
10. In order to create a markup image of the nuclei that have been quantified, first decide whether you would like to create a markup on the original image or the inverted image (Figs. 1 and 2).
11. To create a markup image, open an image that you would like to create markups (same image for which Data Files have been created). In order to create a visual representation of algorithm performance, create a "heatmap" macro algorithm using CellMap™ and save it. To create a "heatmap" macro, scroll to the bottom of the algorithm platform and select "Load CSV data files" under Load Data File setting. Again, specify the location

of the data file as in step 8. In the Load Data File Name, enter the name of the "Settings" file. Under Load, Lock and Display Data File, select "Execute." Under Display Data File Mode, select "Heatmap." Under Heatmap Data select Cell Density [Nb per mm2]. Parameters for Heatmap Data Min., Heatmap Data Max., Heatmap Measurement Radius [um], Heatmap Display Radius [um], and Heatmap Color [Hue] should be selected to result in markups that are visually interpretable. For example on a Cy3/FITC image, blue markups are generally visually interpretable. To achieve these, the following parameters can be set: Heatmap Data Min. = 0, Heatmap Data Max. = 5,000, Heatmap Measurement Radius [μm] = 10, Heatmap Display Radius [μm] =4, and Heatmap Color [Hue] = 200.

12. This analysis can then be performed on multiple images (*see* **Note 12**). Once they are stored on the server, multiple images can be run using the same algorithm and analysis can be performed simultaneously. In this manner, one can export the data for multiple parameters in an excel format for data analysis (*see* **Note 13**).

3.5 Potential Problems and Challenges

One of the potential issues in performing image analysis on EB aggregates is that sometimes individual nuclei are superimposed on each other. This creates a significant challenge for the algorithm in interpreting and identifying individual nuclei, especially if they display identical signal or combinations of signals. This is less of an issue when image analysis are performed on single cells as these cells are generally some distance from each other making cellular resolution greater and making it easier for the algorithm to interpret individual nuclei. Image quality is also an important component of image analysis. The higher the image resolution, the easier it becomes to quantify individual images as low resolution images are difficult to use due to our inability to zoom in on cellular substrates and successfully identify color settings with precision.

4 Notes

1. EBs can be placed in PBS and left overnight at 4 °C for next day processing. Leaving EBs in PBS for longer than 48 h may result in poor tissue architecture and suboptimal tissue integrity during processing.
2. 30 % sucrose is best when prepared fresh. While the solution can be stored long-term (~1 month) at 4 °C, one needs to avoid contamination. Use sterile technique and clean pipette tips when using 30 % sucrose solution. Filtering solutions before use is recommended.

3. Abrupt addition of 30 % sucrose to the Eppendorf tube containing EBs will result in disruption of the EB pellet and intermixing of sucrose solution with PBS. It is important to do this step slowly and to add sucrose solution in a manner that will result in little or no mixing of the two solutions.
4. Best method for inserting EBs into the embedding mold blocks consists of going to the bottom of the embedding block with the pipette tip and gently releasing the contents within. As soon as first EBs leave the pipette tip, slowly move the pipette upward thus making a column of EB substrates in the embedding block. This technique can be done either with or without the microscope. Finally, once EBs are embedded into the embedding blocks, it is important to move relatively quickly at this point as EBs will settle to the bottom. Place the block as evenly as possible onto dry ice and let it solidify. Four to six different EB samples can be placed within a single embedding mold (22 mm × 22 mm).
5. Single cells can be stored at 4 °C for maximum of 48 h in PBS. Importantly, the 4 well dishes are not completely sealed to allow for gas exchange during culture and thus need to be sealed if they are to be kept in the fridge. For sealing, please use parafilm and seal the culture wells around the edges.
6. For preparation of slides containing EB sections, Immunopen can be used to delineate the boundary around the tissue sections in order to prevent the applied liquid from spilling. Slides are stored in humidified chambers during immunostaining while parafilm needs to be used to seal the 4 wells containing single cells during ICC in order to prevent evaporation.
7. Aquamount solution can be used to mount slides containing EBs and coverslips containing single cells. After washes, put Aquamount on the slide containing EBs and gently place the coverslip over it ensuring that there are no bubbles present. Similar is done to coverslips containing single cells except that the coverslips are inverted onto a tissue slide and mounted in that fashion. Allow to dry at 4 °C in the dark at room temperature overnight prior to imaging.
8. Annotations need to be performed in the case of embryoid bodies and not when analyzing single cell images. For dissociated single motor neurons, the whole image can be analyzed and for EBs, the region which corresponds to the EB proper should be identified as a region of analysis.
9. It is essential to do this step correctly. Proper color settings for each individual signal and signal combinations will make the image analysis less prone to error and lead to quality in identification and quantification of individual cellular substrates. One potential alternative to analysis of all signal combinations on a

single slide is to export individual channels for analysis and then using a similar algorithm-based approach for quantification of single signals. However, such an approach cannot be used to assess and analyze colocalization but could aid in identification of Color Settings for individual signals after image inversion. One of the potential problems is having low resolution images. As higher resolution images make it easier to zoom into the individual nuclei and annotating them, low resolutions cannot be zoomed into and thus provide a more error-prone annotations from which color settings are defined. It is recommended to select several regions containing an identical signal or combinations of signals prior to color settings as this will ensure less variability in color settings for analysis.

10. While Cell Selection Criteria offers a plethora of settings which can be used to exclude unwanted cellular substrates, we found that we did not need to utilize it for proper identification of nuclear substrates in the context of fluorescently labeled nuclei.
11. An easy method to do this is: go to Start, select "Computer," select "Desktop," select "Data Folder." Then click on the browser and select C:\Users\YourName\Desktop\data files and copy and paste under "Save Data Folder Name."
12. In general sense, markup images are designed to show visually an example of algorithm performance on selected tissues of interest. For the most part, it is sufficient to perform algorithm analysis without actually creating data files to get the desired results. In a sense—one could only do steps 1–8 and running algorithm analysis on entire images without creating markup images. Generally, markup images are useful to create the visual representations of algorithm-based quantifications, but the bulk of analysis and quantifications can be performed without creating data files and desired data can be obtained in this way as well.
13. To export data for multiple images on the same excel file, server hosted images should be accessed by opening the "Project Specimen" and selecting individual images of interest by clicking on the boxes next to the image. Next click "Export Data" immediately above the images. Select which data parameters you would like to export and select "Include All Analysis data." Click "Export" and save the .csv file as an .xls (or .xlsx) file.

References

1. Dasen JS, De Camilli A, Wang B, Tucker PW, Jessell TM (2008) Hox repertoires for motor neuron diversity and connectivity gated by a single accessory factor, FoxP1. Cell 134:304–316
2. Jessell TM (2000) Neuronal specification in the spinal cord: inductive signals and transcriptional codes. Nat Rev Genet 1:20–29
3. Mazzoni EO, Mahony S, Iacovino M, Morrison CA, Mountoufaris G, Closser M, Whyte WA,

Young RA, Kyba M, Gifford DK et al (2011) Embryonic stem cell-based mapping of developmental transcriptional programs. Nat Methods 8:1056–1058

4. Peljto M, Dasen JS, Mazzoni EO, Jessell TM, Wichterle H (2010) Functional diversity of ESC-derived motor neuron subtypes revealed through intraspinal transplantation. Cell Stem Cell 7:355–366
5. Wichterle H, Lieberam I, Porter JA, Jessell TM (2002) Directed differentiation of embryonic stem cells into motor neurons. Cell 110:385–397
6. Yang YM, Gupta SK, Kim KJ, Powers BE, Cerqueira A, Wainger BJ, Ngo HD, Rosowski KA, Schein PA, Ackeifi CA et al (2013) A small molecule screen in stem-cell-derived motor neurons identifies a kinase inhibitor as a candidate therapeutic for ALS. Cell Stem Cell 12 (6):713–726
7. Wichterle H, Peljto M, Nedelec S (2009) Xenotransplantation of embryonic stem cell-derived motor neurons into the developing chick spinal cord. In: Audet J, Stanford WM (eds) Stem cells in regenerative medicine. Methods Mol Biol. vol 482, Springer Protocols, pp 171–183

Methods in Pharmacology and Toxicology (2015): 141–152
DOI 10.1007/7653_2014_25

Published online: 25 October 2014

Development of a Tissue Image Analysis Algorithm for Celiac Drug Development

Erik Hagendorn, Christa Whitney-Miller, Aaron Huber, and Steven J. Potts

Abstract

Celiac disease, an immune-mediated condition related to gluten sensitivity, is gaining pharmaceutical development interest. Recent conversations with the US Food and Drug Administration (FDA) indicate pathology readouts from intestinal biopsies will continue to be a primary clinical trial endpoint. The existing methodology, the Marsh-Oberhuber score, is a qualitative assessment of celiac severity, combining a morphological criterion known as villous height to crypt depth ratio (VC), with an assessment of localized immune response, manually estimating intraepithelial lymphocyte (IEL) counts. A stereology and image analysis based whole slide imaging methodology was developed for use in CLIA-based clinical trials. Experimental Design: A series of ten normal and ten abnormal patient small bowel biopsies were manually evaluated by two pathologists to determine celiac disease (CD) state using the standard Marsh score. Two quantitative methods were developed—an automated stereological methodology was used to evaluate surface area on whole slide images and an image analysis complementary approach. Methods: Stereology line probes were used to count one-dimensional "hits" on points at the distal ends of the lines which exist over reference tissue area, and "cuts" through the two-dimensional range of the line as it passes through the epithelium of the reference tissue to background, or vice versa. Results: There was strong concordance between the pathologist scores, and the automated stereology analysis, with the automated approaches able to sufficiently delineate intermediate grades of disease, normally more difficult in visual assessments. Conclusion: The quantitative methodology is a valuable addition to CLIA-based clinical trials. Quantitation provides reproducible and unbiased endpoints that can evaluate both the morphological and immune response in therapeutic clinical studies.

Key words Celiac disease, Tissue image analysis, Stereology, Morphometry, Villous atrophy, Crypt hyperplasia

1 Introduction

Celiac disease illustrates both the herd mentality of pharmaceutical drug development as well as a prime example of the difficulties in quantifying morphology in clinical tissue biopsies. Approximately 1 % of the United States population has celiac disease, and in Western Europe the numbers range from 2.4 % in Finland to 0.3 % in Germany [1]. Of the 1.8 million Americans with celiac disease, 1.4 million of them are not aware they have the digestive disorder [2]. Partly this low diagnosis rate is due to the complexity of symptoms. Celiac is an immune reaction to the gliadin in gluten, a complex glycoprotein rich in proline and glutamine, and not entirely degradable by intestinal enzymes. The clinical symptoms are variable; more common presentations include diarrhea, malnutrition, anemia, and/or joint pain. Other presentations include constipation, depression, fatigue, osteoporosis, acid reflux, infertility, dermatological conditions, as well as others. The average time to diagnosis can be years, and many medical practitioners, particularly in the United States, remain highly ignorant of the complexity of potential symptoms. Patients are generally diagnosed by meeting four of five rules: (1) typical clinical symptoms of celiac disease, (2) positive serological markers such as serum anti-transglutaminase (TTG) antibodies or anti-gliadin antibodies, (3) small intestinal biopsy showing absent or blunted villi and increased numbers of intraepithelial cells, (4) positive genetic screening for HLA-DQ2 or DQ8, and (5) improvement of symptoms on a gluten-free diet [3].

Despite the availability of serologic tests, the small intestinal biopsy remains the gold standard for diagnosis. Histology scoring is based on the Marsh-Oberhuber classification, focused on increased intraepithelial lymphocytes (IELs), crypt hyperplasia, and villous atrophy (Table 1) [4].

Until recently, celiac has received scant attention from the pharmaceutical industry, primarily because of the perceived competition of an available low-cost cure, the strict lifelong adoption of a gluten-free diet. But compliance with this diet is not simple, with gluten almost ubiquitous in restaurants, food products, and even drug prescriptions. Along with patients who have extremely high sensitivities to even trace levels of gluten is a substantial subset of celiacs who do not respond to a gluten-free diet, termed refractory celiac disease (RCD). It may be that the combination of RCD patients and patients with extremely high sensitivities to gluten

Table 1
Marsh-Oberhuber classification of celiac disease

Marsh class	Type of lesion	Villous architecture	Crypts	IELs
Marsh I	Infiltrative	Normal	Normal	>30/100 enterocytes
Marsh II	Infiltrative-hyperplastic	Normal	Hyperplasia	>30/100 enterocytes
Marsh III				
3A	Flat destructive	Mild villous atrophy	Hyperplasia	>30/100 enterocytes
3B	Flat destructive	Moderate villous atrophy	Hyperplasia	>30/100 enterocytes
3C	Flat destructive	Total villous atrophy	Hyperplasia	>30/100 enterocytes
Marsh IV	Atrophic-hypoplastic	Total villous atrophy	Hyperplasia	>30/100 enterocytes

will be enough to demonstrate to pharmaceutical executives that a market does indeed exist and demand is growing.

Patients with celiac disease are at risk for a number of long-term complications, including osteoporosis, small intestinal lymphoma, type 1 diabetes, thyroid and liver disorders, psoriasis, and lupus [5]. In children, early detection and compliance with a gluten-free diet can lead to risk profiles equivalent to the general population; however, adults who were identified with celiac late in life or have difficulty with gluten-free compliance, the risk of complications is substantially higher.

In the last several years, several celiac drug programs have emerged, primarily driven by small innovative firms. Alvine Pharmaceuticals ALV003 recently published Phase 2 trials results with a glutenase that breaks down gluten and is designed to be part of a gluten-free diet for individuals with high gluten sensitivity [6]. Biopsies from subjects in the placebo group showed evidence of mucosal injury after gluten challenge, with a mean villous height to crypt depth ratio changing from 2.8 before challenge to 2.0 afterward, and the density of CD3+ intraepithelial lymphocytes changing from 61 to 91 cells/mm after challenge. No significant mucosal deterioration was observed in biopsies from the ALV003 group. The study highlights the difficulties of attempting to measure the villous height to crypt depth ratio, given the variable geometries of the villi.

ImmunusanT is pursuing a vaccine with Nexvax2 in Phase I, with the attempt to introduce immune tolerance to gluten in individuals with the DQ2 gene. Alba Therapeutics partnered with Shire Pharmaceuticals on AT-1001, a drug that attempts to close the tight junctions between endocytes, lowering leaky gut symptoms. In 2009, early phase I trials were unsuccessful, and Cephalon acquired rights to the compound in 2011, and recently initiated Phase 3 trials [7].

BioLineRx's BL-7010 binds directly to gluten, and has been shown to decrease toxicity in mice in nonclinical testing, and recently completed phase 1 safety studies.

The FDA has been clear that one of the primary endpoints for clinical trials in celiac disease will be the biopsy [8]. The Marsh-Oberhuber system was designed as a research tool for staging during diagnosis, not as a scoring scheme for response to therapy. Another difficulty is that the Marsh system includes both immunologic response (the presence of IELs) as well as villous morphology (villi height to crypt depth ratio). While the manual measurement of villous height to crypt depth ratio has been used in some clinical trials, the villi do not orient perfectly, making measurements difficult as a line needs to be drawn from the top of the villi to the depth of the crypt each time.

There is a need during pharmaceutical trials for more reproducible, accurate methods for evaluating morphological changes and immune response in biopsy samples. In this chapter we describe a novel approach to quantifying villous morphology using both image analysis and automated stereology techniques. These two approaches are compared with manual pathology grading to determine their suitability for use in pharmaceutical clinical trials.

2 Methodology

H&E stained sections of 20 human duodenal biopsies were reviewed by two pathologists. Each section contained 1–6 tissue fragments. The pathologists were blinded to the reported diagnosis and any laboratory results. The pathologists used the Marsh-Oberhuber classification to assign a score to each tissue fragment as well as an overall score for each patient (*see* Table 2). The histological characteristics of interest for this study are tissue surface morphometry, or more specifically, the severity of crypt hyperplasia and villous atrophy from celiac disease [9]. The Marsh classification system is used to score the severity of celiac progression, a scheme

Table 2
Overall pathologist Marsh scores (20 patients)

ID	Age	Gender	Marsh grade	Tissue transglutaminase
Celiac patients				
1	25	F	1	Positive
2	29	F	3a	Positive
3	19	F	3c	Positive
4	17	F	3b	Positive
5	56	F	1	Positive
6	42	F	3b	Positive
7	12	F	3b	Positive
8	33	M	3c	Positive
9	58	F	1	Positive
10	28	F	3b	Positive
Healthy control patients				
1N	28	F	0	NA
2N	30	F	0	NA
3N	22	M	0	NA
4N	9	M	0	NA
5N	45	F	0	NA
6N	62	M	0	NA
7N	60	M	0	NA
8N	73	F	0	NA
9N	50	M	0	NA
10N	56	F	0	NA

which scores prognosis from 0 to 4, with the third stage broken into A, B, and C subclasses.

The slides were scanned digitally into high-resolution whole slide images at 20× magnification. Two proprietary tissue image analysis (tIA) algorithms were designed to analyze the surface morphometry of the sections (Flagship Biosciences, Westminster, CO). The first algorithm utilizes automated stereology, a method of using traditional stereology techniques whereby manual operation is limited to observational review of post-analytical markups. The second algorithm used is a derivation of the same stereology design, but is used to calculate two-dimensional surface morphometry features rather than three-dimensional.

Automated stereology utilizes the principles of linear dipole probes, where lines of a fixed length and certain orientation are overlaid atop of the tissue [10]. Each line provides a vector of estimation for surrounding area by acting as a framework for quantifying surface area and volume [11]. Surface area is estimated by counting the changes in phases as the line passes through two-dimensional space. Simply, the line is followed from one end to another, and "cuts" into or out of the epithelium of the tissue are counted. The volume is estimated by an enumeration of "hits" on the reference tissue from either end of the line probe, and a maximum of 2 hits per line. To calculate the surface area to volume ratio, the sum of cuts (c) are divided by the product of the line probe length (l) and the sum of reference tissue hits (h) [12].

Concurrent to the automated stereology analysis, a secondary calculative algorithm measures the perimeter and area of the tissue. Visually, the analysis markups will display a thin line surrounding the tissue, which is the outline of the perimeter measurement. The area of reference tissue should be considered as all areas within the perimeter outlines; no markup pseudo coloration of the tissue was performed. The perimeter to area measurement is calculated by dividing the perimeter (p) by the area (a).

$$\textit{perimeter}\text{: area ratio } (P : A) = \frac{p}{a}$$

$$\text{surface area : volume ratio } (\text{SA} : V) = \frac{\sum_{i=1}^{n} ci}{l * \sum_{i=1}^{n} hi}$$

The measurable covariance between surface area and volume, or perimeter and area are directly related to the amount of exposure the villus has to the outside environment, an ideal method for quantifying celiac disease. A good example of this is observed by analyzing the elliptical eccentricity of a circle and a star. A circle, in comparison to a 100-sided star (Fig. 1), should have a lower SA:V and P:A value. Although the likelihood of a measurement outcome of exactly 0 is nearly impossible, it can be assumed that the farther the value positively deviates from 0, the more eccentric or clinically normal the tissue is.

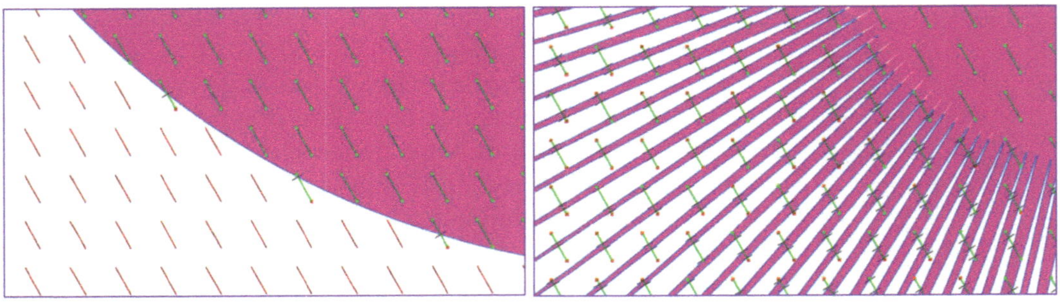

Ellipse	SA:V	P:A
Circle	.0024	.001
Star	.0891	.029

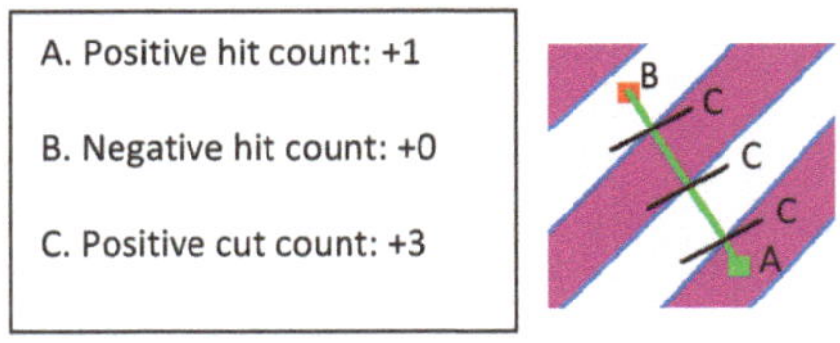

Fig. 1 Synthetic images of a circle (*left*) and a 100-point star (*right*) to demonstrate the use of elliptical eccentricity in quantifying surface morphometry

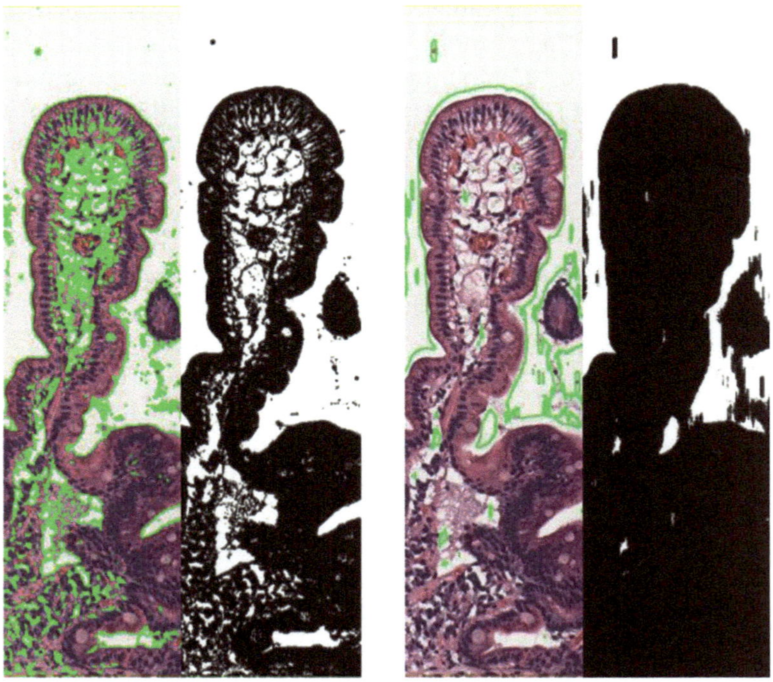

Fig. 2 Low-pass filter (*left*), StereoMap™ tissue completeness correction (*right*)

One of the major hurdles in developing an algorithm for quantifying celiac disease state is the ability to account for tissue completeness, or the lack thereof. A cross-sectional view of the villous presents a solid outline of the outer epithelium, but as one observes more centrally to the lamina propria, tissue density can become very sparse and is difficult for an algorithm to distinguish from background whitespace. As demonstrated in Fig. 2, the analysis markup with a typical low-pass filter is not enough to complete the tissue internally. Special care was also taken to assure that mucosal regions which are not of interest to the analysis are either configured to be omitted by the algorithm or manually excluded from the analysis. For example, large structures of eosinophilic tissue such as the muscularis or submucosa, which can be easily identified at a macro zoom, are removed. Other nonmucosal regions surrounding the tissue, such as tears in the biopsy or artifacts that may cause a disturbance to the eccentricity of the tissue should be removed from analysis.

3 Results (*See* Fig. 3)

Using linear dipoles of a length of 70 μm estimating areas of 100 μm^2, the range of SA:*V* values were .0013–.0068. Image analysis *P*:*A* values range from .00541 to .02438. As shown in Table 2, ten patients were scored by a pathologist as celiac positive, and another ten for celiac negative. A plot of the automated stereology (SA:*V*) and image analysis-based (*P*:*A*) results (Figs. 4 and 5) display the clear decline in values as the villous disease state progresses. When comparing both outcomes, the calculation of the complete tissue morphometry by image analysis (*P*:*A*) provides a more uniform distribution of score groups. The SA:*V* and *P*:*A* groups 1 and 3A show a distinct separation and precision between the tissue transglutaminase (tTG) positive and negative patients. When plotting the SA:*V* and *P*:*A* values in a column scatterplot (Figs. 6 and 7), the separation between the tTG groups is even more evident. A linear regression model of the SA:*V* and *P*:*A* values (Fig. 8) show a strong correlation ($R^2 = .85$) amongst the two methods.

One of the criticisms of both image analysis and stereological techniques in clinical settings is that when the time for whole slide scanning, region of interest capture, computer-based analysis, and pathologist review are combined, the method is far too time-consuming to be utilized in clinical practice. This method was not designed for use in clinical diagnostic settings; it is oriented towards pharmaceutical clinical trials, where accuracy of measurement is more critical than the fast pace of a diagnostic setting. However, the novel approach to automation of the time-consuming stereological point assessment is a contribution that may help reduce overall timelines in stereology. Eventually, such methods will make their way to clinical usage.

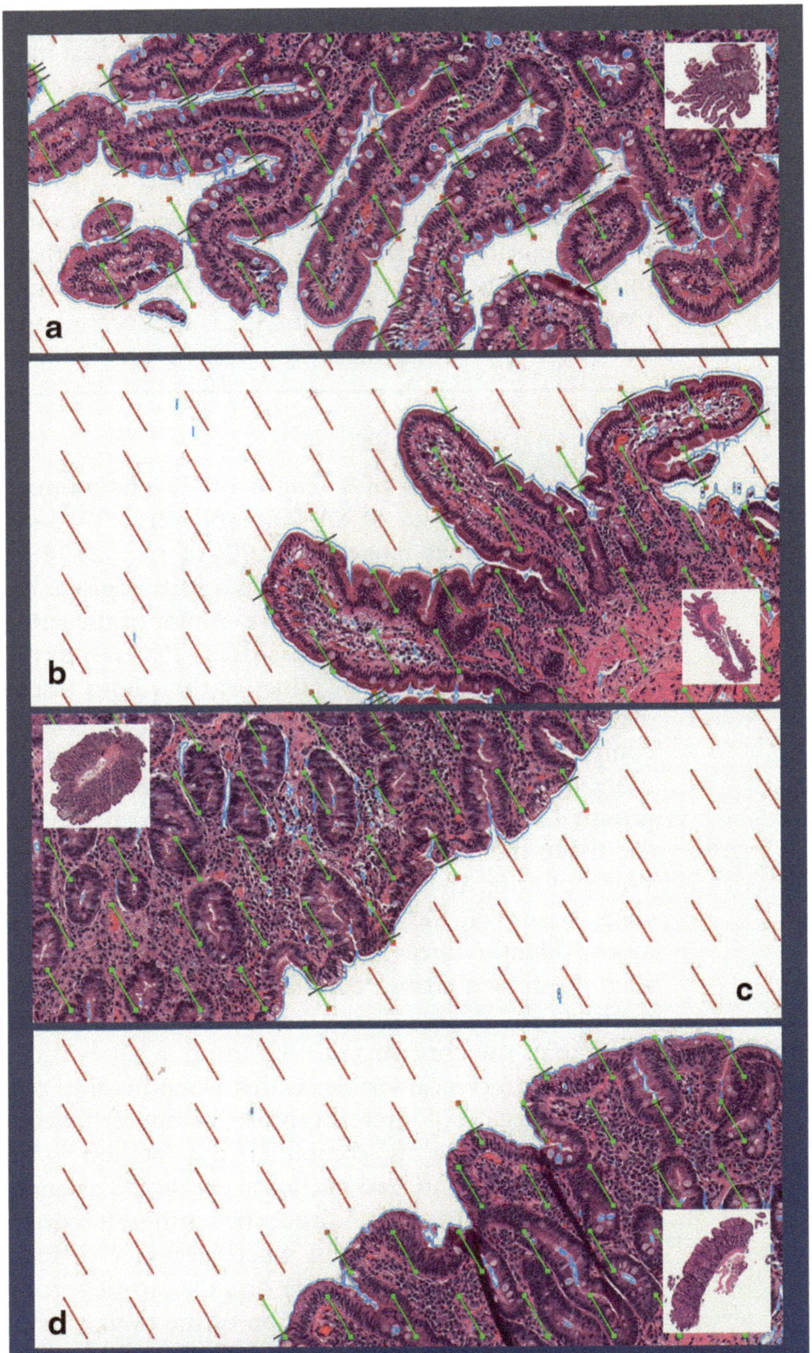

Fig. 3 (a) Normal. (b) Normal. (c) Abnormal. (d) Abnormal

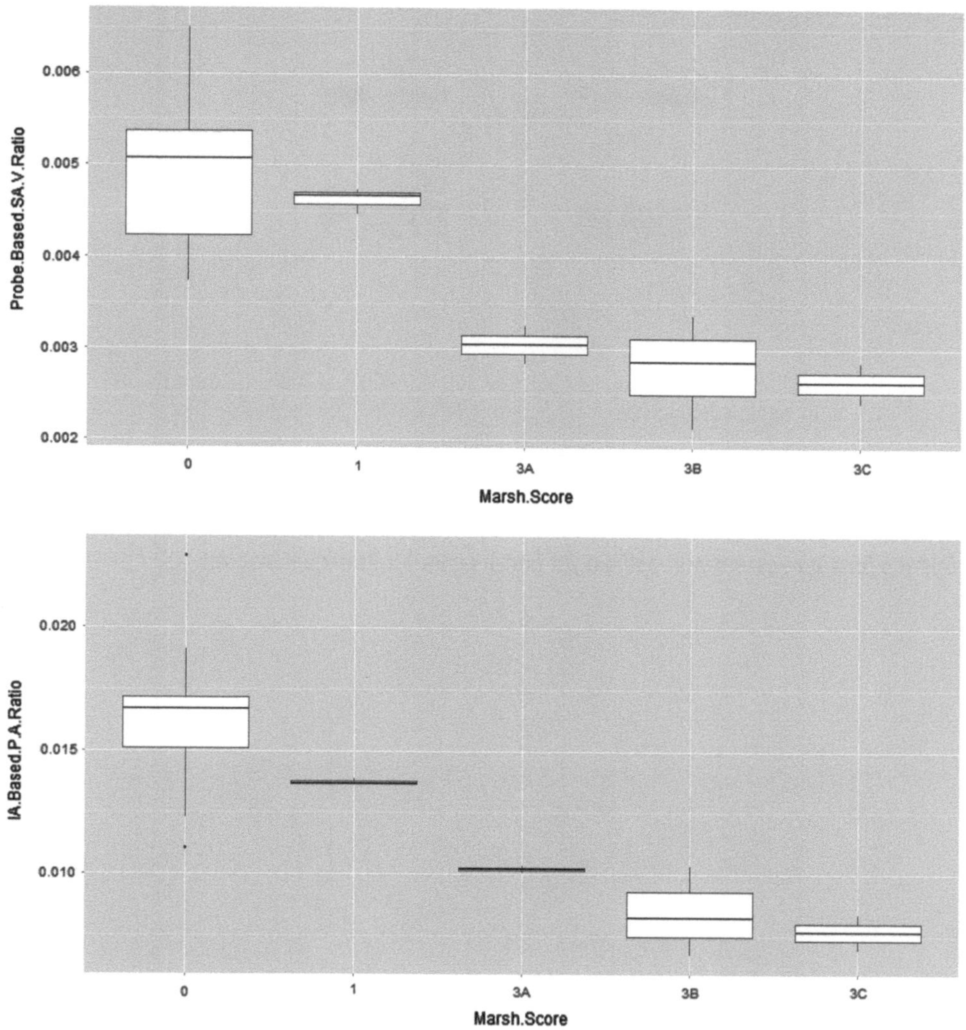

Fig. 4 (*Top*) Estimations of morphometry display the stepwise decrease in values as villus surface exposure falls with patient prognosis. (*Bottom*) Image analysis results show a similar declining trend, but with tighter-result groups and more uniformly spaced score groups

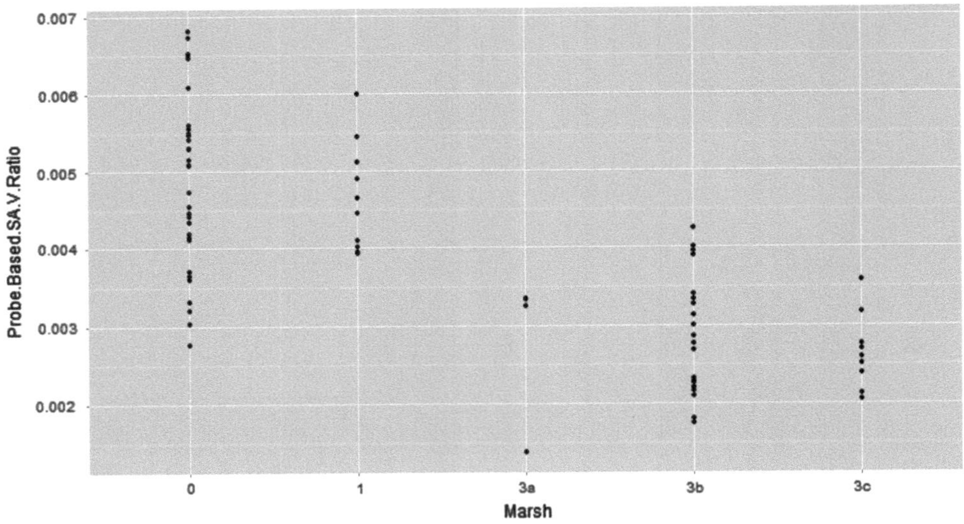

Fig. 5 Surface area agreement with pathologist Marsh score for individual biopsies

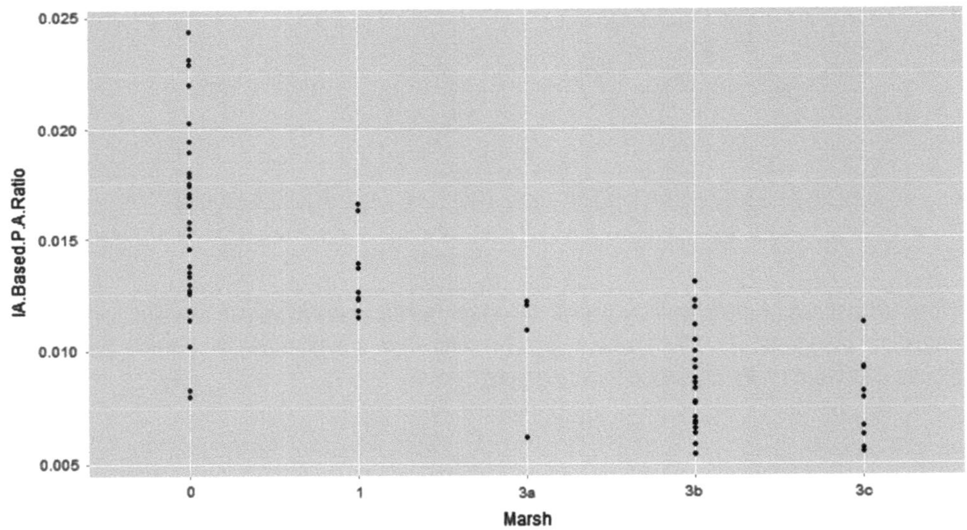

Fig. 6 Image analysis agreement with pathologist Marsh score for individual biopsies

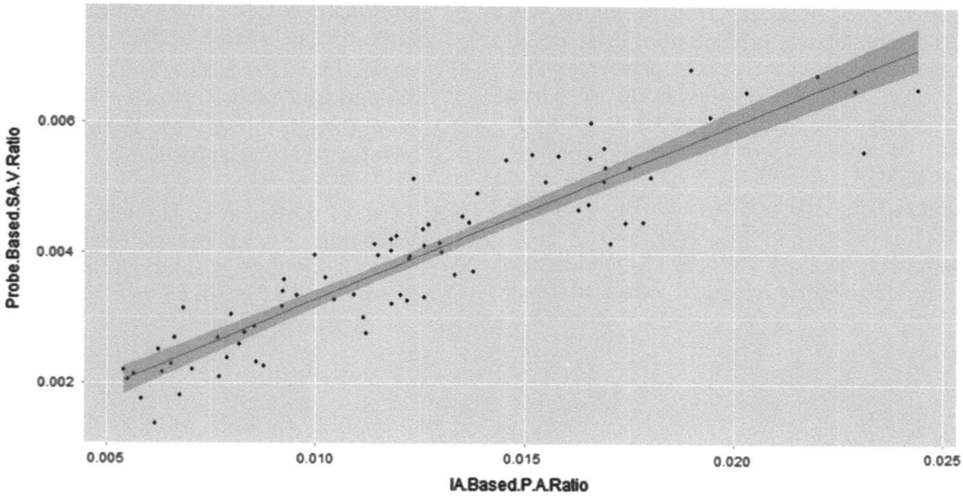

Fig. 7 Agreement between two quantitative measurements of villi geometry

4 Discussion

Agreement between pathologist scores and multisectional patient biopsy measurement outcomes suggest that the use of automated stereology and/or image analysis is an effective tool for the quantification of changes in surface morphometry of gastrointestinal sections. Future work could include comparisons of these methods with manual measurements of villous height to crypt depth ratios [13].

The strong linear correlation between the two measurement techniques demonstrates the robust relationship between stereology and image analysis. Furthermore, the use of one or both of these methods as a tool in screening for drug development or clinical studies, either for Marsh or tTG patient classification, suggests a quantifiable way to develop prognostic profiles.

References

1. Rampertab SD, Mullins GE (eds) (2014) Celiac disease. Humana Press, New York
2. Rubio-Tapia A, Ludvigsson JF (2012) The prevalence of celiac disease in the United States. Am J Gastroenterol 107:1538–1544
3. Catassi C, Fasano A (2010) Celiac disease diagnosis: simple rules are better than complicated algorithms. Am J Med 123(8):691–693
4. Oberhuber G, Granditsch G, Vogelsang H (1999) The histopathology of coeliac disease: time for a standardized report scheme for pathologists. Eur J Gastroenterol Hepatol 11 (10):1185–1194
5. Lewis NR, Holmes GK (2010) Risk of morbidity in contemporary celiac disease. Expert Rev Gastroenterol Hepatol 4(6):767–780
6. Lähdeaho M-L, Kaukinen K, Laurila K et al (2014) Glutenase ALV003 attenuates gluten-induced mucosal injury in patients with celiac disease. Gastroenterology 146 (7):1649–1658
7. Kelly CP, Green PHR (2013) Larazotide acetate in patients with coeliac disease undergoing a gluten challenge: a randomized placebo-controlled study. Aliment Pharmacol Ther 37 (2):252–262

8. United States Food & Drug Administration (2014) Conference presentation: drug development in celiac disease: FDA perspective Jessica Lee – United States Food & Drug Administration, USA. Presented at development of therapies for celiac disease, 20–21 March 2014
9. Corazza GR, Villanacci V (2005) Coeliac disease. J Clin Pathol 58:573–574
10. Risdon RA, Keeling JW (1974) Quantitation of the histological changes found in small intestinal biopsy specimens from children with suspected coeliac disease. Gut 15:9–18
11. Howard CV, Reed MG (2005) Unbiased stereology. Garland Science, New York
12. Wright SG, Tomkins AM (1978) Quantitative histology in giardiasis. J Clin Pathol 31:712–716
13. Taavela J, Koskinen O, Huhtala H et al (2013) Validation of morphometric analyses of small-intestinal biopsy readouts in celiac disease. PLoS One 8(10):e76163

Methods in Pharmacology and Toxicology (2015): 153–162
DOI 10.1007/7653_2014_37

Published online: 22 January 2015

Quantitative Histopathology for Evaluation of In Vivo Biocompatibility Associated with Biomedical Implants

Robert B. Diller, Robert G. Audet, and Robert S. Kellar

Abstract

In the current chapter, digital morphometric analysis (DMA) was used to quantify two markers of biocompatibility around commonly used biomaterials. In the field of biomaterial evaluation for biocompatibility, more sophisticated methods are now being used to precisely characterize the elicited response from the surrounding tissue towards the implanted material. One reason for this is due to the fact that many newer biomaterial innovations are incorporating pharmaceutical agents (e.g., drug eluting stents and drug eluting balloons). Therefore, as described in many of the other chapters in this book, components of toxicology and pharmacology are being evaluated along with biocompatibility.

In this chapter, expanded polytetrafluoroethylene (ePTFE) was compared to polypropylene (PP) for inflammatory and foreign body response. Each material was implanted into dorsal subcutaneous spaces and evaluated after 2, 4, and 12 weeks. Each sample was reacted with an antibody to cluster of differentiation-68 (CD-68). The resulting slides were scanned and evaluated using DMA in order to obtain accurate, reproducible, and consistent results. Expanded PTFE demonstrated a lower overall weighted inflammatory score when compared to PP across all timepoints. This chapter describes the use of DMA as a novel approach to measure the inflammatory score that is associated with a specific biomaterial. Current and future medical devices will need to use various analytical tools to comprehensively assess device, biomaterial, or a combination therapy's biocompatibility. The next chapter further describes how quantitative data from histology and immunohistochemistry assessments can be coupled with quantitative polymerase chain reactions (PCR) as assessment tools for product development.

Key words Quantitative histopathology, Digital morphometric analysis (DMA), Biocompatibility, Biomedical devices, Medical devices, Biomedical implants, Medical implants, Expanded polytetrafluoroethylene (ePTFE), Polypropylene (PP), Inflammation, Inflammatory score, Foreign body response

1 Introduction

All materials elicit a tissue response when implanted into the body; therefore, when designing and evaluating new medical devices, the materials must undergo extensive biocompatibility testing. Biocompatibility is defined as the "ability of a material to perform with an appropriate response in a specific application" (1). The host tissue receiving an implant experiences a wound healing process that includes inflammation, foreign body reactions, and fibrous encapsulation (2). When normal tissue is disrupted, a healthy organism must be able to repair itself through the process of wound healing. The normal wound healing model is characterized by four phases; *hemostasis*, *inflammation*, *proliferation*, and *remodeling*. These phases are not mutually exclusive, overlapping to various extents.

When biomaterials, either synthetic or biological, are implanted into the body, there is an altered response to wound healing (3, 4). A polymer-induced healing response initiates inflammation and a modified wound healing process through the initial implantation surgical procedure. It is understood that different polymers evoke varied wound healing responses that depend on the biocompatibility of each of these materials. However varied these responses are, there are some similarities in their healing characteristics and their deviations from normal wound healing.

Differences between various polymers and the elicited healing response first occur in the inflammatory phase of wound healing. The primary goal of inflammation is to neutralize or destroy an injurious or foreign agent as well as provide a fluid medium for the migration of repair cells (leukocytes and fibroblasts) to the area. Acute inflammation is relatively short lived, lasting minutes to days, and is characterized by polymorphonuclear leukocytes (PMN) and accompanied edema (2). Chronic inflammation can last much longer and remains localized to the implant site. In chronic inflammation the macrophage may very well be the most important cell based on the number of biologically active products it produces (2). Typically macrophages will persist during the presence of a foreign object, whether it is bacteria or implanted materials (3, 5).

The continuation of the chronic inflammatory response into a normal foreign body reaction is recognized by the chronic presence of foreign body giant cells (FBGC) with granulation tissue (2). With the chronic presence of these macrophages and foreign body giant cells, the late phase of inflammation may never resolve, causing the successive phases of normal wound healing to be hindered or never resolved. In the current study PP and ePTFE have been evaluated for the presence of macrophages and FBGCs. While it has been noted by Kellar et al. 2001 and Kidd et al. 2001, implantable materials need to be tested within the tissue the material is being designed for end use, the most common site for initial implantation during the development of a novel material is the subcutaneous space. Therefore evaluations of materials implanted in the subcutaneous locations were the focus of the current study. The subcutaneous space has been used extensively due to the relatively high-throughput, low-cost screening technique for the initial tissue response (6). This model also provides site-specific evaluation of the material to the biological interface that is often indicative of the healing that would be observed in other anatomical regions (7).

2 Materials and Methods

2.1 Slide Scanner

All glass slides were digitally scanned using the Aperio CS slide scanner with a 20× Olympus objective. At 20× magnification the Aperio scanner provides a digital image with a resolution of 0.5 μm/pixel (Aperio, Vista, CA).

2.2 Implants

The materials used were polypropylene mesh (Bard, Tempe, AZ) and thin-walled expanded polytetrafluoroethylene (Bard, Tempe, AZ). Four millimeter (4 mm) round punches were used for implantation into wild-type mouse models (129S1-Sv1mJ, Jackson Labs, Sacramento, CA). All animal studies were performed after approval of protocols by the Northern Arizona University Institutional Animal Care and Use Committee (IACUC). National Institutes of Health (NIH) Guidelines for the Care and Use of Laboratory Animals were observed. Animals were housed in American Association for the Accreditation of Laboratory Animal Care approved facilities.

2.3 Histology and Immunohistochemistry

All explanted tissue samples were paraformaldehyde fixed, paraffin embedded, sectioned at 5 μm, and subsequently processed for immunohistochemistry. Sections were reacted with an antibody to cluster of differentiation-68 (CD-68) (Serotec, clone ED1, Raleigh, NC) used at a final dilution of 1:200. The primary antibody was visualized using a secondary antibody with a peroxidase reaction product recognition system (Universal mouse kit; Dako Inc., Carpinteria, CA). CD-68 is a protein that is expressed in the cytoplasm of activated macrophages and was selected for this analysis because of the high specificity in the current study because of the high specificity of the antibody resulting in a punctate cytoplasmic staining pattern (8, 9). This provides a distinct and clear positive signal for DMA.

2.4 Digital Algorithm

A commercially available algorithm was used to count the number of CD-68$^+$ cells (IHC Nuclear Image Analysis v9, Aperio, Vista, CA). The nuclear algorithm is a cellular counting algorithm which uses input factors based on cellular profiles. Cell parameters were defined by adjusting digital values including nuclear size, roundness, compactness, and elongation. These parameters are adjustable to assist the user with determining the appropriate amount of cellular segmentation. The user can adjust the color values based on the staining of interest. For example the user can use an "eye dropper" tool which chooses specific colors and gradients of color to use as the "positive stain" being measured as well as the background stain. The "eye dropper" tool then provides the user with a breakdown of the color into its red, blue, and green components. In the algorithm setup the user can also change the threshold method in order to determine how the algorithm identifies the edges of the cell. This uses the colors that the user inputs and changes the way the algorithm segments or defines the cell. There is an "amplitude threshold" which adjusts according to the mean intensity of all the pixels and automatically thresholds to one sigma above the mean. The edge threshold method automatically adjusts the threshold according to the mean of edge pixels, using an edge finding method to identify the edge pixels and averages these values to determine the threshold. The manual threshold method uses an upper and lower limit set by the user to eliminate any unwanted background, but it will not automatically adjust to compensate for any lighter or darker staining between slides. The edge threshold method was used in the current study. The algorithm was adjusted using the parameters identified in the Aperio user's guide: (http://tmalab.jhmi.edu/aperiou/userguides/IHC_Nuclear.pdf).

3 Results

All values reported are averages ± standard error of the mean. All of the implants had been fully incorporated into the surrounding tissues at the time of explant.

3.1 Two Weeks

The ePTFE implants ($n = 5$) had an average CD-68 positive macrophage count of 442 ± 85.9; FBGC count of 21 ± 6.1. PP implants ($n = 4$) had an average CD-68 positive macrophage count of 2008.3 ± 213.8; FBGC count was 73.3 ± 10.7 (Fig. 1).

3.2 Four Weeks

Expanded PTFE implants ($n = 4$) had an average CD-68 positive macrophage count of 487.5 ± 107.9; FBGC count of 3 ± 1.5. PP

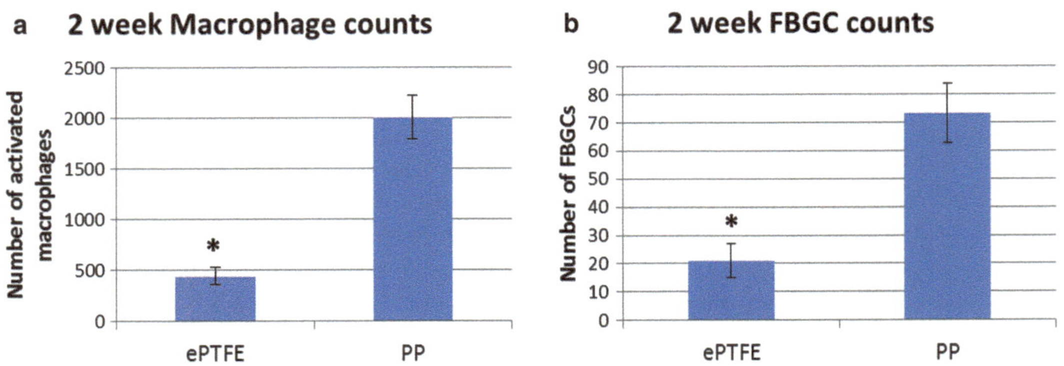

Fig. 1 Graphs depicting the number of cells counted around each material after being implanted for 2 weeks. (**a**) The average number of macrophages surrounding the ePTFE ($n = 5$) and PP ($n = 4$) implants *$p = 0.003$. (**b**) The number of FBGCs counted surrounding each implanted material *$p = 0.006$

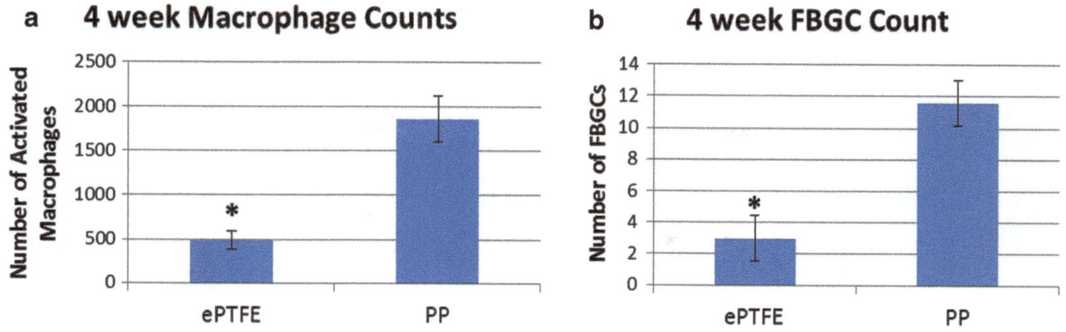

Fig. 2 Graphs depicting the number of cells counted around each material after being implanted for 4 weeks. (**a**) The average number of macrophages found surrounding the ePTFE ($n = 5$) and PP ($n = 4$) implants $*p = 0.004$. (**b**) The number of FBGCs counted surrounding each implanted material $*p = 0.004$

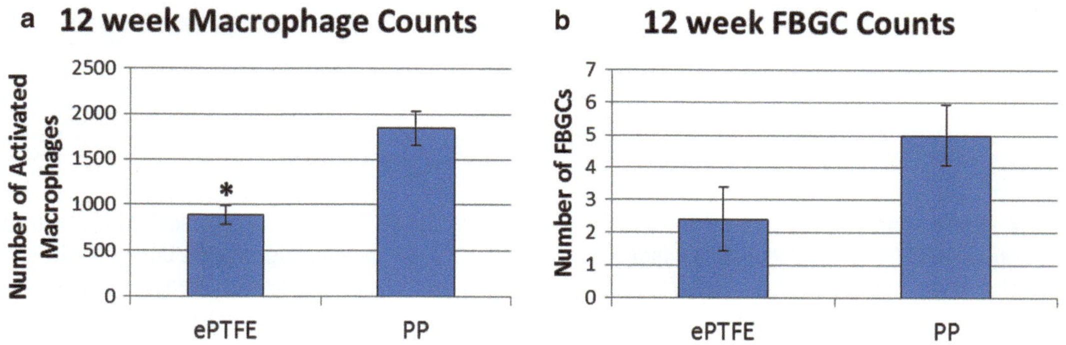

Fig. 3 Graphs depicting the number of cells counted around each material after being implanted for 12 weeks. (**a**) The average number of macrophages found surrounding the ePTFE ($n = 5$) and PP ($n = 5$) implants $*p = 0.002$. (**b**) The number of FBGCs counted surrounding each implanted material. No significant difference

implants ($n = 5$) had an average CD-68 positive macrophage count of 1862.8 ± 259.5; FBGC count of 11.6 ± 1.4 (Fig. 2).

3.3 Twelve Weeks

Expanded PTFE implants ($n = 5$) had an average CD-68 positive macrophage count of 885 ± 102; FBGC count of 2.4 ± 1. PP implants ($n = 5$) had an average CD-68 positive macrophage count of 1844.2 ± 187.7; FBGC count of 5 ± 1 (Fig. 3).

In this study a very porous mesh material (PP) was being compared to a more solid material with less porosity (ePTFE). The mesh has a greater space between the woven material which could allow and possibly encourage macrophages to infiltrate and fill this space. To quantify the inflammatory/foreign body response, an equation was developed to provide weight to various staining intensities and provide a quantitative value to the macrophage and FBGC counts. This equation, the H-score, is

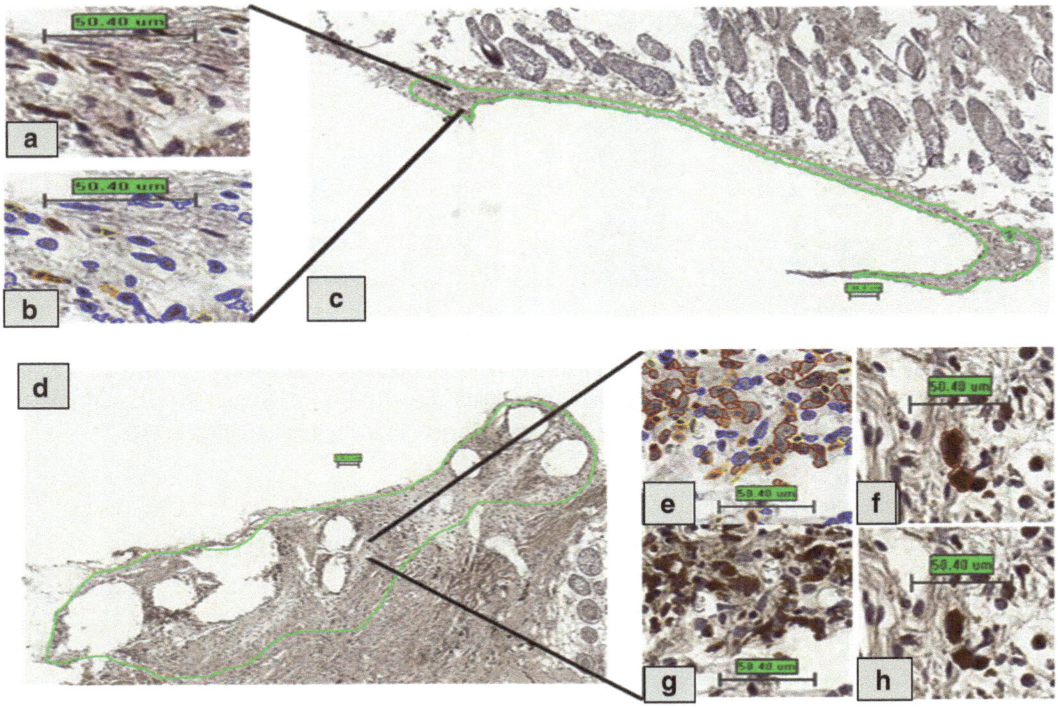

Fig. 4 Representative images of ePTFE and PP reacted with CD-68$^+$ cells, showing the DMA false color markup. (**a**) ePTFE representation of the samples reacted with CD-68$^+$ macrophages (scale bar = 50.40 μm). (**b**) False color markup of the nuclear counting algorithm, *red* = strong positive, *orange* = moderately positive, *yellow* = weak positive, and *blue* = negative (scale bar = 50.40 μm). (**c**) Macro-image of the region of interest around an implant of ePTFE. The material is not present and the majority of the measurement was performed on the superficial surface of the implant (scale bar = 100.8 μm). (**d**) Macro-image of the polypropylene implanted material (scale bar = 100.8 μm). (**e**) False color markup using nuclear counting algorithm to determine inflammatory response. *Red* = strong positive, *orange* = moderately positive, *yellow* = weak positive, and *blue* = negative (scale bar = 50.40 μm). (**f**) Micro-image of CD-68 reacted, activated macrophages (scale bar = 50.40 μm). (**g**) False color markup of FBGC in *red* (scale bar = 50.40 μm). (**h**) Micro-image of FBGC reacted with CD-68 (scale bar = 50.40 μm)

currently used by pathologists (10). The H-score is obtained by the formula:

$$
\begin{aligned}
&(3 \times \text{percentage of strongly staining nuclei}) \\
&\quad + (2 \times \text{percentage of moderately staining nuclei}) \\
&\quad + (\text{percentage of weakly staining nuclei}) \\
&\quad = \text{a range of } 0 \text{ to } 300
\end{aligned}
$$

Strongly staining nuclei were represented by red in the false color markup in the digital algorithm; moderately stained nuclei were represented by orange in the false color markup; and weakly stained nuclei were represented by yellow. Combining the H-score calculations of the counted macrophages and FBGC and dividing by two yields a weighted inflammatory score (Fig. 4).

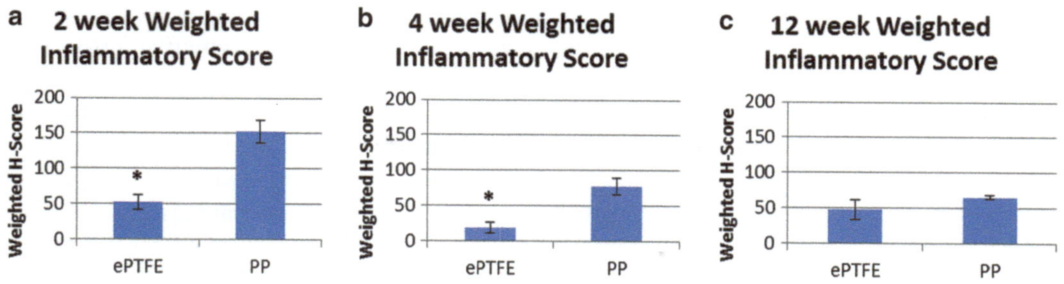

Fig. 5 Graphical representation of the weighted inflammatory score across all three timepoints. As FBGCs diminish over time, the PP weighted inflammatory score also decreases over time. (**a**) Two week weighted inflammatory score $*p = 0.003$. (**b**) Four week weighted inflammatory score $*p = 0.001$. (**c**) Twelve week weighted inflammatory score, no significant difference found

Table 1
Inflammatory index based on the weighted inflammatory score using the weighted H-score

Material	2 Weeks	4 Weeks	12 Weeks
ePTFE	Mildly reactive	Minimally reactive	Mildly reactive
PP	Moderately reactive	Mildly reactive	Mildly reactive

Expanded PTFE is mildly reactive at 2 and 12 weeks and negatively reactive at 4 weeks. PP is moderately reactive at 2 weeks and mildly reactive at 4 and 12 weeks

This will provide weighting to the presence of FBGs as well as a representative overview of the entire inflammatory and foreign body response in a single graphical representation (Fig. 5).

The inflammatory score can then be indexed using the following criteria (10): See Table 1.

$$0 = \text{minimally reactive}\,[0\ \text{to}\ 50],$$
$$1 = \text{mildly reactive}[51\ \text{to}\ 100],$$
$$2 = \text{moderately reactive}\,[101\ \text{to}\ 200],$$
$$3 = \text{strongly reactive}\,[201\ \text{to}\ 300].$$

4 Discussion

The current uses of automated digital analysis have been focused on pharmacological and toxicological effects in histopathology; therefore, much of the literature surrounding digital pathology is driven by cancer and pharmacological research. In these fields automated microscopy and computerized processing have provided increased accuracy, quantification, and standardization (11).

Currently, biocompatibility assessments using histological techniques on explanted materials and associated surrounding tissue are determined utilizing manual methods, including using photomicrographs of a selected number of high-powered fields of view and performing visual or digital measurements across these images (12). This allows bias to enter the analysis because the investigator can be drawn to areas that have a high concentration of staining while possibly ignoring areas with little or no stain. Therefore, the biocompatibility of the entire sample of material is not analyzed, and instead often only a narrow area is evaluated and reported on. Additionally, inter-investigator biasing can be an issue when more than one investigator performs measurements and sample counts. Depending on how these individuals were trained, they may interpret the histological features differently. Investigators may also perform manual evaluations over various periods of time. For example, manual evaluations for large studies may take a single or multiple investigators days or weeks to evaluate, increasing the likelihood of variations and biases that can change from day to day or week to week. Computational whole slide analysis removes these biases by performing measurements with the exact same inputs (and assumptions) across all samples being analyzed, consistently (13).

Digital analysis of histological samples represents a small, but important aspect of biocompatibility testing. By measuring the inflammatory and foreign body response of these devices, the material's biocompatibility can be evaluated. A significant advantage of performing digital analysis around biomaterials is that the investigator receives a more comprehensive overview of the entire material's biocompatibility response versus traditional manual methods that are currently used.

In the current study, two well-characterized and well-used materials in the biomedical industry were evaluated at three time-points to assess the elicited inflammatory response, with each of these materials demonstrating varying tissue-biomaterial responses. Expanded PTFE was found to be mildly reactive at 2 and 12 weeks and minimally reactive at 4 weeks based on a weighted inflammatory response. PP was found to be moderately reactive at 2 weeks and mildly reactive at 4 and 12 weeks based on a weighted inflammatory response. Whole slide digital scans of IHC-reacted slides were created and digital morphometry was used to characterize the tissue-biomaterial interface with respect to inflammation. The results reported in this study are supported by previously published studies where ePTFE elicits a lower inflammatory response when compared to PP. Expanded PTFE has demonstrated a foreign body response present through 21 days (14). Other researchers have found no difference between the inflammatory response between PP and ePTFE in abdominal implants over 28 days (15). At 56 days it has been noted ePTFE has a greater healing response related to granulation tissue formation and the foreign body response (16). In other studies PP has not demonstrated a decrease in macrophage presence between 7 and 90 days (17). The Rosch study used high-powered fields of 100 μm of the mesh; the current study uses DMA to present a more robust analysis of the tissue response surrounding the entire implant.

With an increasing number of new materials being created to support developments in science and medicine, whole slide digital scanning with algorithm-assisted morphometry could help increase the speed and accuracy of biocompatibility testing. Furthermore, these methods could help to reduce or eliminate inter-investigator biases while also providing a whole slide analysis versus limited fields of view analysis which would result in a more accurate assessment of biocompatibility. Finally, these techniques may help to improve the quality, accuracy, and reproducibility of biocompatible testing results, thus allowing a greater ability to directly compare results from different materials.

References

1. Williams DF (1987) Definitions in biomaterials: proceedings of a consensus conference of the European society for biomaterials, Chester, England, 3–5 March 1986
2. Anderson JM (2001) Biological responses to materials. Annu Rev Mater Res 31(1):81–110
3. Anderson JM (1988) Inflammatory response to implants. Am Soc Artif Implant Organs J 34(2):101–107
4. Galante JO, Lemons J, Spector M, Wilson PD, Wright TM (1991) The biologic effects of implant materials. J Orthop Res 9 (5):760–775. doi:10.1002/jor.1100090516
5. Anderson JM, McNally AK (2011) Biocompatibility of implants: lymphocyte/macrophage interactions. In: Seminars in immunopathology, vol 33, no. 3, Springer, pp 221–233. doi:10.1007/s00281-011-0244-1
6. Kidd KR, Dal Ponte DB, Kellar RS, Williams SK (2001) A comparative evaluation of the tissue responses associated with polymeric implants in the rat and mouse. J Biomed Mater Res 59(4):682–689
7. Kellar RS, Landeen LK, Shepherd BR, Naughton GK, Ratcliffe A, Williams SK (2001) Scaffold-based three-dimensional human fibroblast culture provides a structural matrix that supports angiogenesis in infarcted heart tissue. Circulation 104(17):2063–2068
8. Doussis IA, Gatter KC, Mason DY (1993) CD68 reactivity of non-macrophage derived tumours in cytological specimens. J Clin Pathol 46(4):334–336
9. Kellar RS, Lancaster JJ, Thai HM, Juneman E, Johnson NM, Byrne HG, Stansifer M, Arsanjani R, Baer M, Bebbington C, Flashner M, Yarranton G, Goldman S (2011) Antibody to granulocyte macrophage colony-stimulating factor reduces the number of activated tissue macrophages and improves left ventricular function after myocardial infarction in a rat coronary artery ligation model. J Cardiovasc Pharmacol 57(5):568–574
10. Nakopoulou L, Giannopoulou I, Gakiopoulou H, Liapis H, Tzonou A, Davaris PS (1999) Matrix metalloproteinase-1 and -3 in breast cancer: correlation with progesterone receptors and other clinicopathologic features. Hum Pathol 30(4):436–442. doi:10.1016/S0046-8177(99)90120-X
11. Słodkowska J, Filas V, Buszkiewicz E, Trzeciak P, Wojciechowski M, Koktysz R, Garcia Rojo M (2010) Study on breast carcinoma Her2/neu and hormonal receptors status assessed by automated images analysis systems: ACIS III (dako) and ScanScope (aperio). Folia Histochem Cytobiol 48(1):19–25. doi:10.2478/v10042-010-0015-1
12. Cole B, Gomoll A, Yanke A, Pylawka T, Lewis P, MacGillivray J, Williams J (2007) Biocompatibility of a polymer patch for rotator cuff repair. Knee Surg Sports Traumatol Arthrosc 15(5):632–637. doi:10.1007/s00167-006-0187-6
13. Diller RB, Kellar RS (2014) Validating whole slide digital morphometric analysis as a

microscopy tool. Microsc Microanal 1–7. doi:10.1017/S1431927614013567

14. Zhao S, Pinholt EM, Madsen JE, Donath K (2000) Histological evaluation of different biodegradable and non-biodegradable membranes implanted subcutaneously in rats. J Craniomaxillofac Surg 28(2):116–122
15. Voskerician G, Jin J, White MF, Williams CP, Rosen MJ (2010) Effect of biomaterial design criteria on the performance of surgical meshes for abdominal hernia repair: a pre-clinical evaluation in a chronic rat model. J Mater Sci Mater Med 21(6):1989–1995
16. Voskerician G, Gingras PH, Anderson JM (2006) Macroporous condensed poly (tetrafluoroethylene). I. In vivo inflammatory response and healing characteristics. J Biomed Mater Res A 76(2):234–242
17. Rosch R, Junge K, Schachtrupp A, Klinge U, Klosterhalfen B, Schumpelick V (2003) Mesh implants in hernia repair. Eur Surg Res 35 (3):161–166

Methods in Pharmacology and Toxicology (2015): 163–174
DOI 10.1007/7653_2014_38

Published online: 30 January 2015

Quantitative Histomorphometry and Quantitative Polymerase Chain Reaction (PCR) as Assessment Tools for Product Development

Robert G. Audet, Robert B. Diller, and Robert S. Kellar

Abstract

In the current chapter, 12 normal, healthy subjects were enrolled in a clinical study to assess the efficacy of a topically delivered therapeutic to improve the health and appearance of skin. Clinical and histological assessments along with immunohistochemistry and gene expression results were evaluated using quantitative methods for a comprehensive determination of the therapeutic effect. As described in the previous chapter, coupling of various analytic tools in this way can allow for a more complete assessment of a therapeutic activity, a biomedical device's success, or a combination therapy's clinical benefit where a drug coating may be delivered to a targeted area using a biomedical device as a delivery system (e.g., drug eluting stents).

The therapeutic evaluated in the current study was a topical dissolved oxygen dressing (OxygeneSys™ Continuous, AcryMed, Inc., Beaverton, OR). Clinical evaluations demonstrated that the dressing was well tolerated and several measures of skin health and integrity showed improvements compared to a control dressing site. Quantitative data from histology, immunohistochemistry, and gene expression studies demonstrated a general reduction in inflammatory response markers and transcription products (IL-6, IL-8, TNF-alpha, MMP-1, and MMP-12) while facilitating a general increase in structural skin proteins (collagen I, elastin, and filaggrin). Additionally, p53 signals from biopsy samples support the conclusion that the topical therapeutic presented no safety concerns. In summary, the data from this study demonstrated that the dressing had no deleterious effects and stimulated beneficial effects on intact, nonwounded skin. Additionally, quantitative histomorphometry and quantitative polymerase chain reaction (PCR) techniques provided unique tools to comprehensively assess clinical benefits.

Key words Quantitative histomorphometry, Histology, Immunohistochemistry, Quantitative polymerase chain reaction, qPCR, RT-PCR, Gene expression, Product development

1 Introduction

As an organ, skin serves numerous functions, including protection from external environmental insults such as pathogenic organisms, UV radiation, changes in water, humidity, and temperature, and also plays a significant role in the immune system (1). Skin health is dependent on a number of changing physiologic mechanisms that are often compromised with age. For example, wound healing is significantly compromised in the elderly (1) and these wounds can become chronic in nature and present serious clinical issues if they are left untreated (2). Our elderly population is not the only subject group with issues or concerns about skin health. As the largest and most aesthetically important organ in the body, the skin is a growing area of focus for individuals from all age groups. Geriatric people are interested in curbing the effects of age while younger people are interested in maintaining a youthful, healthy skin condition.

It has become widely accepted in the field of skin care that the nutritional supply of oxygen to the skin is primarily supplied by internal circulation that is widely available in the deeper dermal layers. However, recent data have shown that significant amounts of oxygen may be available via diffusion from the external overlying surface (3). Bioavailability of oxygen in the skin is critically important for a number of reasons. There is a close dependency between tissue oxygenation and wound healing: wounds with a pO_2 less than 30 mmHg are considered to be hypoxic and have more clinically associated challenges such as being slow to heal, having little or no granulation tissue, and having accumulations of necrotic deposits (4). In contrast, those wounds with pO_2 levels greater than 30 mmHg usually have fewer longer term clinical issues and follow a normal course of wound healing (4). Furthermore, wounds deprived of oxygen deposit collagen poorly and are easily infected. Epithelialization represents a final resolution of the wound and its mechanisms are optimized at high oxygen levels (5).

Oxygen is essential for wound healing and normal skin organ function. Since there is limited diffusion across the stratum corneum into the epidermis, the goal of the current study was to evaluate if the topical delivery of a total dissolved oxygen in dressing form on intact human skin would improve clinical and histologic skin functioning. Biopsy samples were taken from subjects at active and control sites following 8 weeks of treatment. Biopsy samples were coronally sectioned, with one half processed for histopathology to assess impact on hydration, oxidative stress, and structural proteins and the second half processed for real-time RT-PCR analysis to assess impact on inflammatory markers. These data were correlated with clinically relevant markers such as desquamation, hydration, and roughness. Results from these evaluations suggest active mechanisms are in play with the use of topical oxygen therapy to intact, healthy skin. No safety issues were seen in the current study and structurally significant and biologically relevant differences were detected as a result of 8 weeks of active treatment.

2 Materials and Methods

2.1 Human Subjects

A total of 50 healthy subjects (men and women ages 50–69 years; mean age 58.4) completed a single-site, randomized, controlled, 8-week study. Of these 50 subjects, 12 were randomly selected for biopsy collection. Subjects had age-appropriate photoaging and stable concomitant medications. Informed consent was obtained from all subjects in the study, which was approved by the Concordia Clinical Research Institutional Review Board, New Jersey.

The semiocclusive, absorbent, oxygen-enriched dressing (Active Group, OxygeneSys™ Continuous, AcryMed, Inc., Beaverton, OR) was affixed to the skin covering the anterior tibia on one limb and the contralateral limb was covered with a Kling® bandage to function as the control. A computer-generated randomization scheme determined which limb (left or right) would receive the experimental dressing. The dressing was wet with an ampule of eye moisturizer and affixed to the shin with a Kling® dressing held together with paper tape. The dressing was applied daily by the subject following bathing and worn for 24 h continuously. The location of the dressing placement was noted by the investigator with black indelible ink. Subjects were permitted to continue using their own skin care, cleansing, and makeup products but were not allowed to begin any new products for the 8-week duration of the study. No skin care products of any kind were used on the shins where the dressing was applied.

2.2 Clinical Assessments

Study subjects evaluated in a blinded manner were assessed by the same investigator (clinician) throughout the study. Dressings were removed prior to clinical grading and all parameters were evaluated at 4 and 8 weeks on a 5-point ordinal scale, from 0 (no signs or symptoms) to 4 (very dramatic signs and symptoms resulting in discomfort, representing an adverse reaction). A compliance check visit was performed at 1 week. Clinical investigator assessed efficacy parameters were desquamation, roughness, erythema, and skin texture; and tolerability parameters were itching, stinging, and burning. Digital images of each shin were collected at baseline, 4 weeks and 8 weeks. Skin hydration and water loss were measured with the appropriate Dermalab instruments and probes (Cortex Technology, Denmark): corneometer, TEWL (transepidermal water loss), elasticity, and skin coloration (6). Sensory monofilament test was performed by drawing a cotton fiber over the skin.

2.3 Biopsy

One 3 mm full thickness skin biopsy was taken from each shin (randomized active and control) of 12 randomly selected subjects at week 8. Each biopsy was coronally sectioned in half (superficial to deep) with one half immediately placed in ice-cold fixative (2 % paraformaldehyde (Electron Microscopy Sciences, Hatfield, PA) in PBS and incubated at 4 °C for 48 h for histologic and immunohistologic analyses. The remaining half was placed in ice-cold RNAlater (Sigma Chemical Company, St. Louis, MO) and incubated overnight; then stored at −80 °C until processed for real-time RT-PCR analysis.

2.4 Histology

Histologic and histopathologic analyses were performed at the 8-week time point, comparing control to active site and assessed from 5 μm serially sectioned paraffin-embedded tissues, and either stained with hematoxylin and eosin for histopathology or primary antibodies for immunohistochemistry (IHC). Primary IHC antibodies were oxidative stress DNA adduct 8-hydroxy-2-deoxyguanosine (8-OHdG) (Abcam, Cambridge, MA), water-glycerol channel aquaporin-3 (AQP3) (Santa Cruz Biotechnology, Santa Cruz, CA), structural proteins filaggrin (Vector Laboratories, Burlingame, CA), collagen I (Abcam, Cambridge, MA), and elastin (Abcam, Cambridge, MA), and processed using standard immunohistochemistry methods.

Digital, whole-slide scans (Aperio ScanScope CS) were used for all evaluations to quantify changes in levels, using established digital pathology algorithms (7). All artifacts were manually excluded prior to digital algorithm. H-score, a widely used pathology method for quantitatively evaluating staining features, was used to determine changes in staining intensity. The H-score is directly related to staining intensity, scored as 0, 1+, 2+, or 3+ of the area, cell, or object and calculated by the formula: $(3 \times \text{percentage of } 3+) + (2 \times \text{percentage of } 2+) + (\text{percentage of } 1+)$. Results range from 0 to 300 (8).

2.5 Real-Time PCR

Nucleic acids were extracted from tissues frozen in RNAlater by homogenizing in RLT lysis buffer on ice (Qiagen, Valencia, CA) with an Omni THq rotor stator (Omni International, Kennesaw, GA) for 30 s per sample. Total RNA was isolated using the RNeasy Mini Kit (Qiagen, Valencia, CA). First strand cDNA was reverse transcribed with the Superscript III synthesis system (Invitrogen, Carlsbad, CA) and was diluted 2:1 prior to addition to 10 μl PCR reactions containing 2× Taqman Universal PCR Master Mix (Applied Biosystems, Carlsbad, CA) and amplified on a 7900 Real-Time PCR System (Applied Biosystems, Carlsbad, CA) (Fig. 1). Real-time analysis was performed with the following Taqman probes: IL-6, IL-8, MMP-1 (collagenase), MMP-12 (elastase), TNF-α, TP53, VEGF and GAPDH. Samples were analyzed using the $2^{-\Delta\Delta Ct}$ method, using GAPDH as the housekeeping gene to normalize for sample to sample variations in RNA/cDNA. Data is presented as "fold change" in active vs. control.

2.6 Statistical Analysis

A paired student's *t* test was used to determine differences between the active group vs. the control group. A p value less than 0.05 was considered to be statistically significant.

Fig. 1 Applied Biosystems, Carlsbad, CA. 7900 Real-Time PCR System

3 Results

3.1 Human Subjects

A total of 50 subjects completed the study, 12 of which participated in the biopsy collection for this paper, without any major adverse events or deviations from the study design.

3.2 Clinical Measurements

There was a statistically significant increase in stratum corneum hydration from baseline to 4 weeks, 4–8 weeks, and baseline to 8 weeks ($p < 0.05$ at each time point) with an average increase in hydration of 41 μS (micro Siemens) in active vs. control from baseline to 8 weeks (Table 1) (6). There was a statistically significant decrease in the treatment efficacy parameters of desquamation, roughness and skin texture. There were no statistically significant changes in transepidermal water loss (TEWL), skin coloration monofilament sensorial measurement, or elasticity measures across all time points (data not shown).

3.3 Histology

There were no noticeable histopathologic differences between active vs. control sites at the 8-week time point with respect to acanthosis, spongiosis, chronic inflammation, hyperkeratosis, epidermal mononuclear infiltration, focal acantholysis, or dermal edema. Subtle differences were evident in epidermal thickness, vascular prominence, and occasional perivascular mononuclear cells. However, these features were concluded to be consistent with normal human skin and no trend change was observable for any of these characteristics between active and control samples. Representative histopathology is shown in Fig. 2. While there was a slightly lower level of rete pegs in the active vs. control sites, it was not statistically significant (6).

Table 1
Clinical measurements between control and treatment sites (6)

		Control	Treatment	*p* Value
Skin hydration (μS)	0–4 Weeks	−8.18 ± 44.01	15.88 ± 46.17	0.010
	4–8 Weeks	−2.04 ± 32.34	15.22 ± 41.96	0.025
	0–8 Weeks	−10.22 ± 33.91	31.10 ± 44.71	<0.001
Desquamation	0–4 Weeks	−0.27 ± 1.09	−0.49 ± 1.29	0.356
	0–8 Weeks	−0.12 ± 0.99	−1.08 ± 1.19	<0.001
Roughness	0–4 Weeks	−0.33 ± 1.12	−0.53 ± 1.24	0.396
	0–8 Weeks	−0.18 ± 0.99	−1.14 ± 1.15	<0.001
Skin texture	0–4 Weeks	−0.33 ± 1.12	−0.53 ± 1.24	0.396
	0–8 Weeks	−0.18 ± 0.99	−1.14 ± 1.15	<0.001

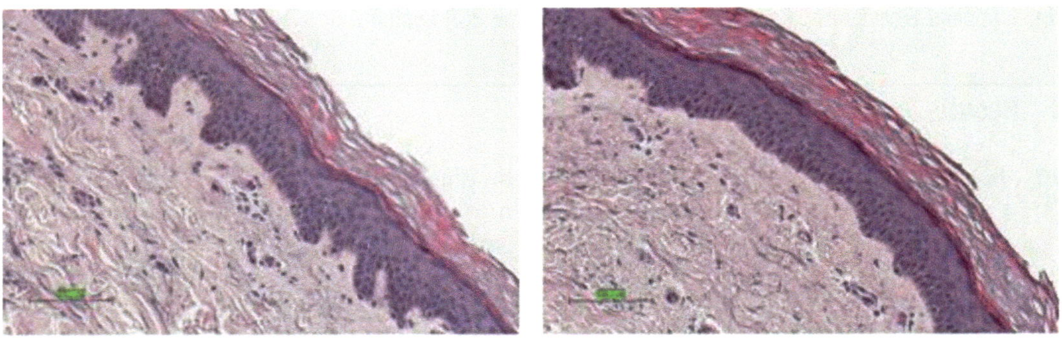

Fig. 2 Representative histopathology (hematoxylin and eosin stain). *Left*, subject #10, control site at the 8-week time point. *Right*, subject #10 active site at the 8-week time point. Scale bars = 100.5 μm (6)

3.4 Immuno histochemistry

Immunohistochemical (IHC) analysis of coronal, serially sectioned biopsies from the 12 subjects revealed a modest increase in 8-OHdG levels in active vs. control sites suggesting increased oxygen was penetrating the epidermis through the dressing, resulting in a measurable increase in DNA adduct formation.

The aquaglyceroporin channel, AQP3, showed a slight decrease in active vs. control sites; however, active sites exhibited a more "circumferential" or membrane-localized staining pattern, suggesting recruitment of AQP3 from the cytoplasm to the membrane to facilitate water and glycerol transport (Fig. 3) (6).

IHC analysis revealed an increase in H-score of filaggrin, collagen I, and elastin in active vs. control sites. While not statistically significant, these key structural protein data collectively show a trend increase, suggesting an influence on the structural organization in the skin (Fig. 4) (6).

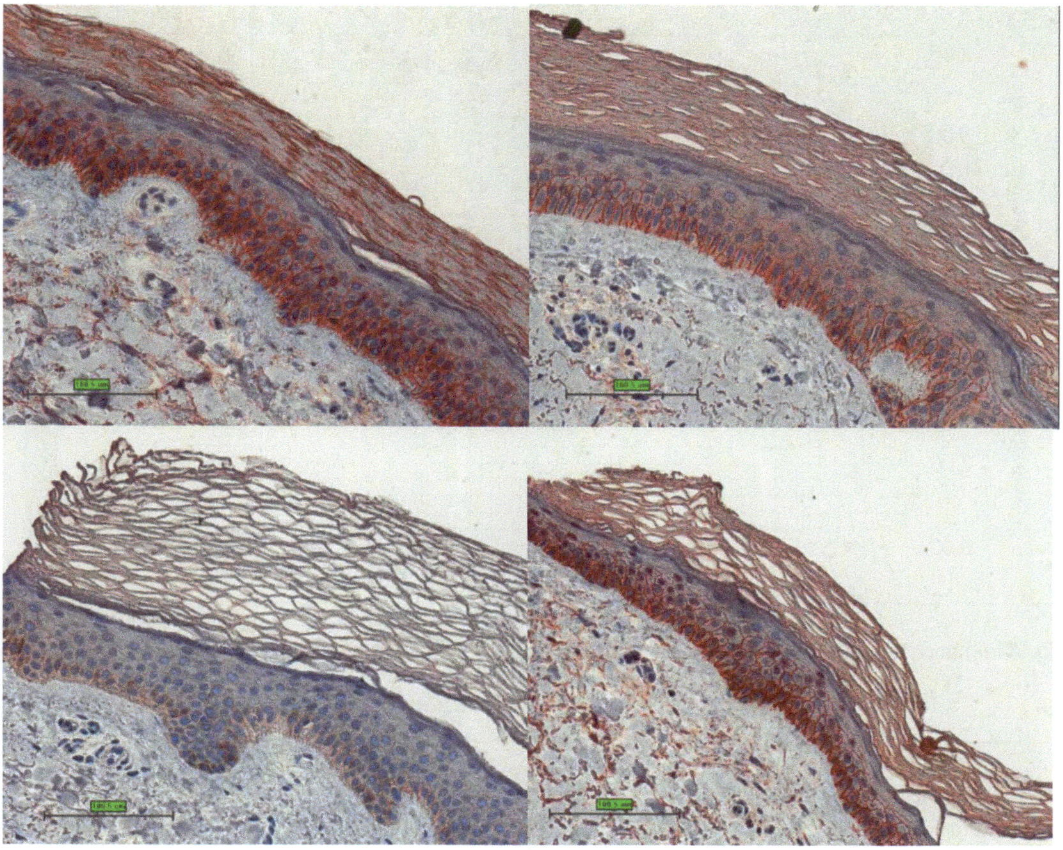

Fig. 3 Representative aquaporin-3 levels at 8 weeks. *Upper left*: subject #10, Control site. *Upper right*: subject #10, Active site. *Lower left*: subject #13, Control site. *Lower right*: subject #13, Active site. Scale bars = 100.5 μm (6)

3.5 Real-Time RT-PCR

GAPDH was used as the housekeeping gene to normalize for sample to sample variations in mRNA. Analysis was performed on a range of inflammatory, structural, angiogenic, and cellular stress genes. Pro-inflammatory cytokines IL-6, IL-8, and TNF-α showed a modest decrease in active vs. control sites. MMP-1 (collagenase) and MMP-12 (elastase) showed a more robust decrease. During inflammation or damage, both MMP-1 and -12 are upregulated to degrade and remodel extracellular matrix (ECM). Trend analysis of these markers suggests consistency in the data and a mechanism of downregulating the expression of inflammatory markers (Fig. 5).

VEGF expression levels were unchanged, suggesting the oxygen dressing neither decreased tissue pO_2 levels to hypoxic levels nor increased pO_2 to hyperoxic levels, both of which would lead to an increase in VEGF expression. TP53 (p53) levels were unchanged. As a central monitor of cellular stress and its environment, including sensing reactive oxygen species (ROS)

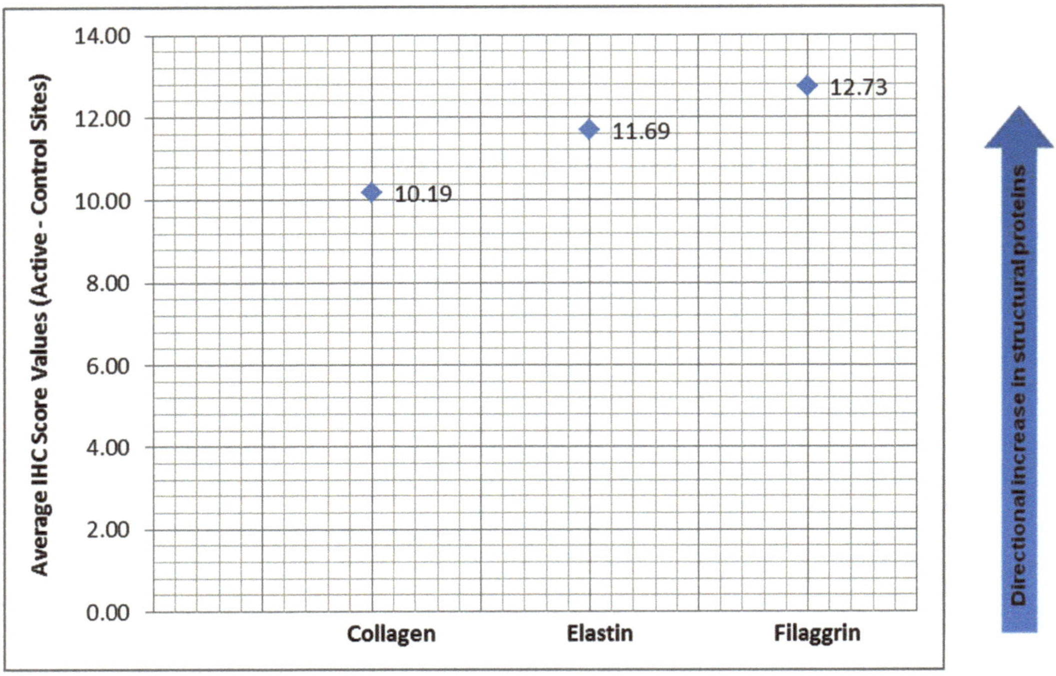

Fig. 4 Immunohistochemistry trend plot of structural proteins collagen I, elastin, and filaggrin at 8 weeks (6)

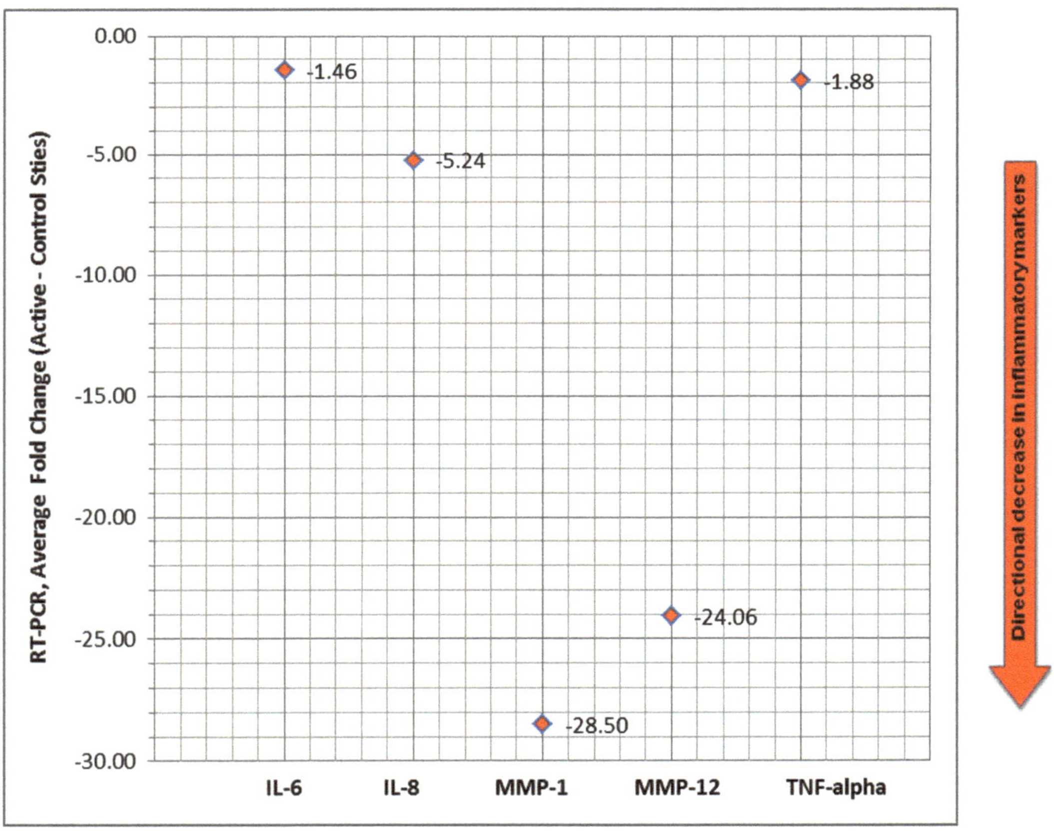

Fig. 5 Real-time RT-PCR trend plot of Inflammatory Markers at 8 weeks (6)

levels, the data suggests the skin covered by the topical oxygen dressing remained healthy with respect to p53 levels, similar to the control.

4 Discussion

In the current study, normal, healthy subjects were enrolled in a pilot clinical evaluation to assess the efficacy and tolerability of a dissolved oxygen dressing for improving skin health and appearance. The dressing provided an oxygen-enriched environment that may promote a favorable milieu for the promotion of healing in patients. Clinical analysis was performed on 50 subjects; histological and gene expression analyses were performed from biopsies on 12 randomly selected subjects. Clinical data indicates that the dressing is well tolerated and several measures of skin health and integrity showed improvement compared to the dressing-only control site. No safety issues were seen during the 8-week study period. This data was supported by the immunohistochemistry and gene expression studies, which showed a general reduction in inflammatory response markers and transcription products concomitant with a general increase in structural proteins. Additionally, there was a significant decrease in investigator-measured desquamation, roughness, and skin texture in active vs. control sites.

The increase in 8-OHdG levels in active vs. control sites suggests the dissolved oxygen dressing increased oxygen levels in the skin. More specifically, there was an increase in 8-OHdG in active vs. control epidermis, suggesting additional oxygen, delivered by the dressing, penetrated the skin and caused an increase in *OH-intermediates. The DNA analogues appear to be managed by normal physiological processes without deleterious consequences. If the DNA damage were beyond the capacity of the cell to manage, we would anticipate seeing an upregulation of p53 mRNA expression.

The levels of both VEGF and p53 were unchanged, suggesting the increased oxygen is within acceptable levels within the cells. Since there was an increase in production of 8-OHdG yet no change in p53 expression, we can infer that the topical oxygen therapy is penetrating the skin and resulting in positive changes that are not causing excessive stress to the integument. This strongly suggests an active therapy mechanism is in effect without negatively impacting cell and tissue health. Furthermore, VEGF levels were unchanged in active vs. control sites, suggesting the dissolved oxygen dressing created no hypoxic or hyperoxic states (9). VEGF expression is extremely sensitive to deleterious physiologic changes in oxygen levels. Therefore, since VEGF expression in the current study was unchanged and p53 levels were effectively unchanged, we can conclude that the therapeutic conditions were safe at the cell and tissue level.

Decreased expression of IL-6, IL-8, TNF-α, MMP-1, and MMP-12 also suggests insignificant stress within the cells and skin. Inflammatory markers IL-6, IL-8, TNF-α, and the matrix metalloproteinases MMP-1 (collagenase) and MMP-12 (elastase) are known to be upregulated when inflammatory processes or mechanisms are in play. They are secreted by leukocytes and regulate a wide range of cellular and tissue responses, recruiting macrophages, neutrophils, inducing angiogenesis, and inducing remodeling of damaged tissue (10). In this study, expression of these markers was downregulated in active vs. control sites, suggesting a decrease in inflammatory pathways or mechanisms and a decrease in breakdown or remodeling of the extracellular matrix within the integument following active treatment.

The increased structural protein levels of collagen I and elastin may be due to reduced protein turnover by a decrease in MMP-1 and MMP-12 mRNA levels. These structural proteins are critically important to the architecture of the skin as an organ. Greater collagen I presence in the dermis may provide skin with greater structural integrity. Further consistency in the data within the current study lies in the fact that three major structural skin proteins (filaggrin, elastin, and collagen I) all increase in their expression in active sites when compared to control sites in the same subject. Furthermore, elastin is a structural protein found in the dermis as well as other critical tissues such as blood vessels, heart, bladder, and ligaments where it provides physiologically relevant elasticity. Elastin levels appear to be higher in active vs. control sites. Greater elastin presence in the dermis may provide skin with greater structural integrity and elasticity. Filaggrin facilitates the organization and condensation of keratinocytes at maturation, contributing to the functional stratum corneum. The increased levels of filaggrin in active vs. control sites, suggests an increased production of filaggrin, likely contributing to more structurally sound barrier and water retention functions. Increased filaggrin levels correlate with the increase in investigator-measured skin hydration and decreases in skin texture, roughness, and desquamation. A trend increase of these three structural proteins suggests consistency in the data and a mechanism of positively influencing the expression of key structural proteins (Fig. 4). Because these proteins trend together, we can conclude with a higher level of confidence that the active treatment is stimulating beneficial structural changes within the underlying skin. A summary of the overall findings in the current study is given in Table 2.

The blinded clinical investigator's measurements of decreased desquamation, roughness, and skin texture in treated vs. control sites correlate with the significant increase in skin hydration values. This data is supported histologically by an increase in filaggrin,

Table 2
Data summary (6)

	Marker	Analysis	Result
Stress monitor	8-OHdG	IHC	Increase
	p53	PCR	No change
	VEGF	PCR	No change
Stratum corneum barrier	AQP-3	IHC	Slight decrease redistributed to membrane
	Hydration	Clinical	Increase
	Desquamation	Clinical	Decrease
	Roughness	Clinical	Decrease
	Skin texture	Clinical	Decrease
Inflammation markers	IL-6	PCR	Decrease
	IL-8	PCR	Decrease
	TNF-α	PCR	Decrease
	MMP-1	PCR	Decrease
	MMP-12	PCR	Decrease
Epidermis structure	Filaggrin	IHC	Increase
	Elastin	IHC	Increase
	Collagen I	IHC	Increase

resulting in increased production of natural moisturizing factors and a redistribution of aquaglyceroporin, AQP3, from the cytoplasm to the membrane.

In summary, the quantitative histomorphometry and quantitative PCR data from this study demonstrate that the dressing has no deleterious effects and appears to stimulate beneficial effects on intact, nonwounded skin. Additionally, clinical assessments, histomorphometry, and RT-PCR data correlate with one another which helps to strengthen study conclusions from the 12 patient subset that was evaluated via skin biopsies. In this way, these methods can be used to help corroborate findings in studies where sample sizes for available biopsies may be limited due to study or clinic restraints.

References

1. Gosain A, DiPietro LA (2004) Aging and wound healing. World J Surg 28:321–326
2. Lazarus GS, Cooper DM, Knighton DR et al (1994) Definitions and guidelines for assessment of wounds and evaluation of healing. Arch Dermatol 130:489–493
3. Roe DF, Gibbons BL, Ladizinsky DA (2010) Topical dissolved oxygen penetrates skin: model and method. J Surg Res 159:e29–e36
4. Heng MC, Harker J, Bardakjian VB, Ayvazian H (2000) Enhanced healing and cost-effectiveness of low-pressure oxygen therapy in healing necrotic wounds: a feasibility study of technology transfer. Ostomy Wound Manage 46:52–60
5. Cianci P, Hunt TK (1993) Adjunctive hyperbaric oxygen therapy in the treatment of diabetic wounds of the foot. In: Levin ME, O'Neal LW, Bowker JH (eds) The diabetic

foot, 5th edn. Mosby Year Book, St. Louis, MO, pp 305–319

6. Kellar RS, Audet RG, Roe DF, Rheins LA, Draelos ZD (2013) Topically delivered dissolved oxygen reduces inflammation and positively influences structural proteins in healthy intact human skin. J Cosmet Dermatol 12:86–95
7. Potts SJ, Young GD, Voelker FA (2010) The role and impact of quantitative discovery pathology. Drug Discov Today 15:943–950
8. McCarty KS Jr, Miller LS, Cox EB et al (1985) Estrogen receptor analysis: correlation of biochemical and immunohistochemical methods using monoclonal antireceptor antibodies. Arch Pathol Lab Med 109:716–721
9. Valko M, Rhodes CJ, Moncol J et al (2006) Free radicals, metals and antioxidants in oxidative stress-induced cancer. Chem Biol Interact 160:1–40
10. Lee Y, Je YJ, Lee SS et al (2012) Changes in transepidermal water loss and skin hydration according to expression of aquaporin-3 in psoriasis. Ann Dermatol 24:168–174

Methods in Pharmacology and Toxicology (2015): 175–182
DOI 10.1007/7653_2014_19

Published online: 28 September 2014

Measuring the Messenger: RNA Histology in Formalin-Fixed Tissues

Steven J. Potts, Mirza Peljto, Mahipal Suraneni, and Joseph S. Krueger

Abstract

RNA in situ hybridization (ISH) complements immunohistochemistry (IHC) by confirming localized gene expression in a tissue. Effective use of ISH requires careful communication and coordination between molecular biologists, histologists, and pathologists. This chapter reviews the literature and experience of the authors in the use of formalin-fixed paraffin-embedded tissues for RNA histology techniques.

Key words RNA in situ hybridization, In situ hybridization (ISH), RNA histology, Chromogenic in situ hybridization (CISH), Formalin-fixed, Paraffin-embedded (FFPE) in situ hybridization, RNA-FISH

1 Introduction

The understanding of the functionality of RNA as both a messenger for protein translation and a self-contained regulatory element has greatly expanded in recent years. There has been long-standing interest in evaluating the contextual expression of RNA in tissue sections or cells, a technique known as in situ hybridization. Over the last four decades, researchers have introduced and refined methodology for evaluating RNA in situ, especially on low-expressing RNA. RNA molecules are difficult to work with in part due to their instability. RNAs can have extremely low half-lives, and the signal time can vary within the same gene [1] (e.g., values from under 10 min half-life to over 3 h have been recorded for different transcripts of c-myc). In addition to overall instability, RNAs are readily degraded by RNases, which are present in the tissue and may begin degrading the RNA signal immediately after surgical removal of the biopsy. While studies investigating the effects of pre-analytical variables (ischemia, fixation time and conditions, etc.) on IHC assays are well documented for clinical markers, little is understood on the effects of pre-analytical variables on RNA ISH. This may explain in part why RNA ISH has limited clinical applications, especially in the context of molecular diagnostics. More common clinical uses include viral infections tests and interphase cytogenetics while currently, the most valuable applications for RNA ISH technology in the biopharma industry are in supplementing IHC studies.

2 Appropriate Applications for RNA ISH in Pharmaceutical Drug Development

RNA in situ hybridization can be used as a stand-alone tool for examination of gene expression within tissue or as a complementary technique to protein expression studies done by IHC. In some cases a reliable antibody reagent may not be available and ISH can be used as a primary method of examining changes in gene expression. Considering the discovery of long noncoding RNAs and miRNAs as well as our understanding of their function bring RNA molecules to the forefront of biology and biomedicine. RNA ISH has also recently gained some ground as a useful technique in molecular histopathology. In the context of cancer diseases, molecular diagnostics (e.g., Her2) are generally based on assessment of protein expression in the context of tumor tissue. One could hypothesize that RNA ISH could be utilized as a complementary strategy for tumor staging, especially in cases where the status based on IHC is equivocal and cannot be resolved.

In the context of molecular pathology studies performed in our laboratory, most work on RNA involves either a discovery program on a target which does not have a suitable tissue-based antibody or mRNA expression assay to confirm expression patterns observed by protein studies with IHC.

RNA ISH also provides a genuine possibility to quantify RNA expression in the context of normal and diseased tissues. However, such quantification methods rely on development of novel technological advances that allow for reliable quantification of gene expression in the context of the tissue. The use of proper image analysis-based solutions may provide for a quantifiable platform for RNA ISH in diagnosis and evaluation of human disease.

3 Target and Signal Amplification Approaches

RNA ISH was initially pioneered by developmental biologists as whole embryo mounts provided an excellent platform for evaluation. The last several decades have witnessed a wide array of creative techniques in improving the signal to noise in experiments from low expressing tissue mRNA. Some innovations have become commercially available and heavily marketed techniques, others remain accessible to the creative researcher with time available to reproduce the technique published from an academic lab. As with every approach in tissue, each pro has a con—the better the pharmaceutical researcher understands the underlying methodology, the more able to apply the right amplification technique for a give situation. What follows is a brief history of the evolving techniques. Developments in radiolabeled ISH are not covered.

ISH originates from FISH (fluorescent in situ hybridization), which was originally developed in 1969 [2, 3] for cytological samples. FISH, as its name implies, uses fluorescent signal to reveal DNA or RNA molecules, utilizing a large probe directly coupled to fluorophores. As FISH evolved, probes were coupled to either biotin or digoxigenin (DIG), and then conjugated to either a fluorescence or colorimetric amplification assay (e.g., alkaline phosphatase or horseradish peroxidase).

RNA ISH is a balance between stringent specificity (i.e., identifying the transcript of interest) and strength of signal with tissue penetration (i.e., short probes tend to penetrate better). A major step forward came in 1998 with the introduction of using up to five 50 bp probes per gene—e.g., each probe coupled to five fluorochromes [4]. This achieved a 25-fold signal increase, and individual RNA molecules could then be viewed as dots with a fluorescent microscope. A further improvement in 2008 modified this technique to utilize up to 50 short 17-22mer probes with a single fluorophore label—*50 RNA-fluor*, with a minimum three nucleotide spacing between them [5]. In larger RNA targets, the spacing and location of probes could be chosen based on optimum GC content (45 %). In theory, this technique would amplify signal 50-fold for an experiment using 50 probes.

While this approach has been predominately a fluorescence-based technology, in 2012, an RNA-enzyme$_n$ conjugate approach with large (up to 938 bp) probes was published called TransISH [6] that returns to the original FISH techniques but with enzyme rather than fluorochrome directly coupled. While it will have the same disadvantages of FISH, it may improve signal to noise and is a technique that will likely continue to evolve for brightfield interpretation of RNA.

In 1992, the first paper was published utilizing an alternative amplification approach, TSA (tyramine signal amplification) for ISH [7]. Since then, this has been a useful approach with many applications. HRP deposits labeled tyramine at the sites of probe-target interaction, which is then further amplified with an enzymatic-chromogenic reaction. TSA can potentially amplify signal 100-fold over conventional FISH, although improvements of 10–30 are more likely in practice [8].

Rolling circle amplification (RSA), using padlocked probes, has proved useful for profiling small mutation changes in RNA in situ. The RSA technique mimics the replication of certain viral genomes with circularized oligonucleotides, and combined with padlock probes, provides extreme sensitivity to point mutations in tissue [9, 10].

Branched DNA (bDNA) probes is a technique that has received a recent surge of commercial promotion by two companies with related technologies. Originally developed in 2001 [11], a series of probes are hybridized sequentially, generating either chromogenic or fluorescent signals. The system contains four sets of probes: target RNA-specific probes of multiple (>10) oligonucleotides, pre-amplifier probes hybridizing to the target RNA-specific probes, multiple amplifier probes hybridizing to the pre-amplifier probes, and labeled probes conjugated to the amplifier probes. For a large RNA molecule that allows spacing for 20 oligonucleotide probes, up to 8,000 fluorochromes can be labeled (20 probes × 20 amplifier binding sites × 20 label probe binding sites). Theoretically this would yield 160 times greater sensitivity than the multiple singly labeled probe technique described earlier.

In situ PCR has emerged as another method for detection of RNA transcripts within tissues. It is based on a PCR reaction taking place on tissue and then the subsequent localization of transcript product within tissue. The technique is generally used for amplification of low copy gene signals and is to a great degree used for research in viral and infection diseases (for review, see [12]).

Digoxigenin (DIG) is a small molecule derived from plants that can be used to label nucleic acids but also serves as an immuno-tag, thus specifically capable of labeling probes and detection during RNA ISH. Depending on the ultimate use of the probe being generated, DIG can be incorporated into individual probes in several distinct ways that include DIG-modified nucleotides incorporated during PCR reaction by a polymerase and nick translation labeling. Since the early 1990s, DIG-labeled probes have been utilized during ISH studies [13]. DIG probes have some significant advantages over radiolabeled probes [13].

ISH (for DIG-labeled probes) can be subdivided into several specific steps: (1) Probe design, probe preparation, tissue preparation, permeabilization of tissue (usually using proteinase K), hybridization of probe to target mRNA, antibody-phosphatase binding to RNA probe, and antibody staining.

The detection of microRNA (miRNA)—small 18–22 base pair nucleic acids with many known and unknown function—is challenging because their small size precludes usage of many techniques that are useful with larger RNA targets. Researchers have been successful using DIG-labeled locked nucleic acid (LNA) probes for miRNA in frozen [14] and FFPE [15] tissue.

4 Pre-analytic Tissue Factors

Formalin fixation should ideally be less than 24 h. Long periods of formalin storage cause covalent linkages between mRNA and proteins, resulting in less accessible targets.

5 Assay Optimization

Others have looked at optimal conditions for tissue preparation and hybridization steps [16]. In breast cancer formalin-fixed sections, the following conditions serve as useful guidelines for commercial DIG-labeled DNA probes for CISH RNA analysis in tissue [17].

Dewaxing: Three 5-min washes with xylene at 55 °C.

Proteolytic digestion. SPoT-Light Tissue Pretreatment Kit (Zymed) for between 4 and 6 min, and should be optimized for each tissue type (e.g., kidney is typically shorter).

Probe concentrations. Highly dependent on the protocol used, with inconsistent weak signals with too low concentrations and nonspecific binding and fuzzy background with too high concentrations. Typical ranges between 50 and 100 ng/ml.

Hybridization buffer: Sigma hybridization solution (H-7782, Sigma-Aldrich), although others can give equivalent results.

Post-hybridization wash: Nonformamide wash in SSC, but use of formamide should be tested based on the length of the probes. Higher percentage of formamide (e.g., 20 %) can be used to increase stringency of binding, especially for higher GC content mRNA.

6 Use of Controls

The advice from the ISH literature from two decades ago is as relevant then as now—common sense is the best control for ISH interpretation [18]. The interpreter of the slides should be asking, "Is this the tissue where I would expect to see expression?," "Would I expect expression in these types of cells in this tissue?," and "where did I expect to see expression in these cells—the nucleus, cytoplasm, or secreted?" Knowledge of RNA function is still incomplete, so surprises should be noted and evaluated, and further tests explored to determine whether the result is biology or due to experimental error. We recommend that those working on an RNA target first generate a hypothesis regarding expression in the organ architecture, the specific cells, and the localization within those cells. Approaching mRNA work with an a priori tissue hypothesis will better highlight surprises, which can then lead to more systematic methodology for testing reproducibility. With IHC development, there are multiple online resources from both academic groups and antibody vendors for confirming observed nucleus, cytoplasm, or membrane expression patterns with known results, but RNA ISH has yet to develop such sources for the community.

In our experience, the best approach is to utilize two fundamentally different ISH analytical techniques, to determine if similar results are reached. These should not share the same target or signal amplification approaches (e.g., avoid using two commercial approaches that both stem from branched DNA).

Controls transcribed in the sense direction are useful, but one should rule out control errors due to hybridizing of regulatory antisense elements. Alternative negative controls include antisense experiments on genes known for certain to be absent in the tissue of interest and testing for the loss of signal after RNase digestion. Good positive controls include evaluating the identical experiment with a probe just upstream or downstream of the probe of interest. For example, with methodology utilizing multiple probes on a single RNA, one can use software to alternate the probes arranged against a gene into different experiments, preferably with the same overall GC content. Additional positive controls should be run on tissues known to contain the target of interest, although this can be difficult when working on pharmaceutical discovery projects with novel targets.

RNA degradation will continue to be one of the most difficult areas to evaluate effectively. There is active work in the industry on better controls for RNA degradation. One should match a control for RNA degradation with a similar expression level as the gene of interest (e.g., commonly used control approaches look at highly expressed RNA, which will not be helpful information on a low expressor).

A number of cell types will give false positive results (Table 1). When lower levels of these cell types are consistently in the tissues of interest, they can be used as controls. Another common error, in both ISH and IHC, is the mistaking of infiltrating immune cells for tumor. Many of these immune cells will exhibit different binding patterns of ISH than the surrounding tissue, and a pathologist must be involved in the study to exclude these cell types. One also needs to be aware of edge effects in tissues; differential ISH binding patterns around the exterior regions of a tissue should be avoided during analysis.

Table 1
Cell types with tendencies towards false positive RNA expression in tissue experiments

Cell type	Comment	Reference
Eosinophilic granulocytes	Good control in most tissues, but high numbers in some bone marrows and lymphomas	[20]
Paneth cell	Gastrointestinal tract	[21]
Lipofuscin	Brain sections	[22]
Various infiltrating immune cells	These can be mistaken for tumor or target tissue, and often have distinctive binding patterns to RNA	

7 IHC and ISH

When IHC and ISH are conducted for multiplexed evaluation, extreme care should be taken. Formamide, dextran sulfates, and some prehybridization treatments can impact antigen detection. Shorter probes need less stringent conditions for both hybridization and post-hybridization washings. For example, a 20mer DNA probe can utilize 28–37 °C and 10 % formamide, while a 50mer DNA probe may require 30–37 °C and 50 % formamide. The more mild conditions will have less effect on antigen-antibody bonds. When using IHC prior to ISH, it is important to include controls for ISH that did not have IHC performed previously and to double-check that increased RNA degradation is not occurring during the IHC steps. There are trade-offs between whether to perform ISH or IHC first. Alternatively, they can be performed simultaneously, a technique that has become known as immuno-FISH [19].

8 Automation Approaches

The ability to run large numbers of ISH samples reproducibly is still in its infancy. Previous automation approaches have included use of Little Dipper (SciGene, Sunnyvale, CA) and automated workstations like Tekan (Mannedorf, Switzerland). The major autostainer companies (e.g., Dako, Leica, Ventana, Biocare) all offer ISH automation on their platforms. The user must navigate both the SOPs provided by the automation vendor as well as other directions from the probe manufacturers. While reproducible automation results can be achieved, in our experience each new probe approach requires its own optimization and automation strategy, and adequate time and materials should be budgeted for these activities.

9 Conclusion

Histology techniques for evaluation of RNA expression with ISH continue to evolve, and have a rich history of innovation over the last several decades. Improvements will continue in areas like controls for RNA degradation, improved signal amplification of low expressing RNA molecules, and automation platforms. RNA ISH serves as a useful companion to protein-based IHC development when interrogating the local tissue environment.

References

1. Eick D et al (1985) Aberrant c-myc RNAs of Burkitt's lymphoma cells have longer half-lives. EMBO J 4(13B):3717
2. Gall JG, Pardue ML (1969) Formation and detection of RNA-DNA hybrid molecules in cytological preparations. Proc Natl Acad Sci 63(2):378–383
3. John H, Birnstiel M, Jones K (1969) RNA-DNA hybrids at the cytological level. Nature 223(5206):582–587

4. Femino AM et al (1998) Visualization of single RNA transcripts in situ. Science 280 (5363):585–590
5. Raj A et al (2008) Imaging individual mRNA molecules using multiple singly labeled probes. Nat Methods 5(10):877–879
6. Kitaoka M et al (2012) Transglutaminase-mediated in situ hybridization (TransISH) system: a new methodology for simplified mRNA detection. Anal Chem 84(14):5885–5891
7. Adams J (1992) Biotin amplification of biotin and horseradish peroxidase signals in histochemical stains. J Histochem Cytochem 40 (10):1457–1463
8. Chen CH et al (2011) Specific sorting of single bacterial cells with microfabricated fluorescence-activated cell sorting and tyramide signal amplification fluorescence in situ hybridization. Anal Chem 83(19):7269–7275
9. Larsson C et al (2010) In situ detection and genotyping of individual mRNA molecules. Nat Methods 7(5):395–397
10. Nilsson M (2006) Lock and roll: single-molecule genotyping in situ using padlock probes and rolling-circle amplification. Histochem Cell Biol 126(2):159–164
11. Player AN et al (2001) Single-copy gene detection using branched DNA (bDNA) in situ hybridization. J Histochem Cytochem 49 (5):603–611
12. Bagasra O (2007) Protocols for the in situ PCR-amplification and detection of mRNA and DNA sequences. Nat Protoc 2 (11):2782–2795
13. Komminoth P (1992) Digoxigenin as an alternative probe labeling for in situ hybridization. Diagn Mol Pathol 1(2):142–150
14. Song R, Ro S, Yan W (2010) In situ hybridization detection of microRNAs. In: RNA therapeutics. Springer, pp 285–292
15. Jørgensen S et al (2010) Robust one-day in situ hybridization protocol for detection of microRNAs in paraffin samples using LNA probes. Methods 52(4):375–381
16. Evans MF, Aliesky HA, Cooper K (2003) Optimization of biotinyl-tyramide-based in situ hybridization for sensitive background-free applications on formalin-fixed, paraffin-embedded tissue specimens. BMC Clin Pathol 3(1):2
17. Lambros MB et al (2006) Unlocking pathology archives for molecular genetic studies: a reliable method to generate probes for chromogenic and fluorescent in situ hybridization. Lab Invest 86(4):398–408
18. Wilcox JN (1993) Fundamental principles of in situ hybridization. J Histochem Cytochem 41 (12):1725–1733
19. Zirbel R et al (1993) Evidence for a nuclear compartment of transcription and splicing located at chromosome domain boundaries. Chromosome Res 1(2):93–106
20. Niedobitek G et al (1989) Detection of viral genomes in the liver by in situ hybridization using 35S-, bromodeoxyuridine-, and biotin-labeled probes. Am J Pathol 134(3):633
21. Garrett KL, Grounds MD, Beilharz MW (1992) Nonspecific binding of nucleic acid probes to Paneth cells in the gastrointestinal tract with in situ hybridization. J Histochem Cytochem 40(10):1613–1618
22. Steiner I et al (1989) Effects of lipofuscin on in situ hybridization in human neuronal tissue. J Virol Methods 24(1):1–9

Methods in Pharmacology and Toxicology (2015): 183–189
DOI 10.1007/7653_2014_21

Published online: 30 September 2014

Algorithm-Driven Image Analysis Solutions for RNA ISH Quantification in Human Clinical Tissues

Mirza Peljto, Joseph S. Krueger, Nicholas D. Landis, G. David Young, Steven J. Potts, and Holger Lange

Abstract

RNA in situ hybridization (ISH) offers an unprecedented advantage of detecting RNA transcript expression in tissues at cellular resolution while maintaining critical information on tissue morphology and architecture. Lack of standardized scoring paradigms for RNA ISH poses a challenge to translation of RNA ISH into clinical settings. Thus, development of novel tools for RNA ISH quantification and analysis is of significant interest. Here, we demonstrate a novel and proprietary algorithm-driven image analysis-based approach for quantification and analysis of RNA ISH in human clinical tissue samples. This approach allows RNA ISH analysis on cell by cell basis for whole tissues. We discuss in detail methods by which image analysis can be utilized for successful detection and interpretation of ISH signal, and provision of pertinent biological endpoints. Strategies discussed herein highlight the potential of combining RNA ISH and image analysis tools to build scoring paradigms and high-throughput approaches for the analysis and quantification of gene expression in tissues. Once such methods are adopted, RNA ISH may become a frequently utilized technique in clinical settings and biomarker and drug discovery.

Key words Digital image analysis, Tissue image analysis, RNA in situ hybridization, Chromogenic in situ hybridization (CISH), RNA-FISH, Digital pathology

1 Introduction

The ability to visualize RNA molecules in tissues has led to fascinating discoveries over the past few decades. RNA in situ hybridization is a molecular biology technique that has facilitated many important biological discoveries. RNA in situ hybridization (ISH) employs a biochemically labeled complementary RNA or DNA strand probe for detection of RNA transcripts in the context of tissues and individual cells. RNA signal can be revealed by using chromogenic (CISH) or fluorescence (FISH) in situ hybridization. RNA CISH gives one unprecedented advantage over RNA FISH in that the transcript can be monitored in the context of native tissue morphology. Theoretically, this would allow for utilization of RNA CISH in clinical settings as a diagnostic tool but due to lack of any standardized scoring paradigms RNA CISH is not widely used as such. Thus, development of RNA CISH scoring paradigms would likely facilitate the use of RNA CISH as a diagnostic tool alone or in combination with IHC assays.

When coupled to whole-slide imaging, image analysis (IA) tools provide a robust platform for quantitative analysis of RNA CISH in human clinical samples to evaluate gene expression for a wide variety of clinically relevant biomarkers. New image analysis algorithms enable quantification of RNA signal in individual cells across the entire tissue sample with a high degree of precision and accuracy. This poses a unique opportunity for the development of standardized scoring paradigms for RNA CISH and could be the driving factor in efficient use of RNA CISH in the clinical settings.

As a proof of principle study, here we discuss the utilization of CellMap™ (Flagship Biosciences, Boulder, CO) for quantification of Her2 RNA in human breast tissues using a combination of image analysis, whole slide imaging, and RNA CISH. We discuss in great detail parameters that allow for appropriate detection of cells and interpretation of RNA CISH signal by the algorithm on cell-by-cell basis. Several distinct ways for quantification of RNA CISH are presented and discussed. Finally, IA of RNA CISH is compared to Her2 IHC IA scores in order to test the degree of concordance between the two IA-driven scoring paradigms.

2 Image Analysis of Her2 CISH and FISH in Human Breast Cancer Tissues

High-magnification (40×) imaging of the tissues provides for analysis and quantification of RNA ISH on cell-by-cell basis. In this context, RNA signal is generally revealed as individual dots within or surrounding the nuclear area of the cell (see Fig. 1). In combination with nuclear marker stain RNA ISH signal can be deduced for each individual cell within the tissue (i.e., millions of cells/tissue section). This leads to a comprehensive evaluation of cellular substrates for expression of a single gene or combinations of genes in individual cells. Algorithm-based image analysis provides for a series of inputs that drive the identification of RNA signal, detection of individual nuclei, and quantitative analysis of RNA signal in each cell.

Several distinct parameters must be defined for an algorithm to drive appropriate identification and detection of cells and RNA signal. As a basic parameter for cellular detection, CellMap™ algorithm identifies individual cells based on the nuclear marker stain (DAPI for FISH and hematoxylin for CISH; Fig. 2) as well as nuclear size and shape parameters, and overall sensitivity imposed by nuclear detection settings. Marker stain is defined by specific color setting parameters that can be identified for each color specifically. These parameters can be defined separately from parameters that define individual RNA dots. Cellular space can be defined in the context of nuclei (i.e., area 2 μm surrounding the nucleus).

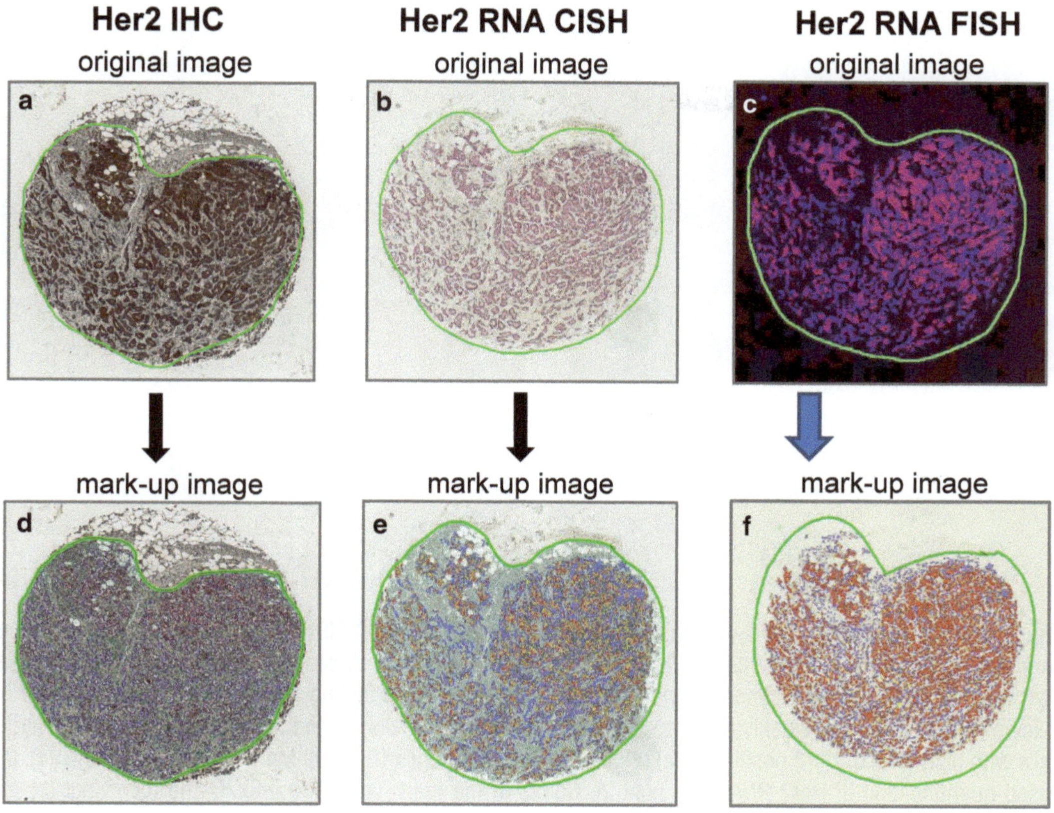

Fig. 1 Whole-slide image analysis of Her2 IHC and RNA CISH. (**a**) Original Her2 IHC image; (**b**) original Her2 CISH image; (**c**) Original Her2 Fish image; (**d**) Her2 IHC image mark-up post-image analysis; (**e**) Her2 RNA CISH image mark-up post-image analysis. *Green line* in all images reflects ROI manually defined by the user prior to image analysis; (**f**) Her2 RNA Fish mark-up post-image analysis (1 × magnification)

Individual dots too are defined in part by color separation from or combination with the nuclear marker stain as well as dot size and shape parameters. For example, user can specify a reference dot size (which can be empirically determined) in this way creating a platform by which the algorithm can separate dot aggregates into appropriate numbers of individual dots (see Fig. 2). Once the detection settings and inputs are specified for individual cells and RNA dots, algorithm analysis can be validated on smaller tissue sections. Once confirmed, algorithm analysis can be run as a high-throughput approach on an unlimited number of whole tissue sections (Fig. 1).

While the general principles of algorithm approaches for RNA FISH and CISH are similar, several specific differences must be considered when constructing algorithms for tissue image analysis. While CISH signal is generally composed of identifiable dots, FISH signal can be more diffuse and of various intensities (Figs. 2 and 3). In addition, CISH provides analysis in the context of tissue morphology, a feature that is not recapitulated by FISH. Finally, DAPI

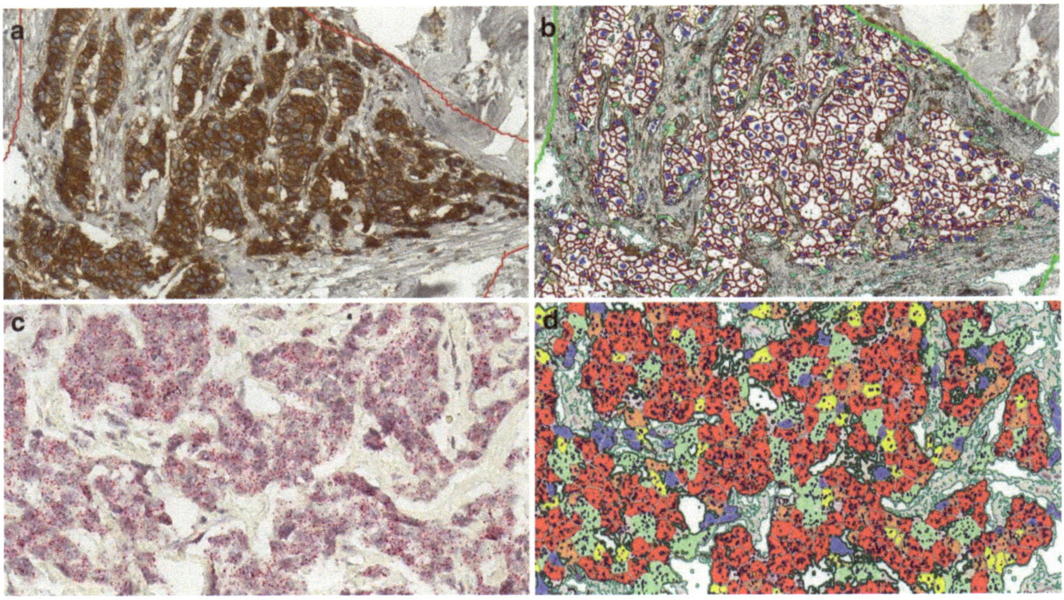

Fig. 2 Examples of algorithm-driven image analysis of Her2 IHC and RNA CISH in human clinical breast cancer samples. (**a**) Original IHC image (20×); (**b**) Mark-up image of (**a**) panel: IA-based quantification of Her2 protein post IA (20×). *Blue color* outlines individual nuclei, while colors pertaining to membrane mark-ups are as follows: *yellow* (refers to weakly positive cells), *orange* (refers to cells of medium positivity), and *red* (marks cells with high level of Her2 positivity); (**c**) original Her2 CISH image (40×); (**d**) Mark-up image of panel (**c**): IA-based quantification of Her2 RNA CISH post IA. Colors in the mark-up image refer to the following: (1) *red* (marks cells containing more than 7 Her2 RNA dots), (2) *orange* (marks cells containing 5–7 Her2 RNA dots), (3) *yellow* (marks cells containing 3–5 Her2 RNA dots), and (4) *blue* (marks cells with fewer than 3 dots)

nuclear marker signal can provide some difficulty for delineation of individual nuclear features apparent in tissue morphology as revealed by CISH (Fig. 3). Algorithm inputs for FISH and CISH must thus reflect these differences in order for image analysis to be effective in each case. Finally, an algorithm can be fine-tuned to only select cells of choice. For example, Her2 CISH was quantified only for tumor cells, while omitting the stromal contribution (Fig. 2).

CellMap™ algorithm outputs provide distinct biological endpoints for RNA ISH quantification on cell-by-cell basis. These can be user-defined depending on the assay or study and include but are not limited to: (1) localization of RNA dots; (2) intensity of individual dots; and (3) number of dots per cell. Such outputs can provide for replicable scoring paradigms. For example, bins of positivity can be established using numbers of dots as thresholds. As an example, cells with <2 dots could be identified as weakly positive, cells with 2–4 dots as medium positive, and cells with >4 dots as strongly positive. Some pertinent endpoints are demonstrated in Fig. 4. Percentages of these subpopulations can be deduced and various replicable scoring paradigms established (e.g., percentage of positive cells or a weighted score similar to an H-score used in pathology).

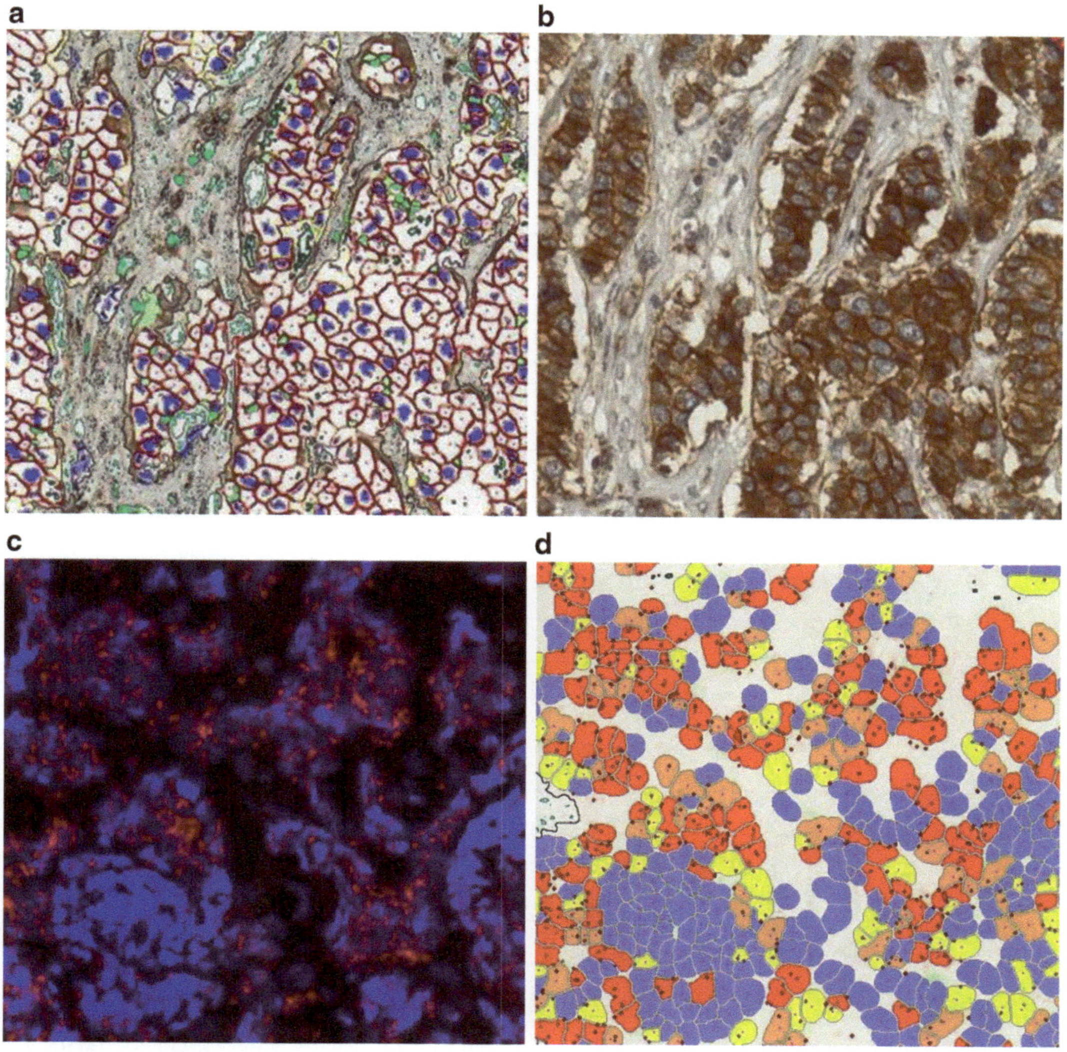

Fig. 3 Examples of algorithm-driven image analysis of Her2 IHC and RNA FISH in human clinical breast cancer samples. (**a**) Original IHC image (20×); (**b**) Mark-up image of (**a**) panel: IA-based quantification of Her2 protein post IA (40×); (**c**) original Her2 FISH image (40×); (**d**) Mark-up image of panel (**c**): IA-based quantification of Her2 RNA FISH post IA. For explanation of color schemes in mark-up images, please refer to Fig. 2 legend

For an algorithm-driven image analysis to be used as a tool in quantification of RNA transcript in tissue, the algorithm must provide the following capabilities: (1) identification of individual cells (e.g., dependent on nuclear parameters); (2) identification of RNA ISH signal (i.e., dots); (3) assigning of dots to individual cells; (4) individual cell-based outputs (i.e., RNA ISH signal intensity/cell or RNA ISH dots quantity/cell). To illustrate the algorithm-based image analysis approach, we utilized CellMap0.6™ algorithm to analyze and quantify Her2 CISH and FISH in the context of human breast cancer tissues. The algorithm was tuned to only

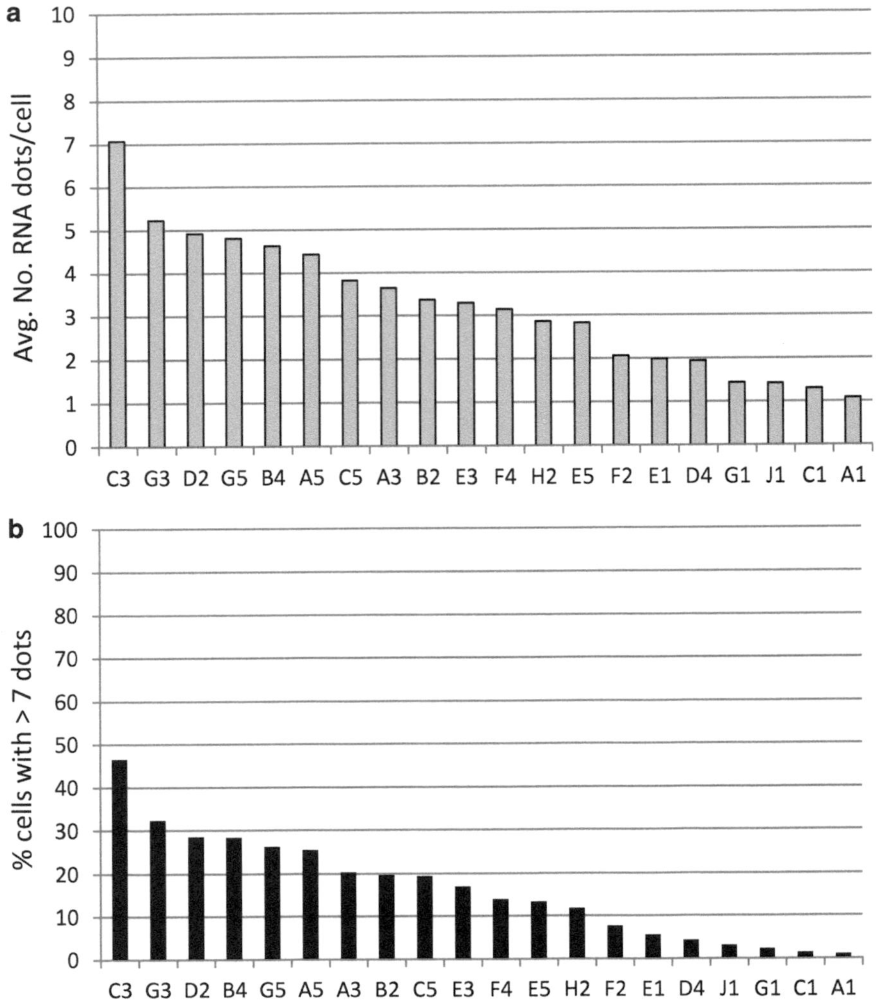

Fig. 4 Examples of quantifiable endpoints from algorithm-driven image analysis of Her2 RNA CISH. (**a**) Ranking of human clinical breast cancer samples based on IA-derived RNA CISH score (average number of RNA dots/cell); (**b**) ranking of human clinical breast cancer samples based on IA-derived RNA CISH score (percent of cells with high Her2 RNA expression reflected by greater than 7 RNA dots)

include tumor cells, and identify individual nuclei based on the DAPI and hematoxylin stains for FISH and CISH, respectively. The algorithm appropriately identifies individual RNA dots and drives the quantification of transcript on cell-by-cell basis (Figs. 2 and 3). In addition, Her2 IHC was performed on these tissues to reveal the levels of Her2 protein expression in addition to Her2 RNA. Standardized ASCO/CAP Her2 scoring paradigm was used to assess Her2 IHC while image analysis derived scoring paradigm was used for assessment of Her2 RNA using CISH. While 79 % concordance was observed between Her2 IHC and CISH across samples, several cases for which Her2 IHC status was determined as "equivocal," RNA CISH score was high (data not shown). This raises the possibility that RNA ISH could be used as a potential complementing strategy to IHC scoring paradigms in clinical and diagnostic settings. Studies illustrated in this example clearly support the idea that image analysis can provide a suitable high-throughput platform for reliable quantification of RNA signal in the context of human clinical tissues.

3 Discussion

RNA ISH is a robust molecular biology technique that can be used to assess gene expression in the tissue while maintaining information on tissue morphology. Importantly, unlike other RNA quantification techniques (e.g., quantitative real-Tree PCR (qRTPCR)), RNA ISH reveals transcript expression levels on cell-by-cell basis. Despite, RNA ISH is not used in clinical settings and as a molecular diagnostic tool likely due to lack of appropriate and standardized scoring paradigms for RNA ISH assessment. As demonstrated herein, image analysis can be used as a robust and effective tool for quantification of RNA ISH in clinical tissues. Several scoring paradigms can be obtained using algorithm-driven image analysis tools. These studies highlight the potential of algorithm-based image analysis tools for RNA analysis and quantification. Future studies will provide further evidence on successful combining of image analysis with RNA ISH in molecular diagnostics, clinical settings, and drug discovery.

Methods in Pharmacology and Toxicology (2015): 191–202
DOI 10.1007/7653_2014_30

Published online: 02 December 2014

Solid Tissue-Based DNA Analysis by FISH in Research and Molecular Diagnostics

Marcus Otte

Abstract

Molecular research in cancer has lead to the identification of many genomic alterations like single point mutations and deletions as well as gene locus amplifications, deletions, and translocations. Fluorescence in situ hybridization (FISH) is state of the art for high-resolution analysis of gene locus amplification, deletion, and translocation and has become of increasing importance in research and clinical diagnosis over the past years. In contrast to biochemical methods like DNA sequencing and PCR, FISH offers the advantage to analyze genomic alterations at cellular level in the context of tissue architecture and cell types. Therefore FISH can analyze intratumor heterogeneity of copy number variations and translocations with highest precision.

Reliable and robust FISH data depend on several key steps including probe design, probe hybridization, FISH signal pattern evaluation, and data analysis, which is discussed in this chapter. Progresses in automated ISH, image scanning, and tissue image analysis promises further improvements and are waiting to be integrated into research and routine diagnosis. Despite the technological progress in FISH, tissue sample quality is still key and the most limiting factor for high-quality ISH data.

Key words FISH, Cancer, Diagnostic, Molecular pathology, Copy number variation, Translocation

Abbreviations

aCGH	Array comparative genomic hybridization
BAC	Bacterial artificial chromosome
CISH	Chromogenic in situ hybridization
CNV	Copy number variation
DAPI	4′,6-Diamidino-2-phenylindole
FDA	Food and Drug Administration
FFPE	Formalin-fixed, paraffin-embedded
FISH	Fluorescence in situ hybridization
FOV	Field of view
ISH	In situ hybridization
NGS	Next generation sequencing
RT	Room temperature
SISH	Silver-enhanced in situ hybridization
TMA	Tissue micro array

1 Introduction

The basic principle of fluorescence in situ hybridization (FISH) is the hybridization of short (25–80 bp) and labeled DNA probes to complementary regions of chromosomal DNA of the interphase nuclei in FFPE or cryo tissue sections (1.5–5 μm). Microscopic inspection of the hybridized tissue slide allows the identification of the hybridization signals, appearing as small fluorescent spots. Since each spot represents one binding event, copy number variations (CNV) can be evaluated by simple enumeration of the fluorescent spots (Fig. 1). Moreover, depending on probe design, changes of the chromosome structure (i.e. translocation) are reflected in the signal pattern and can therefore be analyzed.

Chromosomal alterations like gene locus gains or losses as well as chromosomal translocations have gained an increased interest in research during the last two decades. Some of this interest is driven by the systematic focus on the development of personalized therapy options using functionally well-characterized genetic defects for target selection and subsequent patient stratification. Moreover, several genomic changes have been proved to be useful for disease diagnosis and patient prognosis (Table 1). In the field of hematopoietic malignancies the analysis of chromosomal changes for diagnostic purposes by FISH has been used for many years and has demonstrated clinical utility of FISH diagnostic [1, 2].

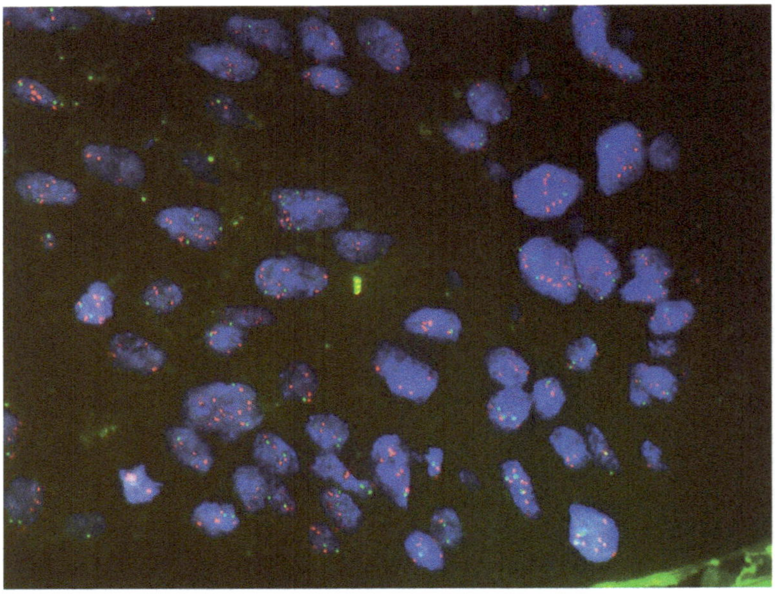

Fig. 1 Dual color FISH assay (*red*: target/allele-specific probe, *green*: control/centromere-specific probe, *blue*: nuclei) in FFPE NSCLC. No z-stacking (100× objective magnification)

Table 1
FDA-approved FISH-based diagnostic solutions

Target(s)	Aberration	Indication	Clinical use
Her-2	Gain	Breast cancer	Companion diagnostic
Alk	Translocation	NSCLC	Companion diagnostic
TOP2A	Gain and loss	Breast cancer	Prognosis
TP53, ATM, d13s319 and d12z3	Loss (TP53, ATM, d13s319); gain (d12z3)	MM	Prognosis
Chromosome 3, 4, 17 and 9p21	Aneuploidy (3,4,17); loss (9p21)	Bladder cancer	Diagnosis
EGR1	Loss	AML	Prognosis

Therefore FISH is not only a highly valuable research tool, but it also has become an increasingly important diagnostic tool for high-resolution analysis of chromosomal alterations.

Compared to conventional cytogenetic analysis, which requires the time-consuming preparation of metaphase cells by cell culture technologies, FISH uses interphase nuclei and can therefore be easily applied for the analysis of tissue sections. However, the power of FISH analysis depends on well-designed FISH probes derived from careful characterized and evaluated chromosomal alterations.

New diagnostic applications and tools especially in the field of cancer have been developed (Table 1) and more are in development, which make use of FISH for patient stratification and diagnosis based on CNV, translocations, and aneusomy.

In addition to the increased availability of FDA-approved FISH-based diagnostic kits (Table 1), a significantly larger number of commercially available research tools have been developed, which are used extensively for characterization of patient cohorts and inter- and intratumor heterogeneity. Several FISH-based diagnostic approaches for patient stratification are in development (i.e. c-Met, FGFR1, FGFR2, EGFR etc.) and clinically utility needs to be demonstrated. Furthermore, several diagnostic and prognostic marker candidates analyzed by FISH are in evaluation stages [3–6].

With the introduction of Herceptin™ treatment for breast cancer, FISH has become a frequent diagnostic tool in the molecular pathology lab for stratification of 2+ scored Her-2 tumors. Hence, FFPE tissue samples have become a sample type for routine

FISH diagnostic which has raised new challenges in the field of sample preparation, pre-analytical conditions, standardization, data recording, and interpretation.

2 Technical Considerations

The FISH procedure can be separated into six main steps: (1) probe design, (2) sample pre-treatment, (3) probe hybridization, (4) posthybridization, (5) hybridization pattern evaluation, and (6) data analysis. All steps have their own limitations and pitfalls, and potential solutions are briefly discussed below. Each probe design and FISH assay requires careful case-by-case decisions and may require protocol adaptations and/or optimization strategies reflecting the desired goal of the assay.

2.1 Probe Design

Probe design is based on the availability of target sequence and genomic mapping data of the targeted allele region(s) including also potential control regions (i.e. centromere regions). A number of empirical rules for probe design, production, and quality control have been developed over the last three decades, making this field an art of its own right. Several commercial service providers offer quality controlled, customized design and production of FISH probes with different options for labeling. Due to the laborious procedure of FISH probe design and production many labs prefer the customized option.

However, several software tools do exist for probe selection and optimization [7–9]. Typically, software-designed probes lack repetitive sequences by design and cover smaller target regions than BAC-based probes. However, FISH probes covering larger target regions (i.e. >200 kb) can be obtained by using more than one BAC for probe preparation. In this case special care has to be taken for blocking cross hybridization of repetitive DNA sequences (i.e. adding Cot-1 DNA during hybridization) or to exclude repetitive DNA sequences from probe production.

For probe labeling the selection of the optimal labeling method depends on the starting material used. Several protocols have been developed using nick translation, PCR-based label incorporation or chemical modification [10]. The different labeling methods vary significantly in the labeling efficiency. However, labeling efficiency has a huge impact on sensitivity and the size of the probe region required for appropriate signal intensity.

Today, state-of-the-art FISH probes are direct-labeled and lack repetitive genomic elements. Hence, using direct-labeled FISH probes improves the efficiency of hybridization and posthybridization procedures. However, the advantages of direct-labeling do not come without a downside. Direct-labeled FISH probes require a relative large target region (~300–800 kb) for optimal signal sensitivity of probe hybridization detection (depending also on the labeling method). Depending on the desired gene locus and probe design strategy, smaller regions might be required and indirect labeling (i.e. biotin) might become the preferred method of choice. Biotin-labeled probes allow the use of signal amplification cascades, by adding several layers of FITC-avidin/biotinylated goat anti-avidin pairs for signal development [11, 12]. However, even target regions as small as 5–10 kb have been visualized successfully in cell lines using the CAtalyzed Reporter Deposition (CARD) method [13].

For routine diagnosis, direct-labeled probes are more efficient in handling. However, improved detection sensitivity due to implementation of more tedious probe development protocols becomes significantly facilitated by staining automates. Reliable detection of small target regions in research and potentially also for clinical routine applications in FFPE tissue sections is becoming technically feasible.

Finally, FISH probes should be carefully validated for hybridization specificity as well as for FISH signal intensity and morphology (optimally small and dense) in FFPE tissue samples. In best case, FISH signal intensity and morphology in FFPE tissue is done directly using the target tissue since signal morphology might vary between tissue/cell types and depends on the size of the nuclei analyzed. However, FFPE thymus tissue samples could be used for initial FISH probe testing especially when various FISH probe designs have to be compared initially site by site.

2.2 Sample Pre-treatment

Sample pretreatment for FFPE sections is crucial for successful hybridization and starts with the removal of the paraffin by an incubation of the tissue sections in an upright position at 65 °C for 35 min. Next, slides are incubated in Hemo-De (or xylene) for 10 min at RT. Deparaffinization is completed by washing the slides for two times in 100 % EtOH for 5 min at RT. Next slides will be immersed in 0.2 N HCl for 20 min at RT, washed with distilled water (3 min, RT) and two times with 2× SSC buffer (3 min, RT).

Furthermore the slides will undergo a heat pretreatment using a microwave (600 W, 20 min). During heat pre-treatment boiling should be avoided. For heat pre-treatment slides should be immersed in a pre-treatment solution. Typically the pre-treatment solution for heat pre-treatments contains a chaotropic agent (i.e. 1 N NaSCN). After heat pre-treatment the slides are shortly washed at RT (distilled water, 1 min; two times 2× SSC, 5 min) and subsequently immersed in a proteinase solution (i.e. proteinase K), typically at 37 °C. The protein digestion might vary between 40 and 70 min depending on tissue type and pre-analytic conditions used (i.e. fixation). Tissue pretreatment kits with heat pre-treatment solution and protease are available from several vendors.

In each case the optimal proteinase incubation time for different tissue types and fixation conditions has to be fine-tuned for each kit.

After proteinase treatment, slides are washed (distilled water, 1 min; two times 2× SSC, 5 min), dehydrated (70 % EtOH, 90 % EtOH, 100 % EtOH, each 1 min), and dried for 10 min before probe hybridization.

In general, weak and insufficient hybridization signals can be improved by increasing the proteinase treatment. However, too lengthy proteinase treatment can result in strong degradation of the nucleus resulting in a DNA loss. Microscopically, degraded nuclei appear as blue rings or so-called doughnuts when stained with DAPI and a reduction of the proteinase incubation time is indicated.

The heat pretreatment can also be performed in a steam cooker (120 °C, 10 min) instead of a microwave, which, in some cases, can improve the hybridization results.

Insufficient deparaffinization can also be a reason for weak or absent hybridization signals. A common cause for insufficient deparaffinization is multiple reuse of Hemo-De (or xylene).

In general, special care should be taken when handling xylene, HCl, and NaSCN, and safety instructions for these reagents should be followed without compromise.

2.3 Probe Hybridization

Commercial FISH probes typically come ready to use in an appropriate hybridization buffer and no further adjustment of probe concentration is required. However, if required, individual probes (i.e. target and control) can be mixed and titrated if required in order to obtain appropriately balanced hybridization signals.

The volume of probe needed to cover the entire tissue section may vary depending on section size and thickness and may be in the range of 7 (i.e. biopsy) to 20 μl. In most cases 12 μl is appropriate. The probe should be added to a dried section (see Section 2.2), covered with a cover slip and sealed with rubber cement to avoid any drying during hybridization.

Finally co-denaturation and hybridization is processed in a humid chamber with adjustable temperature (i.e. ThermoBrite®). Co-denaturation is performed at 80 °C for 5 min followed by hybridization at i.e. 38 °C for 16 h. However, optimal hybridization temperature might vary between probes (36–42 °C) and is provided by the manufacturer. Hybridization longer than 16 h does not improve hybridization efficiency, and good results are frequently achieved at shorter hybridization periods. Hybridization is typically carried out overnight in order to allow sample processing within normal working hours. Most importantly, evaporation of the hybridization buffer should be avoided since the drying artifacts prohibit sample analysis.

Too high probe concentration typically lead to strong FISH signals but are sometimes accompanied by a high background. Hence, probe concentration should be lowered. In multicolor

FISH assays the individual probes may need to be carefully titrated in order to achieve an optimally balanced signal-to-noise ratio for all components. Generally however, commercial probes or FISH assays do typically come ready to use.

If probes are self-made, care should be taken regarding probe length. BAC-derived probes need to be fragmented in order to improve the signal–noise ratio [14].

2.4 Posthybridization

During posthybridization excess probe as well as unspecific bound probe is removed by washing. After removing the rubber cement, the cover slip is carefully removed by immersing the slides in 2× SSC/0.1 % NP-40 for 10 min at RT under gentle shaking. To remove sticky cover slips, carefully lift the cover slip with the blade of scalpel just slightly in order to allow entry of buffer.

When cover slips are removed wash the slides two times for 2 min in 2× SSC/0.1 % NP-40 at RT. For stringent washing incubate the slides in 2× SSC/0.3 % NP-40 at 72 °C for 5 min and agitate gently. If the stringency of the washing has to be increased in order to lower the background binding of the probe, (1) the SSC concentration can be stepwise lowered or (2) the temperature can be increased. Due to the toxicity of formamide, it should be avoided in washing buffers.

After stringency washing the nuclei are stained with DAPI, most elegantly by using an antifade mounting medium with DAPI.

2.5 Hybridization Pattern Evaluation

FISH signals are best analyzed using a 100× objective magnification (oil immersion objective) and a specific filter set depending on the fluorophores used. Either each fluorophore is analyzed separately in a specific filter channel or triple or multipass filters are used allowing the identification of DAPI and two additional fluorophores. Since tissue samples are three dimensional, each sample has to evaluated (or scanned for digital recording) in multiple layers to ensure proper detection and recording of the entire set of FISH signals.

Copy number variations (CNV) are typically analyzed with dual color FISH assays. An allele specific (target) and a centromere-specific probe (control) have been combined for the CNV FISH assays. Gene amplification as well as gene deletion assays use a target and a control region located on the same chromosome. The control region is typically represented by a multiple repeated chromosome-specific satellite DNA sequence. However, chromosome-specific satellite elements are not available for chromosomes 5, 13, 14, 19, 21, and 22, and other regions have to be identified for assay control purposes.

Sometimes the signal morphology between target and control differ, and the fluorescence intensity between target and control might vary. These differences should be small and should have been tackled during probe design and optimization.

Translocation assays can have quite complex signal patterns [15], but most typically two probes are used to produce a simple split or merged signal [16]. The design of the assay defines which signal pattern is indicative for translocation. However, for proper pattern analysis, predefined signal categories, i.e., split signal, merge signal, etc. are analyzed and enumerated for individual nuclei. The definition of signal categories may require additional rules to guide the observer. For example a split signal is typically defined when the distance between two signals is three times the signal diameter. It is obvious that the robustness of pattern analysis as well as the identification of complex diagnostic pattern could be greatly improved by digital image analysis.

In general FISH signal reading has to be limited to the region of interest (i.e. carcinoma cells and only the numbers of red and green FISH signals in nonoverlapping nuclei with complete and intact contours should evaluated and used for analysis). FISH signals should be recorded for individual cancer cells/nuclei and recorded counts should be listed for each field of view (FOV) analyzed. For some tissue types the identification of nonoverlapping nuclei can be challenging, and therefore thinner sections are in general recommended (i.e. 2–4 μm). The identification of proper regions for FISH analysis can be greatly supported by a consecutive H&E section.

For FISH signal analysis at least three FOVs with at least 30 evaluable nuclei per FOV should be analyzed to produce a robust basis for subsequent data analysis. Signal counts and pattern category (for translocation assay) should be noted and recorded for individual nuclei.

Furthermore, new FISH projects should be controlled by site-by-site hybridization and evaluation of appropriate control tissue samples (i.e. nonmalignant/normal control tissue) in order to identify and correct for technical artifacts based on section thickness, hybridization, and reading. This control helps to define proper cut-off values, i.e., for gene locus deletion. It should be kept in mind that these cut-off values reflect the entire procedure for FISH analysis (i.e., section thickness, probe, hybridization conditions, etc.) and can vary significantly when critical aspects of the protocol are changed.

2.6 Data Analysis

Target and control signals are enumerated and recorded per nuclei in CNV assays, in translocation assays type and number of each pattern category are recorded.

However, since various categories of CNV types can be observed (i.e., amplification, deletion, etc.), the observed nuclear hybridization pattern has to be interpreted using categories for possible signal patterns (Table 2). Subsequently the signal classification allows a more detailed data analysis. The standard analysis of overall target gene ratio (Eq. 1) and overall gene dose (Eq. 2)

Table 2
FISH signal categories for CNV analysis

Category	Comment	Rules	Signal pattern [target]:[control]
Normal (N)	No CNV present, only one allowed signal pattern	[target] = [control] = 2 Gene ratio = 1 Target dose = 2	[2]:[2]
Amplified (A)	Target gene locus amplification, multiple signal patterns possible	[target] > [control] Gene ratio ≥ 2.0 Target dose ≥ 3.0	[3]:[2]; [4]:[2]; [6]:[2]; [7]:[3]; etc.
Polysomy (P)	Target gene dose elevation due to polysomy of target homing chromosome, multiple signal patterns possible	[target] = [control] > 2 Gene ratio = 1 Control dose > 2 Target dose > 2 Gene dose	[3]:[3]; [4]:[4]; [5]:[5]; [6]:[6]; etc.
Deletion (D)	Loss of one or both target copies, normal control, only tow signal patterns allowed	[target] < [control] Gene ratio < 1 Control dose = 2 Target dose ≤ 1	[1]:[2]; [0]:[2]
Other (O)	Anything not matched by the before mentioned categories	–	[2]:[1]; [4]:[3]; [5]:[4]; [4]:[5] etc.
No signal (NS)	Complete lack of any detectable hybridization signal	[target] = [control] = 0	[0]:[0]

reflect poorly the tumor heterogeneity, and the overall averages may even be misleading. A more detailed data analysis provides the estimation of the percentage of cells with gene locus amplification, deletion, and polysomy. Furthermore, the categories itself can be characterized based on the observed pattern distribution. Most important well-justified thresholds can be estimated for individual categories (i.e. deletion, amplification) when compared with nonmalignant control hybridizations. Typically the histogram of the frequency of cells of category deletion (hemizygous and/or homozygous) in nonmalignant tissue vs. carcinoma helps to define the threshold for real hemizygous and homozygous deletion events.

In Table 2, the category "other" can be further subdivided in patterns belonging to a signal gain type (i.e. target gene ratio > 1 but <2) or signal loss type (i.e. target gene ratio < 1 but not [1]:[2] or [0]:[2]).

$$\text{Target gene ratio} = \text{Total target counts}/\text{Total control counts} \quad (1)$$

$$\text{Gene dose} = \text{Total target counts}/\text{Total number of nuclei evaluated} \quad (2)$$

Plotting the overall target gene ratio (Eq. 1) vs. the percentage of cells of the category, "amplification" indicates which overall target gene ratio is required to have 50, 75, or more than 90 % of cell with target gene amplification (i.e. gene ratio $\geq$ 2). For many amplified targets we have observed that in tumor samples with an overall gene ratio of 2, only approximately 50 % of the nuclei presented a gene locus amplification pattern with a target gene ratio of at least 2. Typically a significant subgroup (20 %) belonged to the "signal gain type" (see above). A significant degree of the cell population lacked any detectable signs of gene locus amplification. A more detailed data analysis is also of advantage when drug responses are studied in heterogeneous samples.

3 General

Several pitfalls do still exist for FISH analysis. Care should be taken to use appropriate glass slides for preparation of tissue sections (i.e. DAKO FLEX IHC slides). Due to the relative harsh tissue pretreatment procedure, optimal slides should be used to prevent tissue loss.

In some cases, loss of tissue may occur due to the lack of proper tissue fixation. In these cases an increased baking time and a post-fixation step using 4 % paraformaldehyde (5 min, RT) may help to reduce tissue loss.

On the other extreme, over-fixation can limit or completely preclude probe hybridization, which cannot be overcome by increasing protease treatment time. Microwave-assisted hybridization has been shown to improve hybridization efficiency for over-fixed tissue samples [17]. Tissue fixation problems derive primarily from relatively old biorepository samples. Today, standardized and state-of-the-art tissue preparation, fixation, and storage has progressed and its crucial impact for molecular pathology has been recognized and substantial improvements have been made [18, 19].

Autofluorescence can interfere with sample analysis, which might also be due to inadequate pre-analytical conditions [20, 21]. Use of CISH or SISH instead of FISH avoids limitations due to autofluorescence in general since standard light microscopy is applied instead of fluorescence microscopy. In general CISH and SISH may have some advantages over FISH, which might be of higher relevance for routine diagnosis [22, 23].

However, FISH has several advantages in research. Hybridized slides can be simply de-strained by incubation in distilled water for 5 min and reused for new hybridization experiments. Since hybridized slides can be stored for several months at 4 °C when properly sealed, precious tissue samples can be stored and analyzed more efficiently. Typically two to three FISH assays can be performed in series using the same TMA/tissue section. When combined with the preparation of TMAs, tissue consumption of precious tissue samples can be drastically reduced, and the data output per TMA section significantly increased.

4 Final Remarks

ISH is a powerful method for high-resolution analysis of chromosomal alterations and provides the benefit from preservation of tissue structure and cell-based analysis. Intratumor heterogeneities, which can hardly be depicted by "averaging" biochemical methods like aCGH and NGS, are easily derived from ISH data. However, the advantage of FISH comes also with certain disadvantages like the relative time- and labor-consuming sample processing and analysis. Despite the fact of significant protocol improvements [24, 25] and progress in tissue image analysis [26], ISH is in its infancy with regard to automation and tissue image analysis. The present evidence so far suggests that tissue image analysis improves diagnostic accuracy [27]. Given the potential superiority of FISH over IHC for determining the Her-2 status [28], further improvement in FISH seemed to be well reasoned.

Despite the promising progresses made in ISH automation and tissue image analysis, the importance of further improvement and standardization of pre-analytical procedures for tissue processing cannot be over-emphasized.

References

1. Sreekantaiah C (2007) FISH panels for hematologic malignancies. Cytogenet Genome Res 118:284–296
2. Cuneo A, Bigoni R, Roberti MG, Bardi A, Balsamo R, Piva N, Castoldi G (1997) Detection of numerical aberrations in hematologic neoplasias by fluorescence in situ hybridization. Haematologica 82:85–90
3. Dimou A, Non L, Kwang Chae Y, Tester WJ, Syrigos KN (2014) MET gene copy number predicts worse overall survival in patients with non-small cell lung cancer (NSCLC); a systematic review and meta-analysis. PLoS One 9(9): e107677
4. Phi S, Moore MW, Cotter PD (2013) Genomic rearrangements of PTEN in prostate cancer. Front Oncol 3(240):1–9
5. Sanjmyatav J, Matthes S, Muehr M, Sava D, Sternal M, Wunderlich H, Gajda M, Grimm M-O, Junker K (2014) Identification of high-risk patients with clear cell renal cell carcinoma based on interphase-FISH. Br J Cancer 110:2537–2543
6. Aravidis C, Panani AD, Kosmaidou Z, Thomakos N, Rodolakis A, Antsaklis A (2012) Detection of numerical abnormalities of chromosome 9 and p16/CDKN2A gene alterations in ovarian cancer with fish analysis. Anticancer Res 32(12):5309–5313
7. Beliveau BJ, Joyce EF, Apostolopoulos N, Yilmaz F, Fonseka CY, McCole RB, Chang Y, Li JB, Senaratne TN, Williams BR, Rouillard J-M, Wu C-T (2012) Versatile design and synthesis platform for visualizing genomes with Oligopaint FISH probes. Proc Natl Acad Sci U S A 109(52):21301–21306
8. Navin N, Grubor V, Hicks J, Leibu E, Thomas E, Troge J, Riggs M, Lundin P, Månrér S, Sebat J, Zetterberg A, Wigler M (2006) PROBER: oligonucleotide FISH probe design software. Bioinformatics 22(19):2437–2438
9. Nedbal J, Hobson PS, Fear DJ, Heintzmann R, Gould HJ (2012) Comprehensive FISH probe design tool applied to imaging human immunoglobulin class switch recombination. PLoS One 7(12):e51675

10. Morrison LE, Ramakrishnan R, Ruffalo TM, Wilber KA (2003) Labeling fluorescence in situ hybridization probes for genomic targets, Chapter 2. In: Fan Y-S (ed) Molecular cytogenetics protocols and applications. Humana Press, Totowa, NJ, pp 21–40
11. Korenberg JR, Chen X-N (1995) Human cDNA mapping using a high-resolution R-banding technique and fluorescence in situ hybridization. Cytogenet Cell Genet 69:196–200
12. McQuaid S, Allan GM (1992) Detection protocols for biotinylated probes: optimization using multistep techniques. J Histochem Cytochem 40:569–574
13. Schmidt BF, Chao J, Zhu Z, DeBiasio RL, Fisher G (2014) Signal amplification in the detection of single-copy DNA and RNA by enzyme-catalyzed deposition (CARD) of the novel fluorescent reporter substrate Cy3.29-tyramide. J Histochem Cytochem 45:365
14. Lawrence JB, Singer RH, Villnave CA, Stein JL, Stein GS (1988) Intracellular distribution of histone mRNAs in human fibroblasts studied by in situ hybridization. Proc Natl Acad Sci U S A 85:463–467
15. Attard G, Clark J, Ambroisine L, Fisher G, Kovacs G, Flohr P, Berney D, Foster CS, Fletcher A, Gerald WL, Moller H, Reuter V, De Bono JS, Scardino P, Cuzick J, Cooper CS, Transatlantic prostate group (2008) Duplication of the fusion of TMPRSS2 to ERG sequences identifies fatal human prostate cancer. Oncogene 27(3):253–263
16. Finelli P, Fabris S, Zagano S, Baldini L, Intini D, Nobili L, Lombardi L, Maiolo AT, Neri A (1999) Detection of t(4;14)(p16.3;q32) chromosomal translocation in multiple myeloma by double-color fluorescent in situ hybridization. Blood 94(2):724–732
17. Sugimura H (2008) Detection of chromosome changes in pathology archives: an application of microwave-assisted fluorescence in situ hybridization to human carcinogenesis studies. Carcinogenesis 29(4):681–687
18. Groenen PJTA, Blokx WAM, Diepenbroek C, Burgers L, Visinoni F, Wesseling P, van Krieken JHJM (2011) Preparing pathology for personalized medicine: possibilities for improvement of the pre-analytical phase. Histopathology 59 (1):1–7
19. Hicks DG, Boyce BF (2012) The challenge and importance of standardizing pre-analytical variables in surgical pathology specimens for clinical care and translational research. Biotech Histochem 87(1):14–17
20. Babic A, Loftin IR, Stanislaw S, Wang M, Miller R, Warren SM, Zhang W, Lau A, Miller M, Wu P, Padilla M, Grogan TM, Pestic-Dragovich L, McElhinny AS (2010) The impact of pre-analytical processing on staining quality for H&E, dual hapten, dual color in situ hybridization and fluorescent in situ hybridization assays. Methods 52(4):287–300
21. Tokumasu F, Dvorak J (2003) Development and application of quantum dots for immunocytochemistry of human erythrocytes. J Microsc 211:256–261
22. Poulsen TS, Espersen MLM, Kofoed V, Dabetic T, Høgdall E, Balslev E (2013) Comparison of fluorescence in situ hybridization and chromogenic in situ hybridization for low and high throughput HER2 genetic testing. Int J Breast Cancer 2013:1–5
23. Francis GD, Jones MA, Beadle GF, Stein SRB (2009) Bright-field in situ hybridization for gene amplification in breast cancer using tissue microarrays: correlation between chromogenic (CISH) and automated silver-enhanced (SISH) methods with patient outcome. Diagn Mol Pathol 18(2):88–95
24. Ridderstråle KK, Grushko TA, Kim H-J, Olopade OI (2005) Single-day FISH procedure for paraffin-embedded tissue using a microwave oven. Biotechniques 39(3):316–318
25. Bull JH, Harnden P (1998) Efficient nuclear FISH on paraffin-embedded tissue sections using microwave pretreatment. Biotechniques 26:416–422
26. Laurent C, Guérin M, Frenois FX, Thuries V, Jalabert L, Brousset P, Valmary-Degano S (2013) Whole-slide imaging is a robust alternative to traditional fluorescent microscopy for fluorescence in situ hybridization imaging using break-apart DNA probes. Hum Pathol 44(8):1544–1555
27. Levy MJ, Clain JE, Halling KC, Kipp BR, Rajan E, Roberts LR, Root RM, Sebo TJ, Topazian MD, Wang KK, Wiesema MJ, Gores GJ (2007) Preliminary experience comparing routine cytology results with the composite results of digital image analysis and fluorescence in situ hybridization in patients undergoing EUS-guided FNA. Gastrointest Endosc 66(3):483–490
28. Sauter G, Lee J, Bartlett JMS, Slamon DJ, Press MF (2009) Guidelines for human epidermal growth factor receptor 2 testing: biologic and methodologic considerations. J Clin Oncol 27:1323–1333

Methods in Pharmacology and Toxicology (2015): 203–217
DOI 10.1007/7653_2014_17

Published online: 16 September 2014

Preanalytic Considerations for Molecular Genomic Analyses of Tissue

Maureen Cronin

Abstract

Progress in the era of molecular characterization of patients, diseases, and drugs, relies on access to clinical biospecimens characteristic of the general population and of patients with specific disease states to identify predictive and prognostic biomarkers for individualized therapeutic planning. Biospecimens properly consented for clinical research can be difficult to obtain and often are quite limited in volume and mass, making it incumbent on clinical investigators to use these resources effectively to maximize the learning that comes from their analysis. This includes careful documentation of the specimens themselves along with preservation methods and storage planning to support the study objectives. In addition, careful experimental planning focused on the molecules that will be analyzed and commitment to assuring laboratory processes and methods have been fully developed and pre-tested using nonessential, representative biospecimens as surrogates for the intended study specimens are necessary. Following these steps will result in high yields of the best quality data and will ensure interpretable results that will contribute to advancing knowledge of disease states and applicable therapeutic options.

Key words Molecular pathology, Genomics, FFPE tissue, Molecular profiling, Nucleic acids, Biospecimen

1 Introduction

Some of the newest laboratory tools for characterizing tissue specimens involve molecular genomic methods. These new "omics" methods enable analyses far more powerful and comprehensive than have ever before been possible as they integrate related sets of biological molecules to assess patient disease status and patient prognosis or predict patient-specific therapeutic response. Technologies encompassed by molecular assays include DNA microarray, PCR-based or sequence-based polymorphism (SNP variant) profiling, copy number and structural rearrangement assessments; RNA gene expression profiling by microarray, RT-PCR-based methods or other enzymatic methods [1–10]; and massively parallel next-generation sequencing (NGS) which as a single platform can perform essentially all analyses of interest for both RNA and DNA, including ChIPseq and epigenetic methylation profiling [11–13]. Essentially any tissue, including blood and bone marrow, and surgically resected or biopsied solid tumors can be a source of molecular markers to characterize disease. The most recent source of nucleic acid for molecular analysis is cell-free DNA and RNA circulating in the plasma or serum [14, 15]. The performance of molecular analysis technology platforms reaches maximum utility when the data output from these analytical methods is integrated with standard histopathology data allowing the biology to be interpreted in the context of the additional insight provided by well-established anatomic pathology tissue assessment. Successfully integrating molecular and histopathology analytical results relies on collected tissue being carefully managed and documented; tissue preservation conditions being selected to conserve both anatomic and molecular integrity (to the degree possible); and selecting molecular purification methods that maximize the quality and yield of nucleic acids consistent with the intended downstream assay requirements [16–24].

Careful control and documentation of preanalytic factors affecting the integrity of the biospecimen being collected goes a long way toward preventing analysis failures and providing a sufficient yield of molecular materials to support downstream analytical assays. Managing preanalytic factors will also enable deconvolution of any batch process effects or the resolution of assay outlier results which can often be traced to variability that occurs during tissue collection and processing. Preanalytic considerations affecting immunohistochemistry (IHC) based tests have been addressed elsewhere [25]. Many of the considerations important to gaining high-quality IHC assay results, including minimizing and documenting warm and cold ischemic time prior to tissue fixation or tissue freezing, are equally important to successful, interpretable molecular genomic analyses. In addition, full documentation of biospecimen characteristics, including specimen size, anatomic location, condition at harvest, and microscopic cellular characteristics can aid in the interpretation of gene expression profiles, genetic variant profiles, and methylation patterns generated from nucleic acids isolated from these specimens.

2 Selecting Tissue Specimens Suitable for Intended Study Design

Preanalytic considerations for molecular genomic analyses of tissue begin with selection of the tissue to be analyzed. Success of a molecular analysis study relies both on the molecules of interest being present at measureable levels in tissue that is representative of the disease or condition under study and that tissue is actually accessible for collection from a representative patient cohort. When human tissue collection and analysis are included in a study design, it is important to understand the routine pathology practice followed for treating the patients and disease under study. Often the most effective way to conduct a study using biospecimens is to focus on using specimens that are routinely collected in the course of patient care. Residual tissue not needed for routine diagnostic or test procedures may be available for experimental analysis with proper arrangements and informed consent. In cases where the target tissue required for the study is not routinely collected in the course of patient care, the amount and type of tissue required should be estimated and arrangements made for its collection, transport, and storage with clinical sites where patients of interest are seen. For some types of studies, using tissues commercially available from biobanks annotated with information about the tissue source, processing and storage may be suitable (**Note 1**).

Consulting with a biostatistician in advance of tissue collection to define the appropriate number of samples and appropriate sample distribution relative to the population under study is highly recommended [26]. Tissue processing and molecular assays may be very well executed only to have the study fail due to improper sample selection. Ideally, biospecimen collection should be planned prospectively for a study, ensuring the likelihood that sufficient amounts of the target tissue will be available for the intended molecular analysis from the intended study population. However, this is often not possible. More commonly, eligible, consented, representative patients may be difficult to find and tissue specimens may be limited in quantity and quality. Available patients will be heterogeneous on many levels, including disease status and available mass of target tissue; for example, cancer patients in neoadjuvant, adjuvant, or metastatic settings will vary widely with respect to tissue accessibility for biopsy or resection depending on the tumor location, the disease stage, and the patient status. Some tissues such as breast, skin, or prostate are more accessible for biopsy than others such as brain, liver, pancreas, and lung. Additionally, it is important to plan for collecting a patient-matched blood or epithelial cell source when germline DNA is to be used as a reference specimen for tumor somatic mutation profiling.

Additional points to consider as tissue is collected to answer a scientific or clinical question posed by a study protocol in oncology are that molecular analysis of primary tumors and late stage metastatic tumors offers very different opportunities to study disease at the molecular level and that tumor tissue is heterogeneous. As has already been noted, patients as populations are heterogeneous with respect to disease, demographic, and individual patient physical characteristics. Individual patient tissue heterogeneity must also be taken into account during study planning, tissue collection, tissue processing, and molecular analysis [27–29]. For example, blood sampling is influenced by time of day and patient status for factors such as feeding/fasting, hydration, infection, drug therapy, age, and gender. Blood volume collected and preservatives used in collection tubes must be noted with detailed description of the collection timing and patient status. Tissue specimens are intrinsically heterogeneous with respect to cellular and stromal composition so must be collected with the intended analysis and study design in mind. Breast tumors, for example, are of several molecular subtypes and may vary in multiple tissue components and the presence of normal breast tissue. Metastatic cancer frequently involves multiple anatomic sites in addition to the primary tumor site and it is possible a patient may have multiple primary tumors. Clinical study success depends on planning and arranging collection of adequate, representative tissue relevant to the study design (**Note 2**).

3 Tissue Stabilization and Preservation by Freezing or Chemical Fixation

Selecting a tissue preservation method involves a series of decisions. Choosing fixation conditions, including fixation agent, or selecting which freezing process to employ when tissue is to be stored frozen is highly relevant to recovering high-quality nucleic acids that perform well in later molecular assays. Similarly, as has been described for IHC assays, the degree to which the tissue processing methods are standardized and documented will affect later molecular assay performance and interpretation [25]. There are hundreds of biospecimen preservation methods in use among surgical pathology laboratories, varying by lab and by type of biospecimen [30–34]. Generally, although tissue preservation processes may be quite consistent or well controlled within a laboratory, details of these processes vary significantly among different laboratories leading to a spectrum of performance ranges when samples are processed for molecular assays. Highest quality (maximum length, cross link free, protein free) DNA and RNA is generally obtained from fresh stabilized or rapidly frozen biospecimens. Immediate immersion of fresh solid or liquid biospecimens into nucleic acid stabilization agents such as RNA*later*®(Life Technologies, Grand Island, NY), RNA Retain® (Asuragen, Austin, TX), Trizol® reagent (Life Technologies, Grand Island, NY), or PAXGene® tubes (Qiagen, Germantown, MD) will result in neutralization of nuclease enzymes responsible for nucleic acid degradation and rapid chelation of the ionic cofactors required for most nuclease enzymes to be active [35–40]. Samples treated as directed by commercial product instructions are highly stable at room temperature (22–25 °C) for days and at cold temperature (4 °C or −20 °C) for longer periods allowing storage while arranging for nucleic acid and protein isolation. It is extremely important to observe the recommended ratio of tissue volume to stabilization reagent volume guidelines for successful results. Exceeding the tissue to reagent proportion guideline will result in some degree of nuclease activity in the tissue before nuclease inactivation is complete resulting in partial or full degradation of RNA or DNA.

Among chemical fixation methods for tissues, alcohol-based and buffer-neutralized formalin-based fixatives are the most sparing of DNA and RNA damage. Kilobases long nucleic acids can be recovered from freshly fixed biospecimens that have not been over-fixed or microwave heated during fixation. However, it is important to know the exact fixative formulation that has been used since heavy metals such as Hg^{2+}, Mn^{2+}, Mg^{2+}, Zn^{2+}(Zenker's reagent), and Cu^{2+} (Hollandes reagent) are frequently added to formalin since they alter the staining properties of some tissues and therefore are standard additives for some tissue anatomic pathology preparation protocols. Picric acid-based fixatives such as Bouin's reagent and unbuffered formalin accelerate nucleic acid damage and result in increased fragmentation. Similarly harsh are glycol-based fixatives such as Prefer (Anatech, Ltd., Battle Creek, MI) and Pen-Fix™ (Thermo Fischer Scientific, Waltham, MA). Microwave treatment is often used during fixation to increase the rate of fixative penetration into tissue but results in increased fragmentation of RNA and DNA. Under-fixing tissue can be just as damaging to DNA and RNA integrity as some reagents or over-fixation since incompletely fixed tissues may remain hydrated with active nucleases resulting in nucleic acid hydrolysis. Careful documentation of the reagents and protocols used as well as standardizing tissue volume to fixative volume ratio and time and temperature of fixation will maximize the quality of the recovered nucleic acids and the quality of the molecular analyses that can be performed.

In much the same way that fresh tissue is preserved using nucleic acid stabilization reagents, biospecimens can be quickly frozen upon harvest then maintained in liquid nitrogen, on dry ice or in a −80 °C freezer for extended periods of time (days to years) with minimal damage to the nucleic acids until they are extracted. Biospecimens frozen in this way are more difficult to prepare for nucleic acid extraction than specimens maintained in stabilization reagents since frozen samples must be kept thoroughly frozen while they are pulverized, minced, or homogenized into a state that will allow rapid exposure to inhibitors to stop all nuclease activity. Should partial or complete thawing occur while handling frozen specimens, nucleic acids will likely be exposed to active nuclease enzymes before extraction can be successfully completed, resulting in poorer quality substrates for many analytical assays. It should be noted however, that during the last decade, many commercial nucleic acid extraction methods have been optimized and automated to recover nucleic acids damaged by tissue processing fixation methods or by nucleases. Typical damage may include double-strand breaks in DNA or backbone-strand breaks in RNA that result in short fragments; and chemical cross-linking of proteins with nucleotide bases or cross-linking between nucleotide bases, particularly in DNA (**Note 3**).

4 Nucleic Acid Isolation from Frozen and Fixed Tissue Specimens

Once tissue has been collected, processed, or stabilized and its histopathology characteristics defined, isolation of the molecular fractions of interest can be performed. The first decision to make is whether the entire biospecimen or only part of each specimen will be extracted. For example, if a molecular signal is expected to be detected only in cancer cells and not the surrounding stroma or adjacent normal tissue, tumor tissue enrichment may be performed prior to molecular purification. Relevant details informing the decision to enrich for a minor tissue component may include whether DNA, RNA, or both are required for the study design and how much of each is needed. With the fast pace of technology change occurring in the arena of molecular pathology, conservative use of biospecimens is to be encouraged. However, it is also very difficult to predict the yield of useful RNA or DNA that can be obtained from a biospecimen and pure RNA and DNA can be successfully stored for indefinitely long periods under controlled conditions. Therefore, it may be most practical to extract DNA and/or RNA from all available tissue and store the purified biomolecules rather than the tissue (**Note 4**).

One of the most reliable and therefore widely used methods of extracting nucleic acids from frozen tissue is homogenization in a chaotropic salt reagent, such as guanidine isothiocyanate mixed with phenol (Trizol®). Tissue processed this way results in protein being separated from DNA and RNA fractions that can be further purified independently [35]. Grinding frozen tissue in liquid nitrogen with a mortar and pestle or some other means of pulverization keeps the tissue frozen and the nucleases inactive while preparing it for quick addition into either Trizol or a concentrated proteinase K cocktail. In both cases, as protein is exposed to the chaotrope or proteinase K, it becomes completely denatured or degraded. Specific nucleic acid fractions can be purified from the tissue lysate and high purity DNA, RNA, or both can be obtained by these methods. Proteinase K-based extraction reagents include those provided in commercial nucleic acid extraction kits such as MasterPure® (Epicentre, Madison, WI), HiPure® (Roche Molecular, Indianapolis, IN), RNAeasy® or DNAeasy® (Qiagen, Germantown, MD), and Maxwell® (Promega, Fitchberg, WI), among others. Frozen tissues are often fixed in OCT (optimal cutting temperature) compound making them amenable to thin sectioning using a cryostat for microscopic examination. These tissue sections may be handled for extraction similarly to other frozen biospecimens. However, it should be noted that nucleic acid extraction from OCT-compound prepared tissue may show lower efficiency recovery or lower integrity nucleic acids than those from directly frozen or fresh biospecimens. This is possibly due to extraction reagent performance being compromised by the high PEG (polyethylene glycol) levels in the OCT reagent [36–41].

Nucleic acids can be successfully extracted from most fixed tissues and once isolated, have been shown by many investigators to be successful analysis substrates in downstream molecular assays [36–46]. However, the impact of fixation can be highly variable on the quality of the outcomes depending on the combination of fixation chemistry and method used and extraction method selected [47–49]. The most important factors influencing the quality of DNA and RNA obtained from fixed tissues, and the fitness of these molecules for later assays, aside from the characteristics of the tissue itself, are the choice of fixative, the amount of time the specimen is exposed to fixative, and the storage time elapsed after fixation.

The most successful protocols for isolating DNA and RNA from fixed tissues are generally commercially optimized proteinase K-based kits. After fixed tissue sections have been deparaffinized using xylene or a similar organic or inorganic solvent and rinsed in alcohol to rehydrate the tissue, any of the commercial protocols available for isolation of nucleic acids from biospecimens should work well. One point that should be closely attended is that biospecimen digestion is complete with no visible tissue fragments remaining before proceeding with the purification steps. Often more than one addition of proteinase K is required to digest a tissue sample to completion, depending on the specific activity of the enzyme preparation and the mass of tissue that must be processed. Proteinase K window of activity is also limited by auto-digestion, resulting in the need to make multiple additions of fresh enzyme to a tissue specimen to completely dissolve all visible tissue mass. Under processing at the proteinase K digestion step may have several negative effects on a subsequent nucleic acid isolation procedure, including decreased RNA or DNA yield or residual protein contamination due to decreased efficiency of purification, or increased fragmentation of the nucleic acids due to residual nuclease activity before full neutralization by proteinase K. Should any intact tissue remain following the enzymatic digestion steps, the residual tissue should be pelleted by centrifugation and removed before moving on to the purification steps (**Note 5**).

Prior to selecting a nucleic acid purification method, it is important to consider which molecular assays will be performed downstream of the isolation. The probability of successfully executing these assays can be maximized by assessing their RNA or DNA quantity and quality requirements and selecting the extraction process that will meet these needs. There are several reasons for this. First, the amount of DNA or RNA required for a successful assay must be estimated as well as whether RNA, DNA, or both are required for the full spectrum of desired analyses. Another critical consideration that is sometimes overlooked is the size class of nucleic acid to be targeted for isolation since the purification steps often involve nucleic acid binding to silica/glass beads under salt and ethanol conditions which will be preferentially selective for a particular size range and strandedness (duplex DNA or single-stranded RNA or DNA) of nucleic acid. DNA and RNA coming from stabilized fresh tissue or frozen tissue are generally of high molecular weight and nucleic acid isolation conditions optimized for these large molecules should be selected. DNA and RNA coming from fixed tissues generally represent a shorter size class of molecule and an isolation procedure optimized for shorter fragments should be selected. In all tissue types, miRNA presents a special nucleic acid isolation challenge due to its very small size (~19–26 nucleotides). Optimization for miRNA isolation will generally decrease recovery of larger classes of nucleic acid so if both miRNA and longer DNA and RNA molecular analyses are planned, more than one extraction procedure may need to be applied to more than one sample of the biospecimen for optimal results, although some newer procedures claim capability for capturing all relevant nucleic acid populations [50–52].

Once the required amount of DNA and/or RNA has been estimated, some small overage should be added to account for process inefficiencies and performing quality assessment assays to characterize the isolated RNA and/or DNA. In addition, if the nucleic acid isolation yield is sufficient, the possibility should be considered to collect enough nucleic acid for repeating the desired assays in the event a laboratory or computational process failure results in sample or data loss. Finally, storage of residual sample, either as fixed or frozen tissue specimens or as purified nucleic acids should be decided and arranged before initiating tissue processing.

5 Quality Assessment of Nucleic Acids Isolated from Biospecimens

Sample quality assessment for purified RNA and DNA may be limited to quantifying the total amount recovered or be extensive and include capillary or gel-based electrophoresis assessment of the size distribution and purity of the species isolated. It is also important to note that many DNA and RNA isolation and quantitation methods are optimized to larger size classes of molecules ($\geq$100–200 nucleotides) and may not perform well for DNA and RNA isolated from FFPE, miRNA, or other small RNA species. For these applications, kits and quantitation protocols optimized for small RNA should be used [53–56].

Quantifying DNA and RNA from FFPE tissue sources or somewhat degraded nucleic acids as compared with high-quality, high molecular weight RNA and DNA has some special considerations. The two methods most commonly used to quantify RNA and DNA are: (1) optical density at OD_{260} (absorbance peak for nucleic acids) generally evaluated in ratio with both OD_{280} (absorbance peak for amino acids) and OD_{230} (absorbance peak for phenol and carbohydrates) as additional purity checks; and (2) quantitative assays based on fluorescence emission from DNA-intercalating and RNA-binding dyes. PicoGreen® (Life Technologies, Grand Island, NY) and RiboGreen® (Life Technologies, Grand Island, NY) fluorophores are the large dye molecules used for these quantitative assays. Since PicoGreen® requires at least 7–8 nucleotides of dsDNA for intercalation and RiboGreen® requires 5–7 nucleotide phosphates in the backbone to bind, these dyes do not detect free nucleotides. Both dyes show a Stokes shift and increased fluorescence emission of characteristic wave lengths as they intercalate or stack along long nucleic acid molecules. Therefore PicoGreen® and RiboGreen® measurement detection signal is biased toward longer nucleic acid molecules. By contrast, using OD_{260} measurement is sensitive to all nucleotides regardless of length and will often overestimate total nucleic acid quantity relative to fluorescence-based assays [57]. This quantitative bias may translate to assay performance that is uncorrelated with the amount of DNA or RNA used since equal quantities of high and low molecular weight RNA or DNA may perform quite differently in an assay. A high molecular weight sample may perform as expected and give a reliable analytical result, whereas a highly degraded sample will be largely composed of short nucleic acid molecules unable to be measured in PCR, qPCR, or NGS library construction and result in loss of sensitivity and assay failure. For this reason, PicoGreen® and RiboGreen® fluorescence assays are preferred over OD_{260} RNA and DNA sample quantitation when preparing for molecular pathology assays [58–60].

6 Storing Purified RNA and DNA

Once biospecimen RNA and/or DNA have been purified, quantified, and quality checked, arrangements should be made to store the samples in optimal conditions to preserve their integrity and associated data. Both DNA and RNA are stable over the long term (decades) when well purified, stored in neutral buffer with a chelating agent such as EDTA and kept cold. Keeping RNA and DNA in concentrated solution is also recommended. Milligram or microgram quantities per milliliter of solution prevent passive loss to storage container surfaces and are generally more amendable to direct use in molecular assays without having to be concentrated, which can cause sample damage or precipitation. It is also to be recommended that collections of purified RNA and DNA be kept at a standard concentration and volume in racked storage tubes compatible with robotic handling. This makes it easy to quickly and uniformly thaw specimens, remove the amount of analyte needed for the planned assay, and promptly return of the residual specimens to the long-term storage condition. Standard laboratory freezer conditions (−20 °C with no auto-defrost cycle) are sufficient for near-term storage when samples are being actively accessed for assay purposes. However, for longer-term storage, ultralow temperature storage (−80 °C) is generally recommended. Some labs chose to use liquid nitrogen storage. Selecting high-quality, well-sealed freezer tubes offers the benefit of minimizing sample evaporation during extended storage. Using multi-well plates is a more complex decision since samples in plates may not be uniformly accessed and plates need to be re-sealed after each access. Tracking residual sample quantities in plates is also more technically challenging than when tubes are used and may result in sample waste if only accessed using robotic pipetting. Room temperature dry storage conditions are also available for nucleic acids and have been shown, especially for DNA, to offer a stable alternative to freezing for example, the DNAStable product line (Biomatrica, San Diego, CA). Selecting dry storage offers the convenience of room temperature conditions but has the disadvantage of being limited in sample concentration and quantity that can be accommodated, ultimately limiting access options. RNA and DNA stored in this way may not prove suitable for all downstream assays and is most relevant to limited PCR-based, targeted assays performed on DNA.

7 Conclusions

There is a broad and increasing range of molecular assays available for interrogating RNA and DNA from biospecimens. Most assays historically have been limited to specific targeted analyses based on some permutation of PCR technology or dideoxy (Sanger) sequencing. With the advent of massively parallel, next-generation sequencing (NGS) technology, comprehensive analysis of whole genomes, whole exomes, and whole transcriptomes is possible. Whether targeted or comprehensive molecular analysis is planned, it is important that the preanalytical process begins with experimental study design and patient selection, before tissue collection and processing are initiated. This attention to detailed planning will permit maximizing the utility of the biospecimens collected and may continue to inform translational research over time when the samples are carefully documented and stored in stable conditions. Achieving progress in translational research to inform precision medicine is dependent on access to human tissue specimens. Careful planning for the judicious use of this limited resource will pay off in acquisition of high-quality data that will move our understanding of disease forward and lead to increasingly effective therapeutic management.

8 Notes

1. Opportunistic, rather than deliberately planned tissue collection, is a highly risky approach to carrying out high-quality patient molecular profiling studies. Commercial sources of tissue opportunistically collected at postmortem, from hospital pathology laboratories discarding residual nondiagnostic specimens or tissue brokers who pay donors for tissue, will generally offer minimal detailed information about the patient and disease status and will generally not be able to provide specimens representative of the population under study. These tissues do, however, provide a valuable source of tissue for methods and assay development or assay control purposes.
2. Disease heterogeneity at the level of the anatomic tumor location and characteristics is key information necessary for interpreting tumor molecular profiles. Often when more than one primary tumor exists, only one may be collected and molecularly characterized. This is also commonly true for metastatic disease. The study design should include instructions for the management of multiple primary and multiple metastatic tumors in order to make sure the study conduct is aligned with the study objectives. Incomplete or inconsistent tissue harvesting may result in uncorrectable biases among the patient/tumor molecular profiles.
3. In general, careful consideration should be paid to keeping nuclease enzyme activity inhibited until these proteins can be destroyed by chemical fixation, proteinase K digestion, or chaotrope denaturation and phenol extraction. Planning tissue exposure to neutralizing agents should include surface to volume considerations, rates of reagent penetration into tissue, temperature, and mixing. Carelessness at this stage could result in loss of valuable tissue specimens.
4. Consulting with a surgical pathologist about the characteristics of the tissue to be analyzed is highly recommended when planning a clinical molecular profiling study. Tissue heterogeneity is best assessed by a qualified pathologist who can recommend tissue enrichment strategies and indicate what components of each specimen are to be retained and which discarded. It is also wise to store a high-resolution image of an H&E stained section of each specimen at low and high magnification with a calibration scale in the image for measuring the dimensions of the tissues.
5. Many commercial kits for proteinase K tissue digestion provide a single-standard protocol with a single recommended digestion time, temperature, and reagent addition. However, pathology tissues vary quite widely in composition and often

residual, undigested tissue will remain after the recommended digestion process. Discarding this residual tissue will decrease the quantity of nucleic acid recovered and may decrease the quality of purification in subsequent protocol steps unless it is completely digested or removed by centrifugation. It is recommended that test digestions be carried out with the tissue type (s) to be included in the study and adjustment of digestion time and proteinase K addition be made before initiating the study sample extractions.

References

1. Golub TR, Slonim DK, Tamayo P, Huard C, Gaasenbeek M, Mesirov JP, Coller H, Loh ML, Downing JR, Caligiuri MA et al (1999) Molecular classification of cancer: class discovery and class prediction by gene expression monitoring. Science 286:531–537
2. Chang J, Powles TJ, Allred DC, Ashley SE, Clark GM, Makris A, Assersohn L, Gregory RK, Osborne CK, Dowsett M (1999) Biologic markers as predictors of clinical outcome from systemic therapy for primary operable breast cancer. J Clin Oncol 17:3058–3063
3. Sorlie T, Perou CM, Tibshirani R, Aas T, Geisler S, Johnsen H, Hastie T, Eisen MB, van de RM, Jeffrey SS et al (2001) Gene expression patterns of breast carcinomas distinguish tumor subclasses with clinical implications. Proc Natl Acad Sci U S A 98:10869–10874
4. Davis RE, Staudt LM (2002) Molecular diagnosis of lymphoid malignancies by gene expression profiling. Curr Opin Hematol 9:333–338
5. van de Vijver MJ, He YD, van't Veer LJ, Dai H, Hart AA, Voskuil DW, Schreiber GJ, Peterse JL, Roberts C, Marton MJ et al (2002) A gene-expression signature as a predictor of survival in breast cancer. N Engl J Med 347:1999–2009
6. Rosenwald A, Wright G, Wiestner A, Chan WC, Connors JM, Campo E, Gascoyne RD, Grogan TM, Muller-Hermelink HK, Smeland EB et al (2003) The proliferation gene expression signature is a quantitative integrator of oncogenic events that predicts survival in mantle cell lymphoma. Cancer Cell 3:185–197
7. Paik S, Shak S, Tang G, Kim C, Baker J, Cronin M, Baehner FL, Walker MG, Watson D, Park T et al (2004) A multigene assay to predict recurrence of tamoxifen-treated, node-negative breast cancer. N Engl J Med 351:2817–2826
8. Cobleigh MA, Tabesh B, Bitterman P, Baker J, Cronin M, Liu M-L, Borchik R, Mosquera J-M, Walker MG, Shak S (2005) Tumor gene expression and prognosis in breast cancer patients with 10 or more positive lymph nodes. Clin Cancer Res 11:8623–8631
9. Gianni L, Zambetti M, Clark K, Baker J, Cronin M, Wu J, Mariani G, Rodriguez J, Carcangio M, Watson D et al (2005) Gene expression profiles in paraffin-embedded core biopsy tissue predict response to chemotherapy in women with locally advanced breast cancer. J Clin Oncol 23(29):7265–7277
10. Paik S, Tang G, Shak S, Kim C, Baker J, Kim W, Cronin M, Baehner FL, Watson D, Bryant J et al (2006) Gene expression and benefit of chemotherapy in women with node-negative, estrogen receptor-positive breast cancer. J Clin Oncol 24(23):3726–3734
11. Ross JS, Cronin M (2011) Whole cancer genome sequencing by next-generation methods. Am J Clin Pathol 136:527–539
12. Kerick M, Isau M, Timmermann B, Sültmann H, Herwig R, Krobitsch S, Schaefer G, Irmgard Verdorfer I, Bartsch G, Klocker H, Lehrach H, Schweiger MR (2011) Targeted high throughput sequencing in clinical cancer Settings: formaldehyde fixed-paraffin embedded (FFPE) tumor tissues, input amount and tumor heterogeneity. BMC Med Genomics 2011(4):68–80
13. Frampton GM, Alex Fichtenholtz A, Otto GA, Wang K, Downing SR, He J, Schnall-Levin M, White J, Sanford EM, An P, Sun J, Juhn F, Brennan K, Iwanik K, Maillet A, Buell J, White E, Zhao M, Balasubramanian S, Terzic S, Richards T, Banning V, Garcia L, Mahoney K, Zwirko Z, Donahue A, Beltran H, Mosquera JM, Rubin MA, Dogan S, Hedvat CV, Berger MF, Pusztai L, Lechner M, Boshoff C, Jarosz M, Vietz C, Parker A, Miller VA, Ross JS, Curran J, Cronin MT, Stephens PJ, Lipson D, Yelensky R (2013) Development and validation of a clinical cancer genomic profiling test based on massively parallel DNA sequencing. Nat Biotech. doi:10.1038/nbt.2696

14. Murtaza M, Dawson SJ, Tsui DWY, Gale D, Forshew T, Piskorz AM, Parkinson C, Chin SF, Kingsbury Z, Wong ASC, Marass F, Humphray S, Hadfield J, Bentley D, Chin TM, Brenton JD, Caldas C, Rosenfeld N (2013) Non-invasive analysis of acquired resistance to cancer therapy by sequencing of plasma DNA. Nature 497:108–113
15. Chan KCA, Jiang P, Zheng YWL, Liao GJW, Sun H, John Wong J, Siu SSN, Chan WC, Chan SL, Chan ATC, Lai PBS, Chiu RWK, Lo YMD (2013) Cancer genome scanning in plasma: detection of tumor-associated copy number aberrations, single-nucleotide variants, and tumoral heterogeneity by massively parallel sequencing. Clin Chem 59:211–224
16. Moore HM et al (2011) Biospecimen reporting for improved study quality (BRISQ). Cancer Cytopath 119:92–101
17. McShane LM, Cavenagh MM, Lively TG, Eberhard DA, Bigbee WL, Williams PM, Mesirov JP, Polley MC, Kim K, Tricoli JV, Taylor JMG, Shuman DJ, Simon RM, Doroshow JH, Conley BA (2013) Criteria for the use of omics-based predictors in clinical trials: explanation and elaboration. BMC Med 11:220, http://biomedcentral.com/1741–7015/11/220
18. Clark-Langone KM, Wu JY, Sangli C, Chen A, Snable JL, Nguyen A, Hackett JR, Baker J, Yothers G, Kim C, Cronin MT (2007) Biomarker discovery for colon cancer using a 761 gene RT-PCR assay. BMC Genomics 8:279
19. Cronin M, Sangli C, Liu ML, Pho M, Dutta D, Nguyen A, Jeong J, Wu J, Langone KC, Watson D (2007) Analytical validation of the oncotype DX genomic diagnostic test for recurrence prognosis and therapeutic response prediction in node-negative, estrogen receptor-positive breast cancer. Clin Chem 53:1084–1091
20. Coudry RA, Meireles SI, Stoyanova R, Cooper HS, Carpino A, Wang X, Engstrom PF, Clapper ML (2007) Successful application of microarray technology to microdissected formalin-fixed, paraffin-embedded tissue. J Mol Diagn 9(1):70–79
21. Penland SK, Keku TO, Torrice C, He X, Krishnamurthy J, Hoadley KA, Woosley JT, Thomas NE, Perou CM, Sandler RS, Sharpless NE (2007) RNA expression analysis of formalin-fixed paraffin-embedded tumors. Lab Invest 87:383–391
22. Oberli A, Popovici V, Delorenzi M, Baltzer A, Antonov J, Matthey S, Aebi S, Altermatt HJ, Jaggi R (2008) Expression profiling with RNA from formalin-fixed, paraffin-embedded material. BMC Med Genomics 1(9):1–15
23. Rimsza LM, Wright G, Schwartz M, Chan WC, Jaffe ES, Gascoyne RD, Campo E, Rosenwald A, Ott G, Cook JR, Tubbs RR, Brazie RM, Delabie J, Miller TP, Staudt LM (2011) Accurate classification of diffuse large B-cell lymphoma into germinal center and activated B-cell subtypes using a nuclease protection assay on formalin-fixed, paraffin-embedded tissues. Clin Cancer Res 17:3727–3732
24. Gray RG, Quirke P, Handley K, Lopatin M, Magill L, Baehner FL, Beaumont C, Clark-Langone KM, Yoshizawa CN, Lee M, Watson D, Shak S, Kerr DJ (2011) Validation study of a quantitative multigene reversetranscriptase-polymerase chain reaction assay for assessment of recurrence risk in patients with stage II colon cancer. J Clin Oncol 29:4611–4619
25. Hicks D (2013) The impact of pre-analytic variables on tissue quality from clinical samples collected in a routine clinical setting: implications for diagnostic evaluation, drug discovery and translational research. Methods in pharmacology and toxicology, The Humana Press, Springer
26. Mwenifumbo JC, Marra MA (2013) Cancer genome-sequencing study design. Nat Rev Genet 14:322–332
27. Bedard PL, Hansen AR, Ratain MJ, Siu LL (2013) Tumour heterogeneity in the clinic. Nature 501:355–364
28. Ramaswamy S, Ross KN, Lander ES, Golub TR (2003) A molecular signature of metastasis in primary solid tumors. Nat Genet 33:49–54
29. Talantov D, Baden J, Jatkoe T, Hahn K, Yu J, Rajpurohit Y, Jiang Y, Choi C, Ross JS, Atkins D, Wang Y, Mazumder A (2006) A quantitative reverse transcriptase-polymerase chain reaction assay to identify metastatic carcinoma tissue of origin. J Mol Diagn 8:320–329
30. Lewis F, Maughan NJ, Smith V, Hillan K, Quirke P (2001) Unlocking the archive-gene expression in paraffin-embedded tissue. J Pathol 195:66–71
31. Mies C (1994) Molecular biological analysis of paraffin-embedded tissues. Hum Pathol 25 (6):555–560
32. Macabeo-Ong M, Ginzinger DG, Dekker N, McMillan A, Regezi JA, Wong DTW, Jordan RCK (2001) Effect of duration of fixation on quantitative reverse transcription polymerase chain reaction analyses. Mod Pathol 15 (9):979–987
33. Cox ML, Schray CL, Luster CN, Stewart ZS, Korytko PJ, Khan KNM, Paulauskis JD, Dunstan RW (2006) Assessment of fixatives,

fixation, and tissue processing on morphology and RNA integrity. Exp Mol Pathol 80:183–191

34. Hewitt SM, Lewis FA, Cao Y, Conrad RC, Cronin M, Danenberg KD, Goralski TJ, Langmore JP, Raja RH, Williams PM, Palma JF, Warrington JA (2008) Tissue handling and specimen preparation in surgical pathology. Arch Pathol Lab Med 132:1929–1935
35. Chomczynski P, Sacchi N (1987) Single step method of RNA isolation by acid guanidinium thiocyanate-phenol-chloroform extraction. Anal Biochem 162:156–159
36. Cronin MT, Dutta D, Pho M, Nguyen A, Jeong J, Liu M (2008) Tumor marker discovery by expression profiling RNA from formalin fixed paraffin embedded tissues. Methods Mol Biol 520:177–193, Chapter 13
37. Liu ML, Millward C, Baehner F, Sangli C, Dutta D, Pho M, Nguyen A, Jeong J, Cronin MT (2011) RT-PCR gene expression profiling can be performed successfully with RNA from paraffin-embedded tissues prepared using a wide range of different fixatives and conditions. Methods Mol Biol 724:205–237
38. Scott A, Ambannavar R, Jeong J, Liu ML, Cronin MT (2011) RT-PCR gene expression profiling for cancer biomarker discovery from fixed, paraffin-embedded tissues. Methods Mol Biol 724:239–257
39. Cronin M, Pho M, Dutta D, Stephans JC, Shak S, Kiefer MC, Esteban JM, Baker JB (2004) Measurement of gene expression in archival paraffin-embedded tissues: development and performance of a 92-gene reverse transcriptase-polymerase chain reaction assay. Am J Pathol 164:35–42
40. Mullins M, Perreard L, Quackenbush JF, Gauthier N, Bayer S, Ellis M, Parker J, Perou CM, Szabo A, Bernard PS (2007) Agreement in breast cancer classification between microarray and quantitative reverse transcription PCR from fresh-frozen and formalin-fixed, paraffin-embedded tissues. Clin Chem 53 (7):1273–1279
41. Scicchitano MS, Dalmas DA, Bertiaux MA, Anderson SM, Turner LR, Thomas RA, Mirable R, Boyce RW (2006) Preliminary comparison of quantity, quality, and microarray performance of RNA extracted from formalin-fixed, paraffin-embedded, and unfixed frozen tissue samples. J Histochem Cytochem 54 (11):1229–1237
42. Rupp GM, Locker J (1988) Purification and analysis of RNA from paraffin-embedded tissues. Biotechniques 6:56–60
43. Krafft AE, Duncan BW, Bijwaard KE, Taubenberger JK, Lichy JH (1997) Optimization of the isolation and amplification of RNA from formalin-fixed, paraffin-embedded tissue: the armed forces institute of pathology experience and literature review. Mol Diagn 2 (3):217–230
44. Gloghini A, Canal B, Klein U, Dal Maso L, Perin T, Dalla-Favera R, Carbone A (2004) RT-PCR analysis of RNA extracted from Bouin-fixed and paraffin-embedded lymphoid tissues. J Mol Diagn 6:290–296
45. Jackson DP, Lewis FA, Taylor GR, Boylston AW, Quirke P (1990) Tissue extraction of DNA and RNA and analysis by the polymerase chain reaction. J Clin Pathol 43:499–504
46. Roberts L, Bowers J, Sensinger K, Lisowski A, Getts R, Anderson MG (2009) Identification of methods for use of formalin-fixed, paraffin-embedded tissue samples in RNA expression profiling. Genomics 94:341–348
47. Beqaj SH, Flesher R, Walker GR, Smith SA (2007) Use of the real-time PCR assay in conjunction with MagNA Pure for the detection of mycobacterial DNA from fixed specimens. Diagn Mol Pathol 16(3):169–173
48. Ribeiro-Silva A, Zhang H, Jeffrey SS (2007) RNA extraction from ten year old formalin-fixed paraffin-embedded breast cancer samples: a comparison of column purification and magnetic bead-based technologies. BMC Mol Biol 8:118
49. Bohmann K, Hennig G, Rogel U, Poremba C, Mueller BM, Fritz P, Stoerkel S, Schaefer KL (2009) RNA extraction from archival formalin-fixed paraffin-embedded tissue: a comparison of manual, semiautomated, and fully automated purification methods. Clin Chem 55 (9):1719–1727
50. Lee J, Hever A, Willhite D, Zlotnik A, Hevezi P (2005) Effects of RNA degradation on gene expression analysis of human postmortem tissues. FASEB J 19:1356–1358
51. Lee KY, Shibutani M, Inoue K, Kuroiwa K, M. U, Woo GH, Hirose M (2006) Methacarn fixation – effects of tissue processing and storage conditions on detection of mRNAs and proteins in paraffin-embedded tissues. Anal Biochem 351:36–43
52. Kotorashvili A, Ramnauth A, Liu C, Lin J, Ye K, Kim R, Hazan R, Rohan T, Fineberg S, Loudig O (2012) Effective DNA/RNA co-extraction for analysis of MicroRNAs, mRNAs, and genomic DNA from formalin-fixed paraffin-embedded specimens. PLoS One 7(4):e34683. doi:10.1371/journal.pone.0034683

53. von Ahlfen S, Missel A, Bendrat K, Schlumpberger M (2007) Determinants of RNA quality from FFPE samples. Plos One 12:e1261
54. Stanta G, Bonin S (1998) RNA quantitative analysis from fixed and paraffin-embedded tissues: membrane hybridization and capillary electrophoresis. Biotechniques 24:271–276
55. Godfrey TE, Kim SH, Chavira M, Ruff DW, Warren RS, Gray JW, Jensen RH (2000) Quantitative mRNA expression analysis from formalin-fixed, paraffin-embedded tissues using 5′ nuclease quantitative reverse transcription-polymerase chain reaction. J Mol Diagn 2:84–91
56. Abrahamsen HN, Steiniche T, Nexo E, Hamilton-Dutoit SJ, Sorensen BS (2003) Towards quantitative mRNA analysis in paraffin-embedded tissues using real-time reverse transcriptase-polymerase chain reaction: a methodological study on lymph nodes from melanoma patients. J Mol Diagn 5:34–41
57. DNA quantitation: methods and recommendations in use at GenVault, application note, Carlsbad, CA, www.genvault.com
58. Linton KM, Hey Y, Saunders E, Jeziorska M, Denton J, Wilson CL, Swindell R, Dibben S, Miller CJ, Pepper SD, Radford JA, Freemont AJ (2008) Acquisition of biologically relevant gene expression data by Affymetrix microarray analysis of archival formalin-fixed paraffin-embedded tumours. Br J Cancer 98:1403–1414
59. Specht K, Richter T, Muller U, Walch A, Werner M, Hofler H (2001) Quantitative gene expression analysis in microdissected archival formalin-fixed and paraffin-embedded tumor tissue. Am J Pathol 158:419–429
60. Haque T, Faury D, Albrecht S, Lopez-Aguilar E, Hauser P, Garami M, Hanzely Z, Bognar L, DelMaestro RF, Atkinson J, Nantel A, Jabado N (2007) Gene expression profiling from formalin-fixed paraffin-embedded tumors of pediatric glioblastoma. Clin Cancer Res 13 (21):6284–6292

Methods in Pharmacology and Toxicology (2015): 219–257
DOI 10.1007/7653_2014_39

Published online: 30 January 2015

Next-Generation Sequencing (NGS) in Anatomic Pathology Discovery and Practice

Matthew J. McGinniss, David A. Eberhard, and Keith A. Wharton Jr.

Abstract

Next-generation sequencing (NGS), also known as massively parallel sequencing (MPS), is a family of emerging and powerful technologies that generate an enormous amount of nucleic acid sequence data on a time scale and at a cost that is becoming compatible with routine clinical diagnostic applications. NGS relies upon selective amplification or capture of a complex population of nucleic acid molecules within a biological sample. Simultaneous sequencing reactions generate a variably broad and deep digital readout of sequences in the sample, representing specific regions of genes such as exons, a class of RNAs, or whole genomes. NGS can be viewed as a special type of multianalyte "omics" test that overturns the "one-question/one-test" paradigm prevalent in pathology and laboratory medicine for over a century. It has the potential to address thousands of clinically relevant questions at once while also serving as a transformative, hypothesis-generating discovery tool. The power of NGS to generate massive datasets, which by their sheer size and complexity can approach the level of biological complexity within the tissue sample itself, qualifies it as a "disruptive technology" that will task the medical establishment, particularly pathology and laboratory medicine professionals who generate and interpret the data, to implement it in clinical settings. There remain numerous challenges to the routine implementation of NGS technologies in drug development and clinical diagnostics, including scientific/technical, medical, financial, legal/regulatory, and ethical hurdles.

Key words Next-generation sequencing, Massively parallel sequencing, Sanger sequencing, Bioinformatics, Regulatory, Diagnostic test, Clinical trial

1 Introduction

Our purpose in this chapter is to introduce NGS to the broad audience of professionals who will help implement and interpret it in a variety of contexts, with a focus on drug discovery and clinical trials, but also including medical diagnosis. First, we provide a brief overview of current anatomic pathology laboratory practice and multianalyte tests as a framework to incorporate NGS into anatomic pathology practice in diagnostic and therapeutic development. Then, we discuss the major factors that are bringing NGS into clinical medicine, followed by a technical introduction to better understand current and (as best as we can predict) future NGS platforms. We discuss practical aspects of implementation in the clinical laboratory, including automation of the bioinformatics analysis and regulatory issues. We provide current examples with which we have personal experience and close with our view of future challenges in both discovery and diagnostic laboratory settings. Since this field is replete with jargon, we will endeavor to define key terms as we progress to make this topic accessible to a broad audience.

2 Anatomic Pathology and Implementation of "Omics" Tests

Microscopic examination of tissue has served a crucial role in the diagnosis of various diseases and in medical care over the past century and a half, and continues to do so today. Nearly all tissues collected for diagnostic purposes are processed by a histology laboratory. Tissue can be frozen and then sectioned for rapid diagnosis near the operating room, or processed in other ways such as smears or touch preparations for special techniques, but for the clearest preservation of cell and tissue morphology it is commonly fixed (i.e., made rigid by introducing covalent cross-links) in formalin (a buffered solution of formaldehyde), then chemically processed and embedded in paraffin wax blocks. "FFPE" (*f*ormalin *f*ixed *p*araffin *e*mbedded) tissue is thus the main form of tissue generated by pathology labs. Next, FFPE tissue, affixed to a plastic base, is thinly sliced on a microtome, mounted on a glass slide, and then stained with the dyes hematoxylin and eosin (H&E). Tissue is viewed through a microscope, and its interpretation by a pathologist serves as the basis of a report that enters the medical record to guide patient care. Similarly processed and examined tissues from animals treated with therapies under investigation serve as a basis for preclinical evaluation of drug toxicity in pharmaceutical development. Residual FFPE tissue is typically archived for long-term storage.

While H&E slide-based histologic diagnosis represents a "gold standard" for many diseases, the information it provides is limited. In particular, neither H&E nor the variety of "special" (histochemical) stains historically used to highlight tissue attributes (e.g., fibrosis, classes of infectious agents) reveal the status of particular molecules crucial in disease processes. As disease-relevant molecules have been identified, a variety of ancillary molecular-based tests have been developed as an adjunct to H&E tissue diagnosis and evaluation. These molecular tests can be divided into two categories: those that preserve, indeed require, intact tissue architecture; and those that destroy it. Immunohistochemistry (IHC), a technique by which specific antigens are visualized in an intact tissue section, is probably the most commonly used of all molecular techniques in anatomic pathology research and clinical diagnosis (1); others include DNA in situ hybridization (ISH) to detect chromosomal translocations or viruses (2), and RNA ISH to label specific RNA species such as mRNA or miRNA (e.g., (3, 4)).

Among tests that destroy tissue in order to interrogate its chemical complexity are nucleic acid profiling and sequencing as well as various bioanalytical assays (e.g., ELISA) that measure specific analytes such as proteins. Standard of care diagnosis and treatment for many diseases, including most types of cancer, relies on a combination of histologic and molecular assays. For example, in 2014, a lung cancer diagnosed as an adenocarcinoma will be tested for EGFR gene mutations by sequencing (5) and ALK gene rearrangement by fluorescence in situ hybridization (FISH) (6), and a colorectal adenocarcinoma will be sequenced to test for KRAS mutations (7), as each of these tests informs prognosis and choice of therapy. In drug discovery and development, molecular tests can determine if a potential drug target is present or altered in a tissue sample, the consequences of drug exposure, and whether the target is localized to specific cell types, subcellular locations, or tissue components such as tumors, stroma, vessels, or immune cells. In these examples, specific "single-analyte" molecular tests can be performed on tissue to answer a particular question that can provide useful information to make decisions such as which therapy might benefit a particular patient or which patients should be included in a clinical trial. Some of these tests are developed as "companion diagnostics" to predict or document benefit from a particular therapy.

The emergence of "omics"-scale technologies that allow the large-scale interrogation of hundreds, thousands, or millions of discrete molecular entities in a tissue sample has revolutionized discovery and translational research into disease mechanisms and candidate drug target identification (8). Omics technologies include microarrays that, by hybridization reactions, measure nucleic acids including RNAs (mRNA, miRNA, etc.) or DNA alterations such as single-nucleotide polymorphisms (SNPs) and mutations, and gene or chromosome copy number changes; mass spectrometry to measure proteins, metabolites, and other small molecules; NGS; and unknowable future technologies. To date, molecular omics assays require tissue destruction and, unlike IHC or ISH, are not directly interpreted in the context of an intact tissue section. However, microscopic examination of the tissue prior to its destruction is a necessary part of preanalytic quality assessment (e.g., ensuring that the disease process is present in the section of tissue analyzed as well as measuring the disease's extent or severity) and sample preparation (e.g., selection of specific regions of tissue or cells by macro- or microdissection). Furthermore, histopathologic interpretation of the source sample can be crucial to interpretation of omics data.

Although many clinical laboratory tests are single-analyte, generally performed to answer a specific question, omics generate large amounts of data, much of it extraneous, and typically only a subset of which is relevant to the most important questions. The goal of adapting omics tests to clinical decision-making is to identify, interpret, and report all the medically relevant data. Paradigms that can guide implementation of NGS and other omics tests already exist in laboratory medicine, including (1) "test panels" composed of simultaneously performed single-analyte tests such as IHC panels on tumors of specific types (e.g., Her2, ER, and PR on breast cancer), or Chem20 panels on blood samples, which enable diagnosis by pattern recognition (combinations of alterations seen in particular conditions or diseases); (2) "reflex testing," in which certain test results trigger automatic performance of confirmatory or ancillary tests, often part of disease diagnosis by diagnostic algorithm; (3) existing molecular pathology and genetics tests including PCR to detect specific mutations; (4) in vitro diagnostic multivariate index assay (IVDMIA, also called MAAA—multianalyte assay with algorithmic analysis), a multianalyte assay either cleared by the US Food and Drug Administration for a specific diagnostic purpose or offered as a laboratory-developed test (LDT) in specific laboratories (9). Some IVDMIAs originate from large omics datasets, but the analytical platform for the diagnostic test often differs from that used to discover and develop the test. Whatever the format, results from many carefully chosen and simultaneously run tests can enable more accurate and efficient clinical decision-making than is possible from single measurement tests. In principle, NGS/omics tests can be viewed in a similar fashion.

How to approach large omics datasets in a diagnostic context depends on the question(s) to be answered and the types of answer(s) sought. When used as a discovery tool, omics can generate a large list of candidate genes whose measurements (typically abundance of a sequence, or specific mutations) might have meaning (e.g., distinguishing healthy vs. diseased tissues, distinct subtypes of diseases, or consequences of therapy), but the challenge is then to discern the small number of causative or predictive events from a far larger number of downstream, possibly less relevant events—i.e., the "drivers" versus the "passengers." Drivers and critical passenger events (or values) represent potential therapeutic targets. But for diagnostic purposes, omics tests like NGS are in essence an extreme type of multianalyte (multiplex) test that aims to establish a diagnosis or prognosis or predict the response to a particular therapy based on a combination of single-analyte results. Like the IVDMIA, the goal is to define an optimal combination of analyte results, sometimes called a "signature," that best correlates with, or reliably predicts, a particular outcome. In the test development stage, both "hypothesis-directed" (i.e., using known biology to propose hypotheses that can be tested with the omics data sets) and purely statistical "hypothesis-free" approaches can generate candidate signatures. Ultimately, the association of a signature with the disease or treatment characteristic is established empirically using statistical methods on independent cohorts. The combination of single-analyte results (gene expression levels, mutations, etc.) that comprises the signature need not have known roles in the disease process, and individual analyte results by themselves need not have any diagnostic or prognostic power independent of the signature; the signature itself can become a feature, sometimes a defining entity, of a particular type of disease or biological phenomenon.

In the discovery and test development phase, data are typically generated from patient samples—perhaps from a clinical trial—that fulfill specific criteria, in order to compare combinations of single-analyte results with a particular clinical parameter, such as response to intervention, functional status, or survival. A statistical model generates a candidate signature; it is then tested and refined on independent cohorts of the same disease, and then possibly other diseases, to determine how broadly the signature can be applied. To be useful in clinical practice, a signature must predict similar features (e.g., an outcome or survival) in independent cohorts with common features. In addition to demonstrating defined clinical utility, clinical (diagnostic) tests must exhibit robust and reproducible performance characteristics and can only be performed in certified environments. There have been a few notable successes (and many more failures) in discovering, marketing, and implementing omics-based multianalyte tests in the clinical arena during the past two decades, and, based on recent experience, best-practice guidelines for test development and implementation have been proposed (10, 11).

The evolving role of gene expression microarrays in disease diagnosis deserves a brief mention. In the late 1990s and early 2000s, mRNA profiling via microarray platforms began to unveil a limited number of molecular subtypes of diseases, such as diffuse large B-cell lymphoma and breast cancer, that could not be recognized by histopathology or existing molecular criteria (12, 13). For these and many subsequently profiled diseases, identified signatures have potential predictive (treatment response) or prognostic significance; for malignant diseases the subtype gene signatures are often similar to a nonmalignant cell type or stage of development (14). While microarrays continue to serve as a valuable profiling tool in research, drug development, and biomarker development for clinical research, their use in clinical diagnostics has lagged for several reasons, including a need for specialized equipment not present in the routine clinical laboratory, the uncertain meaning of extraneous data, and a lack of clinical utility (i.e., medical decision-making) based on the current standard of diagnosis (15). Consequently, there are ongoing efforts to define minimal signatures from larger data sets and measure them with more widespread technologies (e.g., reverse transcriptase polymerase chain reaction (RT-PCR)), or replicate the findings with different methods (e.g., an IHC panel). A notable example is the commercially successful Oncotype DX test, originally developed from microarray data but clinically validated as a 21-gene panel RT-PCR test. It predicts survival and thus the potential impact of chemotherapy, in estrogen receptor positive breast cancers without histologic evidence of lymphatic invasion (16). The clinical benefit of such tests will continue to be scrutinized, as it has been claimed that a major component of the Oncotype DX "risk score" can be inferred by measuring of tumor mitotic rate with routine Ki67 IHC (17).

Given the pace of new discoveries and technology development, we are likely to see many more candidate omics-based signature tests applied to patient material, particularly tissue, in the coming years. RNA-seq (an NGS-based technique to sequence populations of mRNAs) is poised to gain wider acceptance as a means to generate signatures (18), although competition and economies of scale have driven the cost of microarray profiling downward. Omics technologies, such as NGS, that afford a very broad molecular characterization of patient samples have led some scientists to suggest that these technologies may one day supplant the need for microscopic tissue examination. The history of pathology is full of attempts to introduce new technologies into routine diagnosis, but the only ones that survive (at least until a better test is devised and validated) are those that reliably, rapidly, and economically answer specific clinical questions. NGS is poised to address many clinical questions, including many that are not ready to be answered in other useful ways, or even asked in ways that can be answered. The rapidly declining cost of NGS, coupled with the value added to the patient over their lifetime through knowledge of their genome, suggests that a whole-genome sequence of a patient's germline will soon be a routine part of the medical record (19). However, genomic sequences—just like microscopic slides—are complex datasets that require specialized knowledge and expertise in order to extract and interpret all the data components that are relevant to clinical questions that may arise.

We conclude this section by emphasizing three major points:

1. The intrinsic complexity of tissues and the diversity of cell types comprising them, coupled with alterations in cell function characteristic of diseased tissues, make it likely that morphological interpretation of tissue will remain a central component of anatomic pathology practice in drug development, human disease diagnosis, and delivery of medical care for the foreseeable future. One must therefore view with skepticism claims that omics technologies such as NGS will replace H&E-based histopathology diagnosis; rather, where possible, NGS data will need to be interpreted with reference to histopathology.

2. Ancillary techniques or tests used as part of anatomic pathology practice must demonstrate specific, measurable diagnostic or clinical utility; without it, there will be no reimbursement by government or private insurance programs, and—other than for special research applications—the test will fall into disuse. Multianalyte tests such as NGS are no exception.
3. FFPE tissue plays a central role in anatomic pathology, both in drug development and clinical practice. The advantages of FFPE blocks are that tissue architecture and cytologic features are maintained indefinitely, and—depending on the state of the tissue when embedded, and the storage conditions—many classes of analytes including DNA, RNAs, and certain proteins can be preserved for decades (20–22). The relatively short read length (<150 nucleotides) from most NGS platforms allows archival and routinely processed FFPE tissue to be used as a starting material for analysis, and the chemical changes induced by the fixation process that can lead to artifactual sequence alterations in PCR-based assays are less evident with NGS-based tests (23). However, for research, clinical trial, or diagnostic development-related investigations, the use of archival FFPE tissue blocks (that typically originate from diverse sources) can introduce increased variation in preanalytical and analytical test factors that impact the yield and quality of interpretable data; for this reason, for many studies paired snap-frozen tissue is often a preferred starting material for NGS assays. FFPE tissue preparation introduces chemical changes to various molecular analytes, but these have been fairly well studied, and methods to assay them in appropriately prepared and processed FFPE tissue are becoming established and accepted in clinical practice (24, 25). For example, although not an NGS-based test, the hybridization-based Prosigna test (Nanostring) was recently approved by the FDA to assess breast cancer prognosis from FFPE tissue specimens (26). Alternatively, fixation systems such as PAXGene tissue (Qiagen) that simultaneously preserve tissue morphology and many analytes of interest, including nucleic acids that can be assessed by NGS, might gain greater use for specialized applications or studies (27). Nevertheless, we anticipate that FFPE tissues will remain central to anatomic pathology and diagnostic NGS for the foreseeable future, in part because nearly all diagnostic clinical IHC assays have been developed with FFPE tissues (28).

3 Factors Bringing NGS into the Clinical Arena

Large-scale efforts to define the human genome in health and disease, such as the Human Genome Project and The Cancer Genome Atlas, were spurred by the vision that such knowledge would lead to improvements in the diagnosis and treatment of disease. Early reports of NGS revealing mutations that cause mysterious Mendelian disorders (29, 30) and the complex genomic landscapes of various cancer types (31) have demonstrated the tremendous power of NGS to elucidate disease mechanisms. Furthermore, NGS data is helping to provide a framework for the integration of results from a variety of other experiments, including mRNA or miRNA expression profiling, genome-wide association studies (GWAS), and SNP analyses. There is tremendous enthusiasm for the idea that NGS is not just a sophisticated advance in our research and discovery capabilities but will also transform the approach to routine clinical diagnostics. Already, NGS is being used in diverse fields, including fetal and infectious diseases (32), forensics (33), microbiome characterization (34), and medical genetics (29). Fortunately, nucleic acid sequencing has been used in a variety of clinical testing scenarios for many years. The advances afforded by NGS are the same factors bringing it into clinical practice, and they include massive expansion of the scale at which sequencing can be performed; power to deeply probe sample complexity; high diagnostic yield; declining cost per base pair sequenced; and ability to rapidly and economically extract a large amount of information from clinical samples of a limited size.

The vast amount of sequence interrogation capacity in NGS can be exploited in three ways: (1) by generating sequence from very long stretches of DNA, entire chromosomes, or genomes ("breadth of coverage"); (2) by interrogating smaller selected stretches of nucleic acid, such as a limited number of individual genes, many times over, to allow the sensitive detection of minor components in a mixture of sequence variants ("depth of coverage"); and (3) by running many different samples together, often from different patients, in the same assay run ("multiplexing").

Broad coverage affords the ability to sequence many genes of interest at once—the whole exome or "transcriptome" (all coding exons or mRNAs expressed) or even the entire genome of an individual. A broad approach such as whole-genome sequencing (WGS) can be very powerful when the genes/genomic loci that might be related to the disease or condition of interest are unknown, or are highly variable between patients (e.g., identifying the germline mutations that underlie inherited diseases that are rare or atypical), or that have multigenic etiologies (i.e., diseases that result from interactions between several genes, or interactions of specific genes with the environment), or for identifying the gene rearrangements, translocation breakpoints, and chromosome fusion events that underlie most cancers.

Deep coverage is particularly useful for analyzing complex genomic mixtures where a pathogenic sequence variant may represent only a minor fraction of the total copies in the sample. For example, a patient tumor tissue sample will always contain a variable mixture of cancer cells that contain mutations of interest with various normal non-neoplastic tissue cells such as vascular, inflammatory/immune, and stromal cells with largely wild-type genomes (although mutations have been reported in tumor stroma) (35). Even in the tumor cell population, genomic alterations may be complex and heterogeneous, with genomic evolution of driver and passenger and mutations occurring over the course of treatment or as the tumor grows and metastasizes (36, 37). Deep coverage provides the experimental and statistical power needed to identify and confirm the meaning of such information in complex samples. Current guidelines estimate that 300–500× depth of coverage is required in a clinical setting to identify somatic genome variants in cancer with high sensitivity (38), with greater input DNA required to detect the rarest variants that are not due to preanalytic DNA amplification bias or analytic sequencing errors.

Multiplexing—sequencing samples from multiple patients simultaneously—enables near capacity use of a machine for each analytical run, thus reducing costs. Two examples of this type of test are illustrated later in the chapter.

While the sequencing capacity of NGS is tremendous, it is not unlimited. The amount of data produced depends on the platform; the speed with which the data that can be analyzed depends on computational power and time (39); and the ultimate throughput depends on the efficiency of bioinformatics pipelines and clinical workflows. Therefore, for the foreseeable future a trade-off will remain between breadth and depth of coverage; the optimal approach should be dictated by the questions to be answered.

4 Technical Introduction to NGS

Significant advances in capillary electrophoresis technology in the mid-to-late 1990s (4-color fluorescence-based sequence detection, improved fluorophores, polymerases modified for sequencing, and improved instrumentation) coupled with development of PHRED informatics software, allowed rapid, automated production and analysis of sequence data, culminating in the landmark sequencing of the first human genome in 2001 (40, 41). These so-called pioneer genomes, sequenced de novo, then served as the reference, or baseline, genome sequence that enabled the rapid discovery of many important alterations in cancer genomes (42, 43) and identification of causative mutations in many Mendelian disorders (29, 44). By most measures, NGS is an extreme, atypical multianalyte laboratory test, but to facilitate comparisons with other laboratory tests we divide the NGS workflow into the same steps as any clinical laboratory test: preanalytic, analytic, and postanalytic phases.

Table 1
Comparison of PCR-based mutational analysis, Sanger sequencing, and NGS

Attribute	PCR-based mutation detection methods	Sanger sequencing	NGS
Loci	Only known loci	Known and unknown loci	Known and unknown loci
Sensitivity	Very high sensitivity (LOD 1–5 %)	Moderate LOD ~20 %	High sensitivity LOD 5 % (or better)
Accuracy	High	High raw accuracy	High accuracy
Sample multiplexing	Limited	None	High with sample barcoding
Cost per sample	Low	Modest to high per base ($1,500 per Mb) Low cost per run ($96)	Ultimately lower with barcoding and multigene panels, scalable
Limitation	One assay at a time	Limited throughput; can be challenging for insertion/deletion detection	Steep learning curve

NGS technologies offer distinct advantages over conventional sequencing or genotyping techniques (Table 1). First, the cost of NGS sequencing is substantially lower per base compared to Sanger sequencing. The cost of sequencing an individual genome in 2007 by the Sanger method, for example, was on the order of millions of dollars (45). However, the cost to sequence an individual genome by NGS in a commercial or research setting is, as of 2014, only on the order of several thousand dollars. Second, NGS sequencing can be tuned to low levels of sensitivity such as 5 % mutant allele detection (or lower) depending on the desired depth of coverage (46). In contrast, the lower limit of sensitivity for Sanger sequencing is only ~20 % mutant allele detection (47). NGS platforms offer opportunities for multiplexing patient samples that are not possible with conventional sequencing. The combination of the platform selected, detection chemistries, and library preparation methods dictates capabilities: NGS can assess alterations in DNA or RNA including gene rearrangements, translocations and fusions, CpG methylation patterns, transcript expression profiles, and splicing variants. Depending upon the clinical questions being asked, one must consider which approach (e.g., targeted sequencing, whole exome, or whole genome) and platform (in 2014, commonly Life Technologies/Ion Torrent or Illumina are most commonly used for clinical applications) is most suitable given the cost, specimen types analyzed, mutation sensitivity desired, and the specific types of genomic alterations to be assessed. For example, low-pass WGS, as well as appropriately designed whole exome or targeted panels, can provide some information about DNA copy number variations, but more thorough analysis of DNA copy number variations, as well as RNA expression profiles and DNA methylation patterns, requires specialized library preparation with modifications (48, 49).

4.1 Library Preparation (Preanalytical)

Current NGS platforms require the creation of a composite library of DNA fragments suitable for sequencing. First, the nucleic acid population of interest is isolated from the tissue sample using appropriate methods and then size-selected via mechanical or enzymatic means. If the desired analytes are RNAs, then conversion to DNA is generally required. Some platforms rely upon ligation of the population of DNA fragments with special adapter sequences to allow for clonal amplification by PCR. Current imaging systems used in clinical applications are unable to detect single molecule fluorescence events (e.g., nucleotide incorporation during in vitro DNA polymerase-driven replication) and therefore require simultaneous, identical incorporation events from amplified templates to adequately detect signals (50). Note that "single molecule" real time sequencing using DNA polymerization without amplification—so-called "third generation" sequencing [e.g., Pacific Biosciences (PacBio)]—is a promising technique that can generate longer sequence reads than NGS technologies (51), but it requires a greater quantity of input material to generate such long reads. But, given that the typical sequence read length of material extracted from FFPE blocks is ~150 bp, this may not be a critical operation parameter for most pathology applications. Not far behind are "fourth generation" sequencing technologies such as Nanopore (52). Since they are not routinely used in clinical settings as of 2014, we do not discuss third or fourth generation sequencing technologies further.

The two common methods available for producing amplified templates are emulsion PCR and solid-phase amplification. Emulsion PCR is used for the Ion Torrent PGM platform to clonally amplify library DNA fragments linked to microscopic beads (53). In the aqueous water-oil emulsion, a droplet captures a single bead, which then serves as a PCR "microreactor" that clonally amplifies just that single DNA template molecule (50). These beads, or ion sphere particles, are then deposited directly into the wells of the ion semiconductor chip for subsequent sequencing reactions. Solid-phase amplification is used in the Illumina systems, such as the MiSeq instrument, to generate randomly distributed and clonally amplified clusters on the glass surface of a flow cell (54). This process can produce 100–200 million spatially separated clusters on an imaging area of approximately 2 × 12 mm flow cell ready for downstream sequencing reactions.

Newer flow cells amplify DNA in a semiordered or ordered fashion, allowing increased read density. Both of these amplification methods require template DNA molecules to be immobilized onto a specially prepared surface. This spatial separation of DNA molecules is one of the key features that allows for millions of simultaneous sequencing reactions, resulting in truly massive parallel sequencing (45).

The standardized conditions that are used to amplify complex mixtures of DNA fragments by PCR can introduce sequence artifacts and amplification bias (either under- or overrepresentation of AT- or GC-rich regions) of certain loci (55). However, strategies exist to avoid PCR altogether or to optimize PCR for AT-rich genomic regions (56).

There are other methods of library construction, including variant PCR, molecular inversion probes, and hybrid capture techniques. Molecular inversion probes are designed to flank target-specific amplicons (the regions to be PCR-amplified) with universal spacer regions. These probes anneal at either end of the targeted region and are then gap-filled by a polymerase reaction and ligated (57). Inversion probes are particularly useful in assessing small numbers of targets in a large number of samples. However, the main disadvantages of molecular inversion probe approaches, in comparison to hybrid capture methods, are additional cost and variable recovery (57). Hybrid capture methods, which for targeted panels use substantially less starting material than amplicon-based methods, are useful if the sequencing reaction is intended to detect specific copy number variations and chromosomal translocations that might not be readily detected by routine approaches. So-called "mate-pair" sequencing (58) and optical imaging technologies (59) are also capable of detecting translocations but are not commonly used for clinical diagnosis at present.

4.2 Sequencing Runs (Analytical)

The DNA template molecules in the library are then sequenced on the NGS platform. Most common platforms (such as the Illumina MiSeq or Ion Torrent PGM) perform a repetitive series of DNA polymerase-driven sequencing reactions known as "sequencing by synthesis." Library DNA is used as a template to synthesize complementary DNA fragments, with the sequence read based on incorporation of the four relevant nucleotides (A, C, G, or T) into the new strand, not unlike endogenous DNA replication. Different methodologies (optical or electronic) are used for the actual signal detection step. For example, the Ion Torrent PGM (Life Technologies) uses semiconductor chip technology to detect a change in pH as a hydrogen ion is released during the incorporation of each nucleotide within microscopic wells of the sequencing chip (60). The Illumina platform (MiSeq, HiSeq) sequencing reaction involves optical detection using four-color cyclic reversible terminator chemistry (50). Each cycle requires the incorporation of a nucleotide, imaging of the fluorescent signal, and then cleavage of a noncomplementary hanging strand to allow subsequent nucleotides to be incorporated. These sequencing reactions are performed in pico- or nanoscale reactors in a massively parallel fashion, reducing cost and enhancing the speed at which data can be generated (45). Further improvements in speed are expected with a two-color strategy on Illumina instruments, but they are only beginning to be validated for clinical applications.

4.3 Data Interpretation and Reporting (Postanalytical)

There are two distinct components of the postanalytical phase. The first involves mapping and/or assembly, with technical integration of the data based on the specific NGS experiment; and the second involves comparison with a reference sequence (the patient's germline sequence and/or a mutation database) with context-specific interpretation of any differences. In the first stage, the newly identified DNA sequences (also known as reads) are aligned to a known reference sequence for comparison, and, depending on the indication, a nondiseased sample from the same patient may be sequenced (unless already available). Given the ~3×10^6 single-nucleotide polymorphisms (SNPs) and 5×10^5 distinct genomic structural variations characterized in the human population, and that each person harbors ~60 mutations not present in their parents (61), the number of sequence variants can be in the hundreds for nondiseased samples, and for diseases such as cancer with a heavy mutational load the number of variants to classify can easily number into the thousands. Once all sequence variants (somatic or germline) relative to reference sequence(s) have been identified, the individual variants that are relevant to medical decision-making must be identified and classified. This decision-making could include identifying patients eligible for entry into specific clinical trials; deciding if a certain drug therapy might benefit a cancer patient based upon the presence or absence of particular mutations found in their tumor sample; or matching existing or novel therapies to patients with a mysterious disease previously lacking a diagnosis. Any novel mutation detected should be evaluated for potential clinical significance or pathogenicity. For cancer, this requires the potential importance of the mutation for tumor growth or susceptibility to therapy to be inferred either from literature precedent or by comparison with a growing list of sequence abnormalities in specific tumor types. This interpretative process alone can be very time-consuming when a large list of novel variants is generated.

Each sequencing platform has known strengths and limitations that can impact some clinical applications but not others. For example, in one comparison study from 2012, the Illumina MiSeq instruments generated fewer simple single-nucleotide errors than the 454 GS Junior (Roche) and the Ion Torrent PGM

(Life Technologies) platforms (62). The systematic error rate in paired-end sequencing (sequencing inward from each end of a fragment) reads from high-coverage data is ~1 in 1,000 base pairs, with errors more likely towards the end of reads and in the context of certain sequence motifs (63). In this reference, "systematic errors" are defined as the accumulation of many individual base-call errors in different reads from the same genomic location. For example, GGT and GGC sequence motifs can lead to multiple base-call errors in one direction (*see* Fig. 1 of (63)). New technologies and enhancements will strive to eliminate these subtle but important biases, with the goal of generating error-free sequence.

Each sequencing platform requires distinct assessments to judge run quality. Quality control considerations for evaluating sequencing data include the total number of reads, average read length, quality of individual reads, and metrics indicating the efficiency of the specific method of detection (e.g., optimal cluster density on the Illumina flow cell should yield over 15 million sequence reads, or microsphere loading density should be over 50 % on the Ion Torrent chip). For "clinical-grade" sequencing applications, the user will have to establish consistent and strict quality control criteria and will have to monitor results to ensure the quality and reproducibility of reported sequence. As knowledge accumulates, so will the meaning of particular sequencing results. For example, there may be an expectation of a certain depth of coverage for detection of somatic mutations in tumor specimens on the order of ~250–500× to ensure that low frequency mutant alleles can be detected in samples with a mixed population of cell types, such as somatic mutations in tumor samples with nontumorous stroma (38). There may also be an expectation for the quality of reads to meet a particular criterion, such as >90 % of sequence reads meet or exceed a Q30 quality score (corresponding to the probability of an incorrect base call of 1 in 10^3 and an inferred base call accuracy of 99.9 %). In current practice, the lower limit of detection of mutations in tumor samples is ~5–10 % mutant alleles. One recent report suggests that the lower limit of sensitivity for the Roche 454 GS FLX instrument cannot go below 3.5 % since their data showed that the background false positive variant calls were, depending on the locus, ~0.25–3.5 % allele burden (64).

4.4 Limitations of NGS Technologies

The main advantages of NGS are the extremely high throughput and the resultant lower cost per sequenced base as well as improved sensitivity to detect variant sequence or mutant alleles. As shown in Table 2, both the Illumina MiSeq and Ion Torrent PGM significantly outperform Sanger sequencing with regard to number of reads and data output. Specific limitations of the three main types of NGS approaches (targeted sequencing, whole exome, or whole genome) are outlined in Table 3. Whole genome and some targeted sequencing formats can reliably identify copy number alterations

Table 2
Comparison of Sanger sequencing with two NGS platforms

Platform	DNA input	Samples per run	Run time	Read length	Detection method	Number of reads	Data output
Sanger sequencing	1–3 μg	1–96	3 h	900 bp	Fluorescence	1–96	1–80 kb
Illumina MiSeq[a]	0.1–1 μg	1 Flow cell	27 h	2 × 150 bp	Fluorescence	3×10^6	>4 Gb
Ion Torrent PGM[a]	0.1–1 μg	1 Chip	1.5 h	200 bp	Ion sensing by semiconductor chip	10×10^6	≈1 Gb

[a]Both NGS platforms support sample multiplexing by the barcoding technique, which consists of ligating a unique, short (~10 bp) DNA sequences to each source DNA library fragments in order to link each sequence read to the specific source sample

Table 3
Advantages and limitations associated with various approaches to NGS

Approach	Advantages	Limitations
Targeted sequencing	Easy to obtain high coverage and high sensitivity for clinical sequencing applications Requires relatively small amounts of sample input DNA	Might not detect copy number variation; does not detect RNA expression or noncoding variants
Whole-exome sequencing	Viable alternative to stepwise clinical sequencing of suspected candidate genes for unusual clinical presentations Smaller data sets and storage costs than whole-genome sequencing Costs 5× less than whole-genome sequencing	Does not cover potentially functional noncoding regions including promoters, enhancers, introns, and noncoding RNAs (including microRNA) Limited ability to detect translocations and inversions Limited ability to detect trinucleotide repeat expansions or X-linked recessive disease in females due to skewing of X-inactivation
Whole-genome sequencing	Provides a solid foundation for a complete catalog of oncogenic mutations	Most costly approach Depth of coverage may not be suitable for clinical-grade mutation detection Most of the variants obtained may not be actionable or informative at present Challenging to procure server- or cloud-based data storage for large amounts of data

and structural alterations such as chromosome translocations. In addition, whole-genome sequencing can identify noncoding polymorphisms that may be biologically relevant but cannot be detected by the whole-exome approach. However, depending on the purpose of the assay, such findings may not be relevant to the patient at the time of diagnosis. Thus one needs to match the entire workflow—from sample collection to bioinformatics interpretation—to the specific, intended clinical use, keeping in mind that the questions to be answered are distinct in research/discovery, clinical trials, and in routine clinical diagnostic settings. For example, in cancer applications, whole-genome sequencing may not provide sufficient depth of coverage to permit routine clinical-grade mutation detection as compared to targeted sequencing or whole-exome approaches, whereas targeted panels might not provide sufficient breadth of gene coverage for identification of alterations in genes in rare or unusual tumors. Until bioinformatics interpretation becomes more rapid and wholly automated, one must define the clinical questions and anticipated results prior to choosing the platform as well as deciding on panel versus exome versus whole genome, specifying the need for, and types of, controls depending on clinical context.

5 Bioinformatics Considerations

Bioinformatics solutions for NGS data analysis should provide a high diagnostic yield and practical workflow and minimize potential incidental findings (65). Current options for the application of bioinformatics tools include the purchase of commercially developed packages, adapting or applying current open-source tools, or developing custom applications. The development of user-friendly and automated data processing of sequencing data obtained from samples analyzed on NGS platforms remains an unmet need, whether for discovery research, clinical applications such as molecular oncology, or clinical trials (38). A variety of data analysis algorithms exist to perform specific tasks such as sequence alignments, base calling, or assessing the potential clinical impact of identified sequence variants. Some can be purchased from commercial vendors, and many others are freely available as open-source algorithms via the internet [such as Sorting Intolerant from Tolerant (SIFT), Polyphen-2, and Grantham Variation and Grantham Deviation (GVGD)].

The computational challenges of large NGS data sets are myriad and can present a bottleneck that requires specialized bioinformatics skills (66). For current technologies, data output from sequencing platforms are short sequence reads that can range from 50 to 300 base pairs (Illumina MiSeq) or longer (up to 30,000 bp, PacBio) depending upon the platform and the reagent kit chemistries. Raw sequence reads are usually exported as FASTQ (Illumina MiSeq), BAM (*b*inary *a*lignment *m*ap; Ion Torrent Personal Genome Machine), or HDF5 (PacBio and Oxford Nanopore Technologies) file formats. A variety of factors influence how well different sequence mapping programs perform, including relative GC content of sequences, depth of coverage required, single versus paired-end sequencing, and base calling error rates (67). NGS platforms generate millions of short sequence reads that must first be mapped or aligned to one or more reference sequences; alignment is required for nearly all NGS interpretations. To aid in mapping a large number of sequences, a variety of short read sequence alignment algorithms have been developed, including BWA, Bow Tie, STAR, and Novalign (68, 69). A recent review of several alignment programs, considering accuracy and computational run time, concluded that mapping quality scores might not correlate with the likelihood that mapping is accurate (70). Software alignment tools are designed with different approaches and trade-offs with regard to speed and accuracy and need to be evaluated with respect to the individual user's needs regarding discovery research, targeted clinical sequencing, gene expression, and whole-genome analysis. As expected, most alignment tools showed a linear relationship between the depth of coverage/number of reads and total computational runtime (70).

After sequence reads are base called (the process of assigning nucleotide bases to instrument-generated sequencing signals) and mapped to a reference sequence, the composite sequence needs to be annotated for any previously described or novel sequence variants. For analysis of germline mutations in diploid samples, the expected allelic fraction is either 0.5 for a heterozygous variant or 1.0 for homozygosity. However, somatic mutation calling in preneoplastic or neoplastic samples is more complex, since they usually exhibit aneuploidy, heterogeneity of tumor cell ploidy, and variation in the proportion of tumor versus supporting stromal tissues (42). Thus the allelic fraction for somatic mutations is a continuous variable that can range anywhere from >0 to 1.0. For targeted sequencing analysis of cancer specimens, the tumor somatic mutations can be readily identified by comparing the tumor sequence to that of a matched non-neoplastic germline sample taken from the patient. In the absence of a matched normal sample, well-described germline variants can be filtered out from the tumor sequence after interrogating selected, thoroughly annotated databases. For example, the National Center for Biotechnology Information (NCBI) established dbSNP (http://www.ncbi.nlm.nih.gov/SNP/index.html) for both single-base nucleotide substitutions and short deletion and insertion polymorphisms for many organisms including humans. Another useful database is the 1000 genomes project (the 1000 Genomes Project Consortium 2012). Their aim is to characterize over 95 % of the variants in genomic regions that have an

Table 4
Selected computational aids for mutation calling and clinical annotation

Category	Method	URL	Comment	Reference
Mutation calling	CASAVA	http://support.illumina.com/sequencing/sequencing_software/casava.html		(54)
Indel calling	BreakDancer	http://genome.wustl.edu/tools/cancer-genomics/		(93)
	Pindel	http://www.ebi.ac.uk/~kye/pindel		(94)
Copy number assessment	CBS	https://r-forge.r-project.org/R/?group_id=702		(95)
	SegSeq	http://www.broadinstitute.org/cancer/cga/segseq		(96)
Prediction of effect	SIFT	http://sift.jcvi.org/		(97)
	Polyphen-2	http://genetics.bwh.harvard.edu/pph2/dokuwiki/start		(98)
	dbNSFP	https://sites.google.com/site/jpopgen/dbNSFP		(99)
Visualization	CIRCOS	http://circos.ca/	Many use CIRCOS plots to visualize genomic events	(100)

allele frequency of greater than 1 % (the classic definition of a nucleotide polymorphism). Their web site (http://www.1000genomes.org/) can be used to ascertain if a sequence variation observed in a given patient sample is in fact a germline polymorphism. Furthermore, the COSMIC database (catalog of somatic mutations in cancer) is another well-annotated web site (http://cancer.sanger.ac.uk/cancergenome/projects/cosmic/) designed to store and display somatic mutation information and related details including information relating to various human cancers. The COSMIC database can be routinely queried to see if sequence variations identified in patient samples have been described in other human tumor tissues. A listing of selected software packages and web portals available for mutation calling and clinical annotations is provided in Table 4. Automation of the computational steps and workflows to analyze data from NGS platforms can greatly reduce the hands-on analysis time per sample (66); ideally, future software will query all of these databases or periodically download updated versions of these databases to allow comparisons at the local laboratory. Workflow steps in typical NGS data processing pipelines include processing of raw sequence reads, variant calling, and annotation of sequence variants. One representative pipeline, the Graphical Pipeline for Computational Genomics (GPCG), incorporates these factors to perform NGS data analysis (66). In addition, these data processing pipelines can also increase the efficiency and consistency of rules-based filtering and annotation of variants. Tools, models, and methods to discern altered gene functions in cancer each pose special challenges (71).

6 Implementation of NGS in the Clinical Laboratory

The implementation of NGS tests in the clinical laboratory setting follows many of the common elements for any clinical test. For example, analytic validation is required of any new clinical test, and the associated clinical utility and intended uses must be clearly defined. Procedures must be written, and appropriate staff trained on the new assay, associated reagents, and equipment. The content and format of clinical patient reports must be developed, reviewed, and approved by the medical director. Communications about the new NGS test must be sent to existing and potential clients and to regulatory bodies such as the College of American Pathologists (CAP) and New York State Department of Health (NYSDOH) if applicable. Finally, the laboratory must enroll in any relevant testing proficiency program, such as those routinely offered by the CAP. If none currently exists (e.g., as of 2014, a pilot program for NGS is currently available at CAP), then the laboratory must set up an alternate performance assessment system.

NGS tests also introduce levels of test complexity not routinely seen with other clinical molecular tests. For example, bioinformatics can be a challenge for routine analysis of NGS data, requiring investment in information technology (IT) infrastructure (72). Bioinformatics experts are needed to develop and update pipelines for the handling and filtering of large data sets generated by massively parallel sequencing instruments, and IT/database experts are typically required to assist with the establishment and maintenance of databases, addressing issues such as data security, quality management, business continuity, and patient confidentiality. Furthermore, NGS technologies are constantly being improved and updated. As such, the clinical laboratory using NGS platforms is faced with a continuing series of updates from the various commercial vendors of instruments, chemistries, reagents, and software. If an update is implemented in the clinical laboratory setting, it must be validated and quality assured to document that sequencing results generated are comparable to, or are of higher quality than, those obtained prior to the update. Finally, there are also challenges inherent in the interpretation of sequence variations that arise from NGS testing. For example, the American College of Medical Genetics (ACMG) recommends that the clinical and technical validation of NGS sequence variation should be performed in Clinical Laboratory Improvement Amendments (CLIA)-approved laboratories, and sequences should be interpreted by a board-certified clinical molecular geneticist or a board-certified molecular genetic pathologist (73, 74).

The manner in which NGS results are reported is rapidly evolving. Physicians are busy, generally valuing laboratory reports with bottom-line results and bulleted recommendations for their patients. Even the simplest NGS reports from current targeted sequencing panels require many pages and can require a Ph.D. in genetics to be understood. Pathway analyses, in which mutations are statistically associated with terms or phrases that have been linked to alterations in known drugged or druggable pathways, are one option, but there is a risk of oversimplifying or misinterpreting what can be overwhelmingly complex biology. How to integrate the information in the medical record will require standardization (75). While electronic medical records are slowly replacing paper records, it will be important for raw composite data to be available for future retrospective re-analyses of germline sequence, presuming appropriate informed consent agreements are in place. Paraphrasing Einstein, we believe the goal should be to make things as simple as possible for the customer (patients and their physicians), but not simpler; to do so risks losing future opportunities for new discoveries.

7 Regulatory Requirements

As NGS moves from research and discovery applications to clinical trials or routine diagnostics, regulatory requirements increase proportionally. As with any automated laboratory test, certain performance specifications need to be established for test instruments and applications on NGS platforms. Laboratories performing NGS for clinical diagnostic purposes or for human clinical trials testing will be subject to a wide range of regulatory oversight.

For example, in the USA all laboratories testing human samples for diagnostic purposes are subject to the CLIA regulations that are managed by the Centers for Medicare & Medicaid Services (CMS) and may be monitored by CAP inspections. The objective of the CLIA program is to ensure the quality of laboratory testing. CLIA regulations involve a variety of laboratory best practices including performance of twice annual proficiency testing (PT); establishment of personnel qualifications for technologists, supervisors, and directors; criteria for storage of reagents; content of procedure manuals; and establishment of quality assessment programs. Thus,

Table 5
Next-generation sequencing in the clinic: standards and guidelines

Group/agency	Standards or guidelines	Reference
ACMG	Clinical laboratory standards for next-generation sequencing	[74]
New York State	Next-generation sequencing (NGS) guidelines for somatic genetic variant detection 26 Nov 2012	http://www.wadsworth.org/labcert/TestApproval/forms/NextGenSeq_ONCO_Guidelines.pdf
AMP report	Opportunities and challenges associated with clinical diagnostic genome sequencing	[38]
CDC	Assuring the quality of next-generation sequencing in clinical laboratory practice	[101]
CAP	Molecular pathology checklist (now includes checklist items pertaining to NGS) 29 Jul 2013	www.cap.org/apps/docs/laboratory_accreditation
ACMG	ACMG recommendations for standards for interpretation and reporting of sequence variations	[73]

laboratories can develop and offer testing using NGS platforms as long as they comply with known CLIA requirements; however as of 2014 CLIA has no NGS-specific guidelines in contrast to many other types of testing routinely performed in the clinical laboratory.

Since Sanger-based DNA sequencing has been used in clinical laboratory settings for at least two decades, a variety of standards and guidelines have been published relating to nucleic acid sequencing in general, and by extension, to NGS (Table 5). For example, the CAP has published updates to their molecular pathology checklist (http://www.cap.org/apps/docs/laboratory_accreditation/checklists/new/molecular_pathology_checklist.pdf) to include a variety of NGS-related items covering both wet-bench laboratory work and the bioinformatics pipeline. Highlights include the requirement for a quality management plan specifically for the bioinformatics pipeline used to support NGS analysis, interpretation, and reporting. Another requirement is validation of the analytical wet-bench processes, with revalidation required if changes are made in any key sequencing reagents. In addition, the ACMG has published guidelines on sequence interpretation in general (73), and, most recently, specific laboratory standards for NGS (74) that includes establishment of minimum coverage thresholds and documentation of analytical sensitivity and the false negative rate. These guidelines recommend that laboratories reporting variants in disease-targeted panels evaluate and classify these variants according to ACMG variant classification schemes (73). In addition, some states in the USA have specific regulations that apply to laboratories performing high complexity tests within their borders or on specimens received from patients within those states. For example, the state of New York has among the most rigorous standards, requiring submission of validation materials for all new molecular and cytogenetic assays performed on specimens that originate from that state. They review and approve new tests and also provide oversight and issue certificates of qualification for laboratory directors in the areas of both genetic testing and oncology—molecular and cellular tumor markers.

As NGS gains wider adoption, FDA and other regulatory agencies will require instrument manufacturers to obtain approval for their use as medical devices in the clinical laboratory setting. Tests with class I (low) or II (intermediate) level risk to patients typically require submission of a 510(k) premarket notification where a vendor submits data to demonstrate that the device to be marketed is safe, effective, and substantially equivalent to a device that is currently (and legally) marketed. The more stringent type of device approval required by the FDA is the PMA (premarket approval); this is reserved for devices with the highest level of risk (class III), such as "companion diagnostic" tests that determine likelihood of benefit and/or risk from a particular therapy. Unlike premarket notification, PMA is based on a determination by the FDA that there exists sufficient valid scientific evidence to provide a reasonable assurance that the device is safe and effective for its intended use(s). Several factors determine whether a company will pursue a 510(k) clearance route versus the more rigorous PMA, but current FDA draft guidance suggest that all nucleic acid-based tests will require PMA in the future. Both Illumina (MiSeq) and Life Technologies (Ion Torrent PGM) have submitted their sequencing instruments to the FDA for 510(k) clearance. In late 2013, Illumina received FDA 510(k) clearance for their MiSeq instrument and two cystic fibrosis assays along with a universal assay kit allowing for open uses (76).

8 Use of NGS in Pharmaceutical Clinical Trials

If a disease is caused or driven by mutation or inappropriate expression of a molecule, there might be an opportunity to develop a drug or therapy that preferentially affects the altered molecule, while leaving the endogenous counterpart relatively unaffected, thus improving safety and efficacy. For example, the tyrosine kinase inhibitors erlotinib and gefitinib are much more potent inhibitors of receptor molecules bearing certain mutations in the tyrosine kinase domain than of the wild-type receptor (77); and trastuzumab, a monoclonal antibody against Her2, effectively inhibits the growth of breast cancer cells with Her2 gene amplification and Her2 protein overexpression but has little effect on the growth of cells with normal Her2 gene copy number and expression levels (78). In such cases, assays to detect the presence or structure of an altered gene can serve as a "companion diagnostic" to select patients for treatment likely to respond to, or tolerate, the targeted therapy.

The development of clinical-grade NGS and other "omics" capabilities in testing laboratories has fueled new clinical trial designs intended to speed development and approval process of therapies targeted to patients likely to benefit from the therapy. In oncology, a patient's tumor can be subjected to NGS, and if a mutation in a cancer-associated gene is found, the patient could then be assigned to a treatment arm with a drug targeted against the mutant gene product. For example, in 2014, the US National Cancer Institute (NCI) launched the M-PACT clinical trial which relies on NGS and highly multiplexed genotyping of tumor samples to screen patients for nearly 400 different mutations in 20 genes related to targeted therapeutics (http://clinicaltrials.gov/show/NCT01827384). If tumor mutations are found, patients are randomized either to an arm where they receive treatment targeted to the gene mutation or the gene pathway, or to an arm where the treatment is not specifically focused on the mutated gene. Another example is the Lung-MAP trial (http://clinicaltrials.gov/ct2/show/NCT02154490), which also uses a multidrug, targeted-screening method to match specific patients with new treatments. There is also the Signature clinical trial sponsored by Novartis (http://www.signaturetrial.com/) that assigns patients to treatment with one of eight different investigational cancer drugs based on evidence for activation of signaling pathways in tumors that can be attenuated by those drugs. Such evidence of activation includes mutations in pathway-associated oncogenes or tumor suppressor genes.

The introduction of omics technologies such as NGS into clinical trials has spurred stakeholders in the drug development process to bring forth recommendations for appropriate implementation. In 2013, an NCI-sponsored collaborative group with representatives from various disciplines related to omics-based test development developed a detailed checklist of criteria that are important for determining the readiness of omics-based tests for clinical trials involving patient treatment decisions. These criteria address issues related to specimens, assays, mathematical model development and performance evaluation, and clinical trial design, as well as ethical, legal, and regulatory considerations. The checklist is suggested for use by funding bodies and journals to aid in assessing the quality of studies and strength of evidence, and for the evaluation of proposals for NCI-sponsored clinical trials in which omics tests would be used to guide therapy (11, 79).

9 Sample NGS-Based Tests in the Clinical Laboratory Today (2014)

A variety of service laboratories, both private and academic, are now performing clinical testing using NGS, and increasingly these tests will involve specimens also submitted for interpretation by histopathology. One of the first applications of NGS was the use of whole-exome sequencing in the diagnosis of genetic disorders where other routine or target-specific tests had not identified a causative mutation in the affected patient. For example, exome sequencing of four individuals with Miller syndrome led to the identification of causative mutations in the DHODH gene (80). Furthermore, exome sequencing of parent-child trios has been effective in distinguishing inherited mutations from new ones, thus helping to distinguish whether a novel coding variant is pathogenic or not (30). Baylor College of Medicine recently published a series of routine whole-exome sequencing initiated in October 2011 (44). They were able to identify one or more mutated alleles in 62 out of 250 probands (25 %) submitted for clinical whole-exome sequencing, an impressively high diagnostic yield given these patients previously had no definitive diagnosis. These results suggest that whole-exome sequencing is useful as a diagnostic test for patients with nonspecific or unusual disease presentations of suspected genetic disorders, whether due to inherited or new mutations. In addition, they showed that medically actionable incidental findings in a total of 16 genes were also made in 12 % (30/250) of patients. Interestingly, 9 of these 16 genes were among those recently recommended by the ACMG for reporting of incidental findings (81). For example, finding germline or constitutional mutations in the TP53 tumor suppressor gene would be considered medically actionable, since individuals with a rare hereditary cancer syndrome (Li Fraumeni) often harbor TP53 mutations. In the above cases, the diagnosis did not require lesional tissue from patients because the mutations of interest were present throughout the body. Two recent examples of "mosaic" diseases required comparison of gene sequences from lesional and nonlesional tissue to identify somatic activating mutations in the causative genes: AKT1 in Proteus syndrome (82) and GNAQ in Sturge-Weber syndrome (83). NGS thus represents a vital tool to begin to understand mysterious diseases with underlying genetic etiologies.

NGS has served as a crucial discovery tool in cancer biology, revealing constellations of genetic lesions in various cancers while providing insight into the genesis of the lesions. In clinical oncology, targeted NGS panels are currently used for two purposes: (1) NGS as the basis of prescreening for enrollment in a clinical trial (such as the previously mentioned Lung-MAP protocol for squamous lung cancer) (84); and (2) interrogation of a panel of genes whose mutation implies alteration in pathways targeted by marketed therapies (e.g., that offered by Foundation Medicine http://www.foundationmedicine.com/). Consequently, many academic and commercial laboratories have launched NGS-based molecular oncology tests. For example, the Genomics and Pathology Services at the Washington University School of Medicine offers a comprehensive cancer gene panel, containing 25 key cancer genes using target capture and sequencing on the Illumina HiSeq or MiSeq platform (85). Their extensive clinical report lists the variants identified and stratifies these variants by clinical therapy indications as follows: predictive—a sequence that predicts response or safety signal of a particular therapy; prognostic—conferring increased or decreased risk for a particular outcome irrespective of treatment; reported in cancers or other diseases; variant of unknown significance; and, lastly, known polymorphisms. Foundation Medicine, a publicly traded company, offers an NGS panel designed for solid tumor samples that is comprised of hundreds of genes and is analyzed with the Illumina HiSeq platform. In 2012, they identified genomic alterations associated with therapeutic options in over 50 % of colorectal cancer patients and just over 70 % of non-small-cell lung cancer patients (86). Their test report, sent to the referring clinician, outlines potential targeted therapies, including any FDA-approved therapies, that might be indicated based upon the genomic alterations identified in a tumor sample. In addition, their report also provides details on any clinical trials for which a patient may be eligible, based on the tumor's molecular profile. MLL (Munich Leukemia Laboratory, Germany) demonstrated the feasibility of targeted sequencing of several genes (RUNX1, CEBPA, CBL, NRAS, KRAS, DNMT3A, EZH2, and TP53) for hematological disease characterization and monitoring (87). The University of Cambridge and Cambridge Hospital (UK) recently demonstrated that NGS could monitor circulating tumor DNA present in serial plasma samples obtained in a cohort of 30 women with metastatic breast cancer (88). This novel "liquid biopsy" approach, using readily available patient plasma samples, may be advantageous since it is potentially a more sensitive and less invasive method for serial sampling of metastatic breast cancer. Given the competitive diagnostic test landscape, more clinical laboratories will develop novel NGS tests as well as tests comparable to those offered by competitors, but in our uncertain reimbursement climate it will be a combination of clinical utility and added value to the patient that will determine which tests remain available. Next, we describe two NGS-based oncology tests that two of the authors have helped to implement in support of clinical diagnosis.

One author, MJM, recently helped implement mutation profiling by NGS for selected genes (ASXL1, RUNX1, TP53,

ETV6, and EZH2) at Genoptix Medical Laboratory (Carlsbad, CA USA) in order to stratify the prognosis of patients with myelodysplastic syndrome (MDS) (89). This is an example of how research discoveries rapidly led to a new clinical test, the results of which have stimulated new clinical trials based on test results (89). Five genes were ultimately selected for targeted sequencing because their collective status demonstrated the highest level of significance for overall survival in a multivariate analysis (89). In this assay, 43 exons encompassing these five genes are amplified using the Fluidigm PCR microfluidic platform (an integrated fluidic circuit device that allows preparation of 48 simultaneous libraries). Because 43 exons is far below the instrument's single run capacity of up to 15 gigabytes of output with 25 million sequencing reads and 2×300 bp read lengths, samples from multiple patients can be run in parallel (multiplexed). DNA samples from up to 23 patients are amplified, "barcoded" (labeled with specific DNA sequences not found in the starting material for later association with a specific patient sample), pooled, normalized, and then sequenced simultaneously on the flow cell of an Illumina MiSeq instrument. The resulting sequence data is analyzed with the aid of third-party alignment/variant calling software. Variants are filtered and further annotated using bioinformatics tools, interpreted by a board-certified clinical molecular geneticist, and then reviewed and signed out by a hematopathologist. Validated sample types for this MDS molecular profile assay include bone marrow aspirates, peripheral blood, and cell pellets suitable for cytogenetic analysis. An example of the workflow for the MDS assay is shown in Fig. 1.

Another author, DAE, helped to implement a translational NGS protocol at the University of North Carolina (Chapel Hill), UNCseq™, which incorporates features of both discovery research and clinical application. This protocol, for cancer patients, identifies tumor-specific somatic mutations and copy number alterations in approximately 250 cancer-related genes relevant to basic and clinical outcomes research and provides a mechanism to report potentially actionable findings to the referring clinician (90). Such findings may influence the choice of a particular drug or enrollment in a clinical trial or may provide useful prognostic or diagnostic information. The UNCseq™ workflow is also summarized in Fig. 1. UNCseq™ examines matched tumor and normal samples (frozen or FFPE tissue, blood, or buccal swabs) from each patient to identify tumor-specific variations that differ from germline sequence. Prior to DNA isolation, tumor samples are reviewed by a pathologist to assess sample adequacy, estimate the percentage of tumor in the sample, and, if necessary, identify tissue section areas for macrodissection to enrich the sample's tumor content. UNCseq™ employs a target capture technique to prepare libraries including all exons of the >250 target genes. Library preparation involves a minimal PCR amplification to permit accurate assessment

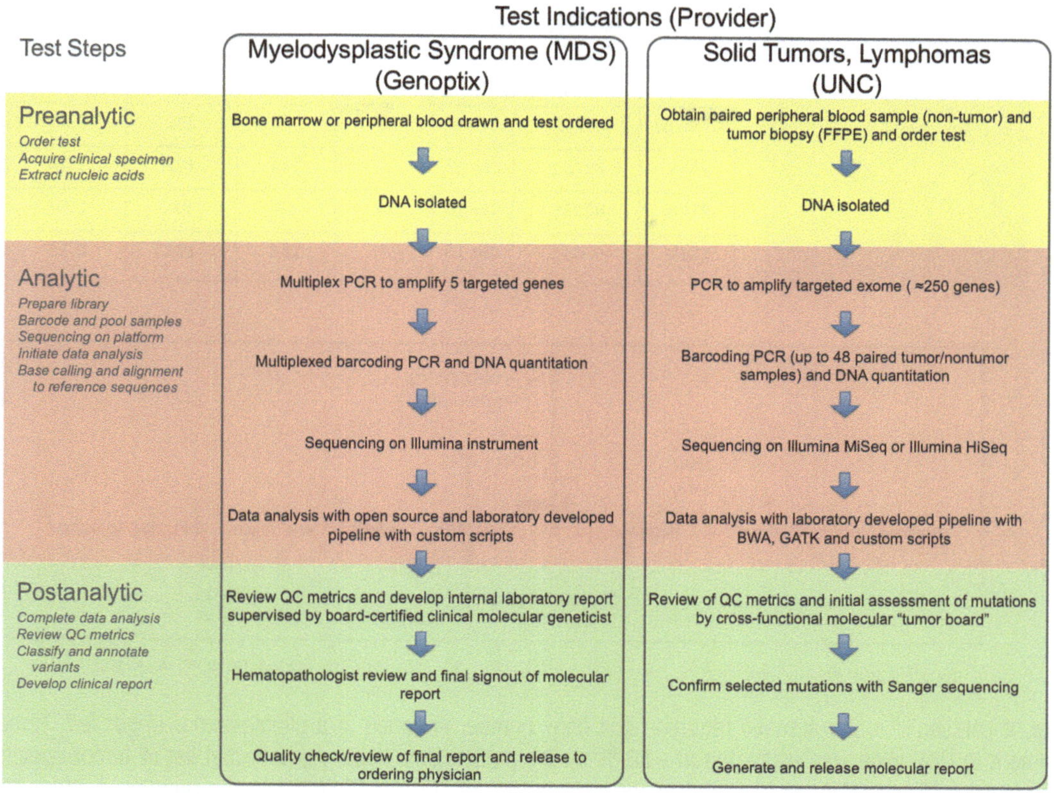

Fig. 1 Workflow of two clinical NGS tests developed with the assistance of two of the authors (MJM and DAE). Preanalytic, analytic, and postanalytic phases of each test are indicated in the *left* column. The *center* column shows the Genoptix MDS mutation profiling assay from the Genoptix Medical Laboratory. The *right* column shows the workflow for the tumor panel UNCseq™ at UNC Chapel Hill

of copy number across all the targeted sequences in the library. Sequencing is performed on Illumina HiSeq or MiSeq platforms at a target coverage depth of 150–500×, permitting reliable detection of mutant alleles at frequencies as low as 5–10 %. The choice of platform is dictated by the desired turnaround time and the number of samples in a multiplexed batch pool. Data are analyzed and summarized using customized applications developed by the UNCseq™ bioinformatics team. An example result is shown in Fig. 2. The UNCseq™ NGS results are reviewed by a Molecular Tumor Board composed of pathologists, oncologists, and bioinformatics experts to determine whether any potentially clinically actionable findings should be confirmed in a CLIA laboratory setting using established methods, such as Sanger sequencing for mutations or fluorescence in situ hybridization (FISH) for copy number alterations. Given the current exploratory nature of the NGS assay, only results from the confirmatory tests are reported in the patient's medical record. The report includes information

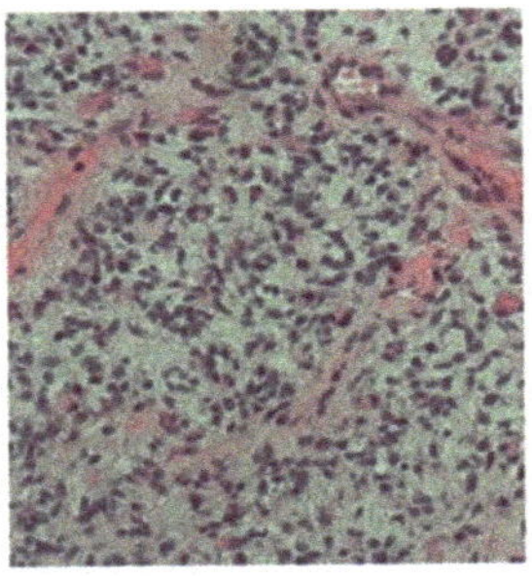

GENE	A.A. MUT	CHR.	No. IN COSMIC	TUMOR DEPTH	NORMAL DEPTH	TUMOR MUT FRQ
EGFR	L658Q	Chr 7	0	6000	177	0.78
CTNNB1	N51S	Chr 3	104	264	288	0.32
JAK3	R175Q	Chr 19	0	144	102	0.44
PTEN	R173C	Chr 10	62	43	97	0.67
CDH1	G382S	Chr 16	3	158	293	0.15

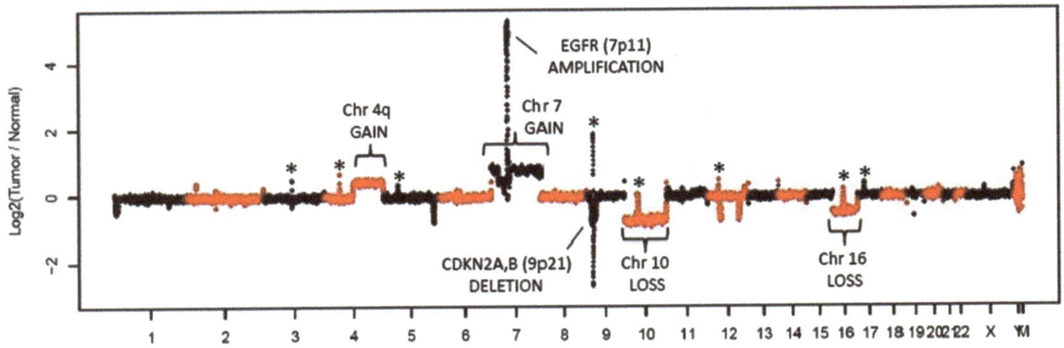

Fig. 2 UNCseq™ tumor somatic mutation and copy number variation in a glioblastoma. *Upper left*: Tumor content in H&E slide was estimated at ~80 % by the pathologist. *Upper right*: Partial list of tumor-specific sequence variants. "A.A. mut" is the amino acid mutation. "No. in COSMIC" is the number of times the sequence variant was already represented in the COSMIC database; those with "0" are novel variants. "Tumor depth" and "Normal depth" are the number of reads for the variant position in tumor and normal samples. "Tumor Mut Frq" is the frequency of the alternate (variant) allele in the tumor sample. *Bottom*: Copy number variation across the genome represented as the log2 ratio of reads in tumor versus normal samples. The *X-axis* indicates each chromosome. *Asterisks* represent centromeric regions where low-depth off-target reads typically result in statistical variations in the T/N ratio. Chromosomal abnormalities inferred from the total NGS read are indicated

about the specific molecular alteration, functional or biological implications if known, and drugs or clinical trials that the oncologist might consider for the patient based on the test results.

10 Challenges and Future Directions

Since completion of the human genome over a decade ago, NGS technologies have made possible significant breakthroughs in understanding of disease, with direct implications for medical care. Technical, diagnostic, regulatory, and bioinformatic precedents are in place from existing pathology and molecular genetics practices to guide NGS implementation in drug development/clinical trial and clinical diagnostic spaces. Technology advances and economies of scale will further reduce the cost of data generation, but we believe the two major challenges remaining are accurate interpretation of the massively complex datasets and defining the limits of the technology. For diseases in which a critical mutation to be identified is the "needle in the haystack," proper experimental design and data aggregation approaches enable identification of the "needle" (i.e., the causative gene mutation). However, for many diseases that are far more complex and heterogeneous across populations—characterized by mutations or variants in dozens to hundreds of genes—including autoimmune and neurodegenerative diseases, and especially cancer, ascertaining the meaning of specific combinations of mutations—a mutational "signature"—can range from difficult to impossible when the signature is expected to be predictive of a particular quality or when each mutation is assessed in isolation. In other words, given that most genes function in pathways with variable activity in different cells and tissues, and that the expression and the ultimate abundance and activities of their encoded proteins in each cell type is what determines their effects in diseased tissue, DNA-based NGS on highly heterogeneous tissue might fall short of providing critical diagnostic or therapeutic insight. For example, CpG DNA methylation and heterochromatic gene silencing—a distinct mechanism of inactivating tumor suppressor genes in cancer—is not detected by routine DNA sequencing, but rather requires library preparation from modified DNA (e.g., bisulfite sequencing) or RNA (e.g., RNA-seq). Similarly, alteration in proteins that do not correspond to measurable abnormalities in nucleic acids will be inaccessible by NGS techniques, although profiling of miRNAs, which regulate protein abundance via mRNA stabilization or degradation, might provide complementary or surrogate insights.

We envision that within our lifetimes, most people, at least in Western industrialized countries with insured healthcare systems, will have germline WGS done at least once to define their "baseline" genome. However, it is unlikely that WGS will be used as a "one-size-fits-all" approach. Amidst increasing enthusiasm for NGS, a prominent report has illustrated the current limitations of WGS, including incomplete coverage of inherited disease genes and uncertainties about clinically reportable findings (39). When a disease such as cancer develops, focused interrogation of gene panels (as in the above examples) might provide greater immediate clinical value, particularly when therapies that target the mutant protein are already approved.

Platforms and applications that are tailored to the particular needs of various clinical areas and diseases will continue to emerge. For example, WGS or more focused NGS approaches could allow assessment of exposure to environmental mutagens as a means to predict the risk of developing cancer. New therapeutic clinical trial designs will be based on genomics and molecular pathology: future trials might rely less on traditional histology-based diagnostic classifications such as lung adenocarcinoma or glioblastoma, moving toward molecularly defined categories of disease that are tied to the mechanisms of action of novel therapies, or combinations thereof, under development.

Genomic DNA sequencing provides only part of the picture. While microarray profiling technology will continue to be used for specific applications, deep sequencing of RNA (RNA-seq) is poised to emerge as an additional or alternative approach for particular research questions and clinical conditions (91). In research settings, NGS approaches are now being used to reveal the importance of epigenetic mechanisms such as DNA methylation and histone modification as critical determinants in various acquired or inherited "genetically based" diseases such as cancer (92).

Given the cost of NGS technologies—including technical, computational, and professional interpretive time—payers will require evidence of patient benefit for every proposed application of NGS in the clinic. For example, in cancer treatment, a broad but targeted gene panel might make the most sense to establish an initial diagnosis, and then a further limited, yet still targeted, approach might be required at tumor relapse based on the known biology of cancer progression for tumors that derive from that site as well as known mechanisms of resistance to therapy.

Reporting of incidental findings to the patient remains a critical, unresolved issue. Hidden in our genomes are perhaps dozens of mutations that singly, or in combination, might predict early onset or susceptibility to a disease that many patients would prefer not to be informed about. As the meaning of these polymorphisms/mutations become apparent, we will need to pay careful attention to informed consent documentation and provide counseling to patients so that they can give or refuse their consent for the reporting of such findings to them.

As NGS reads accumulate, it will be ever more important to be able to query, analyze, and ultimately comprehend the importance of mutational variants and genome rearrangements across diseases and populations, and in response to different treatments. Moreover, as knowledge accumulates, the meaning and possible actions resulting from retrospective data review will increase. For cancer alone, such retrospective analyses could be lifesaving, as the number of approved therapies targeting the diverse but ultimately finite number of cancer pathways increases every year.

Given the current importance of histopathologic interpretation of tissue for medical diagnosis and practice, it will remain important to place NGS data, at least for the foreseeable future, in the context of criteria traditionally used to describe disease, in particular histopathology. For this reason we believe that anatomic pathologists should be involved with clinical interpretation of NGS data, and the pathology profession should continue to help incorporate NGS and other omics training into its repertoire of diagnostic healthcare services, starting in medical school and residency training and continuing with board certification, recertification, and continuing medical education. Everybody on the healthcare team will need to think "molecular" and make it understandable to physicians and their patients. NGS technology has advanced quickly and will continue to do so. But the evidence-based framework we discuss for implementing NGS (or any emerging technology) into clinical practice should apply in the future: data needs to be interpreted in the context of healthcare decision-making and delivery, and tests implemented by professionals.

We are at a pivotal juncture in the history of medicine, poised to develop new ways of diagnosing and treating disease based on what some of the deepest thinkers of our generations consider the ultimate self-knowledge—our genomes. It is incumbent on the community of pathologists and laboratory medicine professionals to help define the appropriate questions to be answered with this powerful technology, and to guide how their answers will best advance knowledge of disease and improve medical care. We are optimistic that these challenges will be met, leading to a not-so-distant future where NGS-based testing will offer dramatic improvements in medical care and clinical outcomes worldwide (19).

Acknowledgements

We'd like to acknowledge Edward Oakeley Ph.D. (Novartis AG) and Wendy Winckler Ph.D. (NIBRI) for critical reviews of the chapter, and Tiffany Liang for preparation of Fig. 1. UNCseq™ is funded in part by the UNC Lineberger Comprehensive Cancer Center's University Cancer Research Fund. We would like to thank H. Shelton Earp III M.D., D. Neil Hayes M.D., M.P.H., Joel Parker Ph.D., Karen Weck M.D., and Nirali Patel M.D., for their leadership in the UNCseq™ program.

References

1. Dunstan RW, Wharton KA Jr, Quigley C, Lowe A (2011) The use of immunohistochemistry for biomarker assessment - can it compete with other technologies? Toxicol Pathol 39(6):988–1002. doi:10.1177/0192623311419163
2. Vaandrager JW, Schuuring E, Zwikstra E, de Boer CJ, Kleiverda KK, van Krieken JH, Kluin-Nelemans HC, van Ommen GJ, Raap AK, Kluin PM (1996) Direct visualization of dispersed 11q13 chromosomal translocations in mantle cell lymphoma by multicolor DNA fiber fluorescence in situ hybridization. Blood 88(4):1177–1182
3. Wang F, Flanagan J, Su N, Wang LC, Bui S, Nielson A, Wu X, Vo HT, Ma XJ, Luo Y (2012) RNAscope: a novel in situ RNA analysis platform for formalin-fixed, paraffin-embedded tissues. J Mol Diagn 14(1):22–29. doi:10.1016/j.jmoldx.2011.08.002
4. Hanna JA, Wimberly H, Kumar S, Slack F, Agarwal S, Rimm DL (2012) Quantitative analysis of microRNAs in tissue microarrays

by in situ hybridization. Biotechniques 52 (4):235–245. doi:10.2144/000113837

5. Gridelli C, de Marinis F, Cappuzzo F, Di Maio M, Hirsch FR, Mok T, Morgillo F, Rosell R, Spigel DR, Yang JC, Ciardiello F (2014) Treatment of advanced non-small-cell lung cancer with epidermal growth factor receptor (EGFR) mutation or ALK gene rearrangement: results of an international expert panel meeting of the Italian Association of Thoracic Oncology. Clin Lung Canc 15(3):173–181. doi:10.1016/j.cllc.2013.12.002
6. Gridelli C, Peters S, Sgambato A, Casaluce F, Adjei AA, Ciardiello F (2014) ALK inhibitors in the treatment of advanced NSCLC. Canc Treat Rev 40(2):300–306. doi:10.1016/j.ctrv.2013.07.002
7. Parsons BL, Myers MB (2013) Personalized cancer treatment and the myth of KRAS wild-type colon tumors. Discov Med 15 (83):259–267
8. Kiechle FL, Zhang X, Holland-Staley CA (2004) The -omics era and its impact. Arch Pathol Lab Med 128(12):1337–1345. doi:10.1043/1543-2165(2004)128<1337:TOEAII>2.0.CO;2
9. Zhang Z (2012) An in vitro diagnostic multivariate index assay (IVDMIA) for ovarian cancer: harvesting the power of multiple biomarkers. Rev Obstet Gynecol 5(1):35–41
10. Micheel CM, Nass SJ, Omenn GS, Committee on the Review of Omics-Based Tests for Predicting Patient Outcomes in Clinical Trials, Board on Health Care Services, Board on Health Sciences Policy, Institute of Medicine (eds) (2012) Evolution of translational omics: lessons learned and the path forward. National Academies Press, Washington, DC, PMID: 24872966
11. McShane LM, Cavenagh MM, Lively TG, Eberhard DA, Bigbee WL, Williams PM, Mesirov JP, Polley MY, Kim KY, Tricoli JV, Taylor JM, Shuman DJ, Simon RM, Doroshow JH, Conley BA (2013) Criteria for the use of omics-based predictors in clinical trials. Nature 502(7471):317–320. doi:10.1038/nature12564
12. Prat A, Perou CM (2011) Deconstructing the molecular portraits of breast cancer. Mol Oncol 5(1):5–23. doi:10.1016/j.molonc.2010.11.003
13. Roschewski M, Staudt LM, Wilson WH (2014) Diffuse large B-cell lymphoma-treatment approaches in the molecular era. Nat Rev Clin Oncol 11(1):12–23. doi:10.1038/nrclinonc.2013.197
14. van de Wetering M, Sancho E, Verweij C, de Lau W, Oving I, Hurlstone A, van der Horn K, Batlle E, Coudreuse D, Haramis AP, Tjon-Pon-Fong M, Moerer P, van den Born M, Soete G, Pals S, Eilers M, Medema R, Clevers H (2002) The beta-catenin/TCF-4 complex imposes a crypt progenitor phenotype on colorectal cancer cells. Cell 111(2):241–250
15. Guarnaccia M, Gentile G, Alessi E, Schneider C, Petralia S, Cavallaro S (2014) Is this the real time for genomics? Genomics 103 (2–3):177–182. doi:10.1016/j.ygeno.2014.02.003
16. Carlson JJ, Roth JA (2013) The impact of the Oncotype Dx breast cancer assay in clinical practice: a systematic review and meta-analysis. Breast Cancer Res Treat 141(1):13–22. doi:10.1007/s10549-013-2666-z
17. Sahebjam S, Aloyz R, Pilavdzic D, Brisson ML, Ferrario C, Bouganim N, Cohen V, Miller WH Jr, Panasci LC (2011) Ki 67 is a major, but not the sole determinant of Oncotype Dx recurrence score. Br J Cancer 105 (9):1342–1345. doi:10.1038/bjc.2011.402
18. Malone JH, Oliver B (2011) Microarrays, deep sequencing and the true measure of the transcriptome. BMC Biol 9:34. doi:10.1186/1741-7007-9-34
19. Kilpivaara O, Aaltonen LA (2013) Diagnostic cancer genome sequencing and the contribution of germline variants. Science 339 (6127):1559–1562. doi:10.1126/science.1233899
20. Maes E, Broeckx V, Mertens I, Sagaert X, Prenen H, Landuyt B, Schoofs L (2013) Analysis of the formalin-fixed paraffin-embedded tissue proteome: pitfalls, challenges, and future prospectives. Amino Acids 45(2):205–218. doi:10.1007/s00726-013-1494-0
21. Peiro-Chova L, Pena-Chilet M, Lopez-Guerrero JA, Garcia-Gimenez JL, Alonso-Yuste E, Burgues O, Lluch A, Ferrer-Lozano J, Ribas G (2013) High stability of microRNAs in tissue samples of compromised quality. Virchows Arch 463(6):765–774. doi:10.1007/s00428-013-1485-2
22. Nam SK, Im J, Kwak Y, Han N, Nam KH, Seo AN, Lee HS (2014) Effects of fixation and storage of human tissue samples on nucleic Acid preservation. Korean J Pathol 48(1):36–42. doi:10.4132/KoreanJPathol.2014.48.1.36
23. Spencer DH, Sehn JK, Abel HJ, Watson MA, Pfeifer JD, Duncavage EJ (2013) Comparison of clinical targeted next-generation sequence

data from formalin-fixed and fresh-frozen tissue specimens. J Mol Diagn 15(5):623–633. doi:10.1016/j.jmoldx.2013.05.004

24. Hewitt SM, Lewis FA, Cao Y, Conrad RC, Cronin M, Danenberg KD, Goralski TJ, Langmore JP, Raja RG, Williams PM, Palma JF, Warrington JA (2008) Tissue handling and specimen preparation in surgical pathology: issues concerning the recovery of nucleic acids from formalin-fixed, paraffin-embedded tissue. Arch Pathol Lab Med 132(12):1929–1935. doi:10.1043/1543-2165-132.12.1929
25. Giusti L, Lucacchini A (2013) Proteomic studies of formalin-fixed paraffin-embedded tissues. Expert Rev Proteomics 10 (2):165–177. doi:10.1586/epr.13.3
26. Nielsen T, Wallden B, Schaper C, Ferree S, Liu S, Gao D, Barry G, Dowidar N, Maysuria M, Storhoff J (2014) Analytical validation of the PAM50-based Prosigna Breast Cancer Prognostic Gene Signature Assay and nCounter Analysis System using formalin-fixed paraffin-embedded breast tumor specimens. BMC Cancer 14:177. doi:10.1186/1471-2407-14-177
27. Kap M, Smedts F, Oosterhuis W, Winther R, Christensen N, Reischauer B, Viertler C, Groelz D, Becker KF, Zatloukal K, Langer R, Slotta-Huspenina J, Bodo K, de Jong B, Oelmuller U, Riegman P (2011) Histological assessment of PAXgene tissue fixation and stabilization reagents. PLoS One 6(11):e27704. doi:10.1371/journal.pone.0027704
28. Belloni B, Lambertini C, Nuciforo P, Phillips J, Bruening E, Wong S, Dummer R (2013) Will PAXgene substitute formalin? A morphological and molecular comparative study using a new fixative system. J Clin Pathol 66 (2):124–135. doi:10.1136/jclinpath-2012-200983
29. Biesecker LG (2010) Exome sequencing makes medical genomics a reality. Nat Genet 42(1):13–14. doi:10.1038/ng0110-13
30. Bamshad MJ, Ng SB, Bigham AW, Tabor HK, Emond MJ, Nickerson DA, Shendure J (2011) Exome sequencing as a tool for Mendelian disease gene discovery. Nat Rev Genet 12(11):745–755. doi:10.1038/nrg3031
31. Cassidy LD, Venkitaraman AR (2012) Genome instability mechanisms and the structure of cancer genomes. Curr Opin Genet Dev 22(1):10–13. doi:10.1016/j.gde.2012.02.003
32. Chiu CY (2013) Viral pathogen discovery. Curr Opin Microbiol 16(4):468–478. doi:10.1016/j.mib.2013.05.001
33. Fortes GG, Speller CF, Hofreiter M, King TE (2013) Phenotypes from ancient DNA: approaches, insights and prospects. Bioessays 35(8):690–695. doi:10.1002/bies.201300036
34. Zaura E (2012) Next-generation sequencing approaches to understanding the oral microbiome. Adv Dent Res 24(2):81–85. doi:10.1177/0022034512449466
35. Madar S, Goldstein I, Rotter V (2013) "Cancer associated fibroblasts" – more than meets the eye. Trends Mol Med 19(8):447–453. doi:10.1016/j.molmed.2013.05.004
36. Lundberg P, Karow A, Nienhold R, Looser R, Hao-Shen H, Nissen I, Girsberger S, Lehmann T, Passweg J, Stern M, Beisel C, Kralovics R, Skoda RC (2014) Clonal evolution and clinical correlates of somatic mutations in myeloproliferative neoplasms. Blood 123 (14):2220–2228. doi:10.1182/blood-2013-11-537167
37. Villamor N, Lopez-Guillermo A, Lopez-Otin C, Campo E (2013) Next-generation sequencing in chronic lymphocytic leukemia. Semin Hematol 50(4):286–295. doi:10.1053/j.seminhematol.2013.09.005
38. Schrijver I, Aziz N, Farkas DH, Furtado M, Gonzalez AF, Greiner TC, Grody WW, Hambuch T, Kalman L, Kant JA, Klein RD, Leonard DG, Lubin IM, Mao R, Nagan N, Pratt VM, Sobel ME, Voelkerding KV, Gibson JS (2012) Opportunities and challenges associated with clinical diagnostic genome sequencing: a report of the Association for Molecular Pathology. J Mol Diagn 14 (6):525–540. doi:10.1016/j.jmoldx.2012.04.006
39. Dewey FE, Grove ME, Pan C, Goldstein BA, Bernstein JA, Chaib H, Merker JD, Goldfeder RL, Enns GM, David SP, Pakdaman N, Ormond KE, Caleshu C, Kingham K, Klein TE, Whirl-Carrillo M, Sakamoto K, Wheeler MT, Butte AJ, Ford JM, Boxer L, Ioannidis JP, Yeung AC, Altman RB, Assimes TL, Snyder M, Ashley EA, Quertermous T (2014) Clinical interpretation and implications of whole-genome sequencing. JAMA 311 (10):1035–1045. doi:10.1001/jama.2014.1717
40. Lander ES, Linton LM, Birren B, Nusbaum C, Zody MC, Baldwin J, Devon K, Dewar K, Doyle M, FitzHugh W, Funke R, Gage D, Harris K, Heaford A, Howland J, Kann L, Lehoczky J, LeVine R, McEwan P, McKernan K, Meldrim J, Mesirov JP, Miranda C, Morris W, Naylor J, Raymond C, Rosetti M, Santos R, Sheridan A, Sougnez C, Stange-Thomann

N, Stojanovic N, Subramanian A, Wyman D, Rogers J, Sulston J, Ainscough R, Beck S, Bentley D, Burton J, Clee C, Carter N, Coulson A, Deadman R, Deloukas P, Dunham A, Dunham I, Durbin R, French L, Grafham D, Gregory S, Hubbard T, Humphray S, Hunt A, Jones M, Lloyd C, McMurray A, Matthews L, Mercer S, Milne S, Mullikin JC, Mungall A, Plumb R, Ross M, Shownkeen R, Sims S, Waterston RH, Wilson RK, Hillier LW, McPherson JD, Marra MA, Mardis ER, Fulton LA, Chinwalla AT, Pepin KH, Gish WR, Chissoe SL, Wendl MC, Delehaunty KD, Miner TL, Delehaunty A, Kramer JB, Cook LL, Fulton RS, Johnson DL, Minx PJ, Clifton SW, Hawkins T, Branscomb E, Predki P, Richardson P, Wenning S, Slezak T, Doggett N, Cheng JF, Olsen A, Lucas S, Elkin C, Uberbacher E, Frazier M, Gibbs RA, Muzny DM, Scherer SE, Bouck JB, Sodergren EJ, Worley KC, Rives CM, Gorrell JH, Metzker ML, Naylor SL, Kucherlapati RS, Nelson DL, Weinstock GM, Sakaki Y, Fujiyama A, Hattori M, Yada T, Toyoda A, Itoh T, Kawagoe C, Watanabe H, Totoki Y, Taylor T, Weissenbach J, Heilig R, Saurin W, Artiguenave F, Brottier P, Bruls T, Pelletier E, Robert C, Wincker P, Smith DR, Doucette-Stamm L, Rubenfield M, Weinstock K, Lee HM, Dubois J, Rosenthal A, Platzer M, Nyakatura G, Taudien S, Rump A, Yang H, Yu J, Wang J, Huang G, Gu J, Hood L, Rowen L, Madan A, Qin S, Davis RW, Federspiel NA, Abola AP, Proctor MJ, Myers RM, Schmutz J, Dickson M, Grimwood J, Cox DR, Olson MV, Kaul R, Raymond C, Shimizu N, Kawasaki K, Minoshima S, Evans GA, Athanasiou M, Schultz R, Roe BA, Chen F, Pan H, Ramser J, Lehrach H, Reinhardt R, McCombie WR, de la Bastide M, Dedhia N, Blocker H, Hornischer K, Nordsiek G, Agarwala R, Aravind L, Bailey JA, Bateman A, Batzoglou S, Birney E, Bork P, Brown DG, Burge CB, Cerutti L, Chen HC, Church D, Clamp M, Copley RR, Doerks T, Eddy SR, Eichler EE, Furey TS, Galagan J, Gilbert JG, Harmon C, Hayashizaki Y, Haussler D, Hermjakob H, Hokamp K, Jang W, Johnson LS, Jones TA, Kasif S, Kaspryzk A, Kennedy S, Kent WJ, Kitts P, Koonin EV, Korf I, Kulp D, Lancet D, Lowe TM, McLysaght A, Mikkelsen T, Moran JV, Mulder N, Pollara VJ, Ponting CP, Schuler G, Schultz J, Slater G, Smit AF, Stupka E, Szustakowski J, Thierry-Mieg D, Thierry-Mieg J, Wagner L, Wallis J, Wheeler R, Williams A, Wolf YI, Wolfe KH, Yang SP, Yeh RF, Collins F, Guyer MS, Peterson J, Felsenfeld A, Wetterstrand KA, Patrinos A, Morgan MJ, de Jong P, Catanese JJ, Osoegawa K, Shizuya H, Choi S, Chen YJ, International Human Genome Sequencing C (2001) Initial sequencing and analysis of the human genome. Nature 409 (6822):860–921. doi: 10.1038/35057062

41. Venter JC, Adams MD, Myers EW, Li PW, Mural RJ, Sutton GG, Smith HO, Yandell M, Evans CA, Holt RA, Gocayne JD, Amanatides P, Ballew RM, Huson DH, Wortman JR, Zhang Q, Kodira CD, Zheng XH, Chen L, Skupski M, Subramanian G, Thomas PD, Zhang J, Gabor Miklos GL, Nelson C, Broder S, Clark AG, Nadeau J, McKusick VA, Zinder N, Levine AJ, Roberts RJ, Simon M, Slayman C, Hunkapiller M, Bolanos R, Delcher A, Dew I, Fasulo D, Flanigan M, Florea L, Halpern A, Hannenhalli S, Kravitz S, Levy S, Mobarry C, Reinert K, Remington K, Abu-Threideh J, Beasley E, Biddick K, Bonazzi V, Brandon R, Cargill M, Chandramouliswaran I, Charlab R, Chaturvedi K, Deng Z, Di Francesco V, Dunn P, Eilbeck K, Evangelista C, Gabrielian AE, Gan W, Ge W, Gong F, Gu Z, Guan P, Heiman TJ, Higgins ME, Ji RR, Ke Z, Ketchum KA, Lai Z, Lei Y, Li Z, Li J, Liang Y, Lin X, Lu F, Merkulov GV, Milshina N, Moore HM, Naik AK, Narayan VA, Neelam B, Nusskern D, Rusch DB, Salzberg S, Shao W, Shue B, Sun J, Wang Z, Wang A, Wang X, Wang J, Wei M, Wides R, Xiao C, Yan C, Yao A, Ye J, Zhan M, Zhang W, Zhang H, Zhao Q, Zheng L, Zhong F, Zhong W, Zhu S, Zhao S, Gilbert D, Baumhueter S, Spier G, Carter C, Cravchik A, Woodage T, Ali F, An H, Awe A, Baldwin D, Baden H, Barnstead M, Barrow I, Beeson K, Busam D, Carver A, Center A, Cheng ML, Curry L, Danaher S, Davenport L, Desilets R, Dietz S, Dodson K, Doup L, Ferriera S, Garg N, Gluecksmann A, Hart B, Haynes J, Haynes C, Heiner C, Hladun S, Hostin D, Houck J, Howland T, Ibegwam C, Johnson J, Kalush F, Kline L, Koduru S, Love A, Mann F, May D, McCawley S, McIntosh T, McMullen I, Moy M, Moy L, Murphy B, Nelson K, Pfannkoch C, Pratts E, Puri V, Qureshi H, Reardon M, Rodriguez R, Rogers YH, Romblad D, Ruhfel B, Scott R, Sitter C, Smallwood M, Stewart E, Strong R, Suh E, Thomas R, Tint NN, Tse S, Vech C, Wang G, Wetter J, Williams S, Williams M, Windsor S, Winn-Deen E, Wolfe K, Zaveri J, Zaveri K, Abril JF, Guigo R, Campbell MJ, Sjolander KV, Karlak B, Kejariwal A, Mi H, Lazareva B, Hatton T, Narechania A, Diemer K, Muruganujan A, Guo N, Sato S, Bafna V, Istrail S, Lippert R, Schwartz R, Walenz B, Yooseph S, Allen D, Basu A, Baxendale J, Blick L,

Caminha M, Carnes-Stine J, Caulk P, Chiang YH, Coyne M, Dahlke C, Mays A, Dombroski M, Donnelly M, Ely D, Esparham S, Fosler C, Gire H, Glanowski S, Glasser K, Glodek A, Gorokhov M, Graham K, Gropman B, Harris M, Heil J, Henderson S, Hoover J, Jennings D, Jordan C, Jordan J, Kasha J, Kagan L, Kraft C, Levitsky A, Lewis M, Liu X, Lopez J, Ma D, Majoros W, McDaniel J, Murphy S, Newman M, Nguyen T, Nguyen N, Nodell M, Pan S, Peck J, Peterson M, Rowe W, Sanders R, Scott J, Simpson M, Smith T, Sprague A, Stockwell T, Turner R, Venter E, Wang M, Wen M, Wu D, Wu M, Xia A, Zandieh A, Zhu X (2001) The sequence of the human genome. Science 291(5507):1304–1351. doi: 10.1126/science.1058040

42. Meyerson M, Gabriel S, Getz G (2010) Advances in understanding cancer genomes through second-generation sequencing. Nat Rev Genet 11(10):685–696. doi:10.1038/nrg2841
43. Yates LR, Campbell PJ (2012) Evolution of the cancer genome. Nat Rev Genet 13 (11):795–806. doi:10.1038/nrg3317
44. Yang Y, Muzny DM, Reid JG, Bainbridge MN, Willis A, Ward PA, Braxton A, Beuten J, Xia F, Niu Z, Hardison M, Person R, Bekheirnia MR, Leduc MS, Kirby A, Pham P, Scull J, Wang M, Ding Y, Plon SE, Lupski JR, Beaudet AL, Gibbs RA, Eng CM (2013) Clinical whole-exome sequencing for the diagnosis of mendelian disorders. N Engl J Med 369(16):1502–1511. doi:10.1056/NEJMoa1306555
45. MacConaill LE (2013) Existing and emerging technologies for tumor genomic profiling. J Clin Oncol 31(15):1815–1824. doi:10.1200/JCO.2012.46.5948
46. Chin EL, da Silva C, Hegde M (2013) Assessment of clinical analytical sensitivity and specificity of next-generation sequencing for detection of simple and complex mutations. BMC Genet 14:6. doi:10.1186/1471-2156-14-6
47. Tsiatis AC, Norris-Kirby A, Rich RG, Hafez MJ, Gocke CD, Eshleman JR, Murphy KM (2010) Comparison of Sanger sequencing, pyrosequencing, and melting curve analysis for the detection of KRAS mutations: diagnostic and clinical implications. J Mol Diagn 12(4):425–432. doi:10.2353/jmoldx.2010.090188
48. Krueger F, Kreck B, Franke A, Andrews SR (2012) DNA methylome analysis using short bisulfite sequencing data. Nat Methods 9(2):145–151. doi:10.1038/nmeth.1828
49. Koboldt DC, Larson DE, Chen K, Ding L, Wilson RK (2012) Massively parallel sequencing approaches for characterization of structural variation. Methods Mol Biol 838:369–384. doi:10.1007/978-1-61779-507-7_18
50. Metzker ML (2010) Sequencing technologies – the next generation. Nat Rev Genet 11(1):31–46. doi:10.1038/nrg2626
51. Ozsolak F (2012) Third-generation sequencing techniques and applications to drug discovery. Expert Opin Drug Discov 7(3):231–243. doi:10.1517/17460441.2012.660145
52. Gut IG (2013) New sequencing technologies. Clin Transl Oncol 15(11):879–881. doi:10.1007/s12094-013-1073-6
53. Xu MY, Aragon AD, Mascarenas MR, Torrez-Martinez N, Edwards JS (2010) Dual primer emulsion PCR for next- generation DNA sequencing. Biotechniques 48(5):409–412. doi:10.2144/000113423
54. Bentley DR, Balasubramanian S, Swerdlow HP, Smith GP, Milton J, Brown CG, Hall KP, Evers DJ, Barnes CL, Bignell HR, Boutell JM, Bryant J, Carter RJ, Keira Cheetham R, Cox AJ, Ellis DJ, Flatbush MR, Gormley NA, Humphray SJ, Irving LJ, Karbelashvili MS, Kirk SM, Li H, Liu X, Maisinger KS, Murray LJ, Obradovic B, Ost T, Parkinson ML, Pratt MR, Rasolonjatovo IM, Reed MT, Rigatti R, Rodighiero C, Ross MT, Sabot A, Sankar SV, Scally A, Schroth GP, Smith ME, Smith VP, Spiridou A, Torrance PE, Tzonev SS, Vermaas EH, Walter K, Wu X, Zhang L, Alam MD, Anastasi C, Aniebo IC, Bailey DM, Bancarz IR, Banerjee S, Barbour SG, Baybayan PA, Benoit VA, Benson KF, Bevis C, Black PJ, Boodhun A, Brennan JS, Bridgham JA, Brown RC, Brown AA, Buermann DH, Bundu AA, Burrows JC, Carter NP, Castillo N, Chiara ECM, Chang S, Neil Cooley R, Crake NR, Dada OO, Diakoumakos KD, Dominguez-Fernandez B, Earnshaw DJ, Egbujor UC, Elmore DW, Etchin SS, Ewan MR, Fedurco M, Fraser LJ, Fuentes Fajardo KV, Scott Furey W, George D, Gietzen KJ, Goddard CP, Golda GS, Granieri PA, Green DE, Gustafson DL, Hansen NF, Harnish K, Haudenschild CD, Heyer NI, Hims MM, Ho JT, Horgan AM, Hoschler K, Hurwitz S, Ivanov DV, Johnson MQ, James T, Huw Jones TA, Kang GD, Kerelska TH, Kersey AD, Khrebtukova I, Kindwall AP, Kingsbury Z, Kokko-Gonzales PI, Kumar A, Laurent MA, Lawley CT, Lee SE, Lee X, Liao AK, Loch JA, Lok M, Luo S, Mammen RM, Martin JW, McCauley PG, McNitt P, Mehta P, Moon

KW, Mullens JW, Newington T, Ning Z, Ling Ng B, Novo SM, O'Neill MJ, Osborne MA, Osnowski A, Ostadan O, Paraschos LL, Pickering L, Pike AC, Pike AC, Chris Pinkard D, Pliskin DP, Podhasky J, Quijano VJ, Raczy C, Rae VH, Rawlings SR, Chiva Rodriguez A, Roe PM, Rogers J, Rogert Bacigalupo MC, Romanov N, Romieu A, Roth RK, Rourke NJ, Ruediger ST, Rusman E, Sanches-Kuiper RM, Schenker MR, Seoane JM, Shaw RJ, Shiver MK, Short SW, Sizto NL, Sluis JP, Smith MA, Ernest Sohna Sohna J, Spence EJ, Stevens K, Sutton N, Szajkowski L, Tregidgo CL, Turcatti G, Vandevondele S, Verhovsky Y, Virk SM, Wakelin S, Walcott GC, Wang J, Worsley GJ, Yan J, Yau L, Zuerlein M, Rogers J, Mullikin JC, Hurles ME, McCooke NJ, West JS, Oaks FL, Lundberg PL, Klenerman D, Durbin R, Smith AJ (2008) Accurate whole human genome sequencing using reversible terminator chemistry. Nature 456 (7218):53–59. doi: 10.1038/nature07517

55. Aird D, Ross MG, Chen WS, Danielsson M, Fennell T, Russ C, Jaffe DB, Nusbaum C, Gnirke A (2011) Analyzing and minimizing PCR amplification bias in Illumina sequencing libraries. Genome Biol 12(2):R18. doi:10.1186/gb-2011-12-2-r18

56. Oyola SO, Otto TD, Gu Y, Maslen G, Manske M, Campino S, Turner DJ, Macinnis B, Kwiatkowski DP, Swerdlow HP, Quail MA (2012) Optimizing Illumina next-generation sequencing library preparation for extremely AT-biased genomes. BMC Genomics 13:1. doi:10.1186/1471-2164-13-1

57. Mamanova L, Coffey AJ, Scott CE, Kozarewa I, Turner EH, Kumar A, Howard E, Shendure J, Turner DJ (2010) Target-enrichment strategies for next-generation sequencing. Nat Methods 7(2):111–118. doi:10.1038/nmeth.1419

58. Lasho T, Johnson SH, Smith DI, Crispino JD, Pardanani A, Vasmatzis G, Tefferi A (2013) Identification of submicroscopic genetic changes and precise breakpoint mapping in myelofibrosis using high resolution mate-pair sequencing. Am J Hematol 88(9):741–746. doi:10.1002/ajh.23495

59. Levy-Sakin M, Ebenstein Y (2013) Beyond sequencing: optical mapping of DNA in the age of nanotechnology and nanoscopy. Curr Opin Biotechnol 24(4):690–698. doi:10.1016/j.copbio.2013.01.009

60. Rothberg JM, Hinz W, Rearick TM, Schultz J, Mileski W, Davey M, Leamon JH, Johnson K, Milgrew MJ, Edwards M, Hoon J, Simons JF, Marran D, Myers JW, Davidson JF, Branting A, Nobile JR, Puc BP, Light D, Clark TA, Huber M, Branciforte JT, Stoner IB, Cawley SE, Lyons M, Fu Y, Homer N, Sedova M, Miao X, Reed B, Sabina J, Feierstein E, Schorn M, Alanjary M, Dimalanta E, Dressman D, Kasinskas R, Sokolsky T, Fidanza JA, Namsaraev E, McKernan KJ, Williams A, Roth GT, Bustillo J (2011) An integrated semiconductor device enabling non-optical genome sequencing. Nature 475 (7356):348–352. doi:10.1038/nature10242

61. Keightley PD (2012) Rates and fitness consequences of new mutations in humans. Genetics 190(2):295–304. doi:10.1534/genetics.111.134668

62. Loman NJ, Misra RV, Dallman TJ, Constantinidou C, Gharbia SE, Wain J, Pallen MJ (2012) Performance comparison of benchtop high-throughput sequencing platforms. Nat Biotechnol 30(5):434–439. doi:10.1038/nbt.2198

63. Meacham F, Boffelli D, Dhahbi J, Martin DI, Singer M, Pachter L (2011) Identification and correction of systematic error in high-throughput sequence data. BMC Bioinformatics 12:451. doi:10.1186/1471-2105-12-451

64. Grossmann V, Roller A, Klein HU, Weissmann S, Kern W, Haferlach C, Dugas M, Haferlach T, Schnittger S, Kohlmann A (2013) Robustness of amplicon deep sequencing underlines its utility in clinical applications. J Mol Diagn 15(4):473–484. doi:10.1016/j.jmoldx.2013.03.003

65. Ellard S, Patrinos GP, Oetting WS (2013) Clinical applications of next-generation sequencing: the 2013 human genome variation society scientific meeting. Hum Mutat 34 (11):1583–1587. doi:10.1002/humu.22400

66. Torri F, Dinov ID, Zamanyan A, Hobel S, Genco A, Petrosyan P, Clark AP, Liu Z, Eggert P, Pierce J, Knowles JA, Ames J, Kesselman C, Toga AW, Potkin SG, Vawter MP, Macciardi F (2012) Next generation sequence analysis and computational genomics using graphical pipeline workflows. Genes 3(3):545–575. doi:10.3390/genes3030545

67. Lin Y, Li J, Shen H, Zhang L, Papasian CJ, Deng HW (2011) Comparative studies of de novo assembly tools for next-generation sequencing technologies. Bioinformatics 27 (15):2031–2037. doi:10.1093/bioinformatics/btr319

68. Li H, Homer N (2010) A survey of sequence alignment algorithms for next-generation sequencing. Brief Bioinform 11(5):473–483. doi:10.1093/bib/bbq015

69. Van Allen EM, Wagle N, Levy MA (2013) Clinical analysis and interpretation of

cancer genome data. J Clin Oncol 31 (15):1825–1833. doi:10.1200/JCO.2013.48.7215

70. Ruffalo M, LaFramboise T, Koyuturk M (2011) Comparative analysis of algorithms for next-generation sequencing read alignment. Bioinformatics 27(20):2790–2796. doi:10.1093/bioinformatics/btr477

71. Chin L, Hahn WC, Getz G, Meyerson M (2011) Making sense of cancer genomic data. Genes Dev 25(6):534–555. doi:10.1101/gad.2017311

72. Gullapalli RR, Desai KV, Santana-Santos L, Kant JA, Becich MJ (2012) Next generation sequencing in clinical medicine: Challenges and lessons for pathology and biomedical informatics. J Pathol Informat 3:40. doi:10.4103/2153-3539.103013

73. Richards CS, Bale S, Bellissimo DB, Das S, Grody WW, Hegde MR, Lyon E, Ward BE, Molecular Subcommittee of the ALQAC (2008) ACMG recommendations for standards for interpretation and reporting of sequence variations: revisions 2007. Genet Med 10(4):294–300. doi:10.1097/GIM.0b013e31816b5cae

74. Rehm HL, Bale SJ, Bayrak-Toydemir P, Berg JS, Brown KK, Deignan JL, Friez MJ, Funke BH, Hegde MR, Lyon E, Working Group of the American College of Medical G, Genomics Laboratory Quality Assurance C (2013) ACMG clinical laboratory standards for next-generation sequencing. Genet Med 15 (9):733–747. doi:10.1038/gim.2013.92

75. Tarczy-Hornoch P, Amendola L, Aronson SJ, Garraway L, Gray S, Grundmeier RW, Hindorff LA, Jarvik G, Karavite D, Lebo M, Plon SE, Van Allen E, Weck KE, White PS, Yang Y (2013) A survey of informatics approaches to whole-exome and whole-genome clinical reporting in the electronic health record. Genet Med 15(10):824–832. doi:10.1038/gim.2013.120

76. Collins FS, Hamburg MA (2013) First FDA authorization for next-generation sequencer. N Engl J Med 369(25):2369–2371. doi:10.1056/NEJMp1314561

77. Gazdar AF (2009) Activating and resistance mutations of EGFR in non-small-cell lung cancer: role in clinical response to EGFR tyrosine kinase inhibitors. Oncogene 28(Suppl 1): S24–31. doi:10.1038/onc.2009.198

78. Stern HM (2012) Improving treatment of HER2-positive cancers: opportunities and challenges. Sci Transl Med 4(127):127rv2. doi:10.1126/scitranslmed.3001539

79. McShane LM, Cavenagh MM, Lively TG, Eberhard DA, Bigbee WL, Williams PM, Mesirov JP, Polley MY, Kim KY, Tricoli JV, Taylor JM, Shuman DJ, Simon RM, Doroshow JH, Conley BA (2013) Criteria for the use of omics-based predictors in clinical trials: explanation and elaboration. BMC Med 11:220. doi:10.1186/1741-7015-11-220

80. Ng SB, Buckingham KJ, Lee C, Bigham AW, Tabor HK, Dent KM, Huff CD, Shannon PT, Jabs EW, Nickerson DA, Shendure J, Bamshad MJ (2010) Exome sequencing identifies the cause of a mendelian disorder. Nat Genet 42(1):30–35. doi:10.1038/ng.499

81. Green RC, Berg JS, Grody WW, Kalia SS, Korf BR, Martin CL, McGuire AL, Nussbaum RL, O'Daniel JM, Ormond KE, Rehm HL, Watson MS, Williams MS, Biesecker LG, American College of Medical Genetics and Genomics (2013) ACMG recommendations for reporting of incidental findings in clinical exome and genome sequencing. Genet Med 15(7):565–574. doi:10.1038/gim.2013.73

82. Lindhurst MJ, Sapp JC, Teer JK, Johnston JJ, Finn EM, Peters K, Turner J, Cannons JL, Bick D, Blakemore L, Blumhorst C, Brockmann K, Calder P, Cherman N, Deardorff MA, Everman DB, Golas G, Greenstein RM, Kato BM, Keppler-Noreuil KM, Kuznetsov SA, Miyamoto RT, Newman K, Ng D, O'Brien K, Rothenberg S, Schwartzentruber DJ, Singhal V, Tirabosco R, Upton J, Wientroub S, Zackai EH, Hoag K, Whitewood-Neal T, Robey PG, Schwartzberg PL, Darling TN, Tosi LL, Mullikin JC, Biesecker LG (2011) A mosaic activating mutation in AKT1 associated with the Proteus syndrome. N Engl J Med 365(7):611–619. doi:10.1056/NEJMoa1104017

83. Shirley MD, Tang H, Gallione CJ, Baugher JD, Frelin LP, Cohen B, North PE, Marchuk DA, Comi AM, Pevsner J (2013) Sturge-Weber syndrome and port-wine stains caused by somatic mutation in GNAQ. N Engl J Med 368(21):1971–1979. doi:10.1056/NEJMoa1213507

84. Ledford H (2013) "Master protocol" aims to revamp cancer trials. Nature 498 (7453):146–147. doi:10.1038/498146a

85. Cottrell CE, Al-Kateb H, Bredemeyer AJ, Duncavage EJ, Spencer DH, Abel HJ, Lockwood CM, Hagemann IS, O'Guin SM, Burcea LC, Sawyer CS, Oschwald DM, Stratman JL, Sher DA, Johnson MR, Brown JT, Cliften PF, George B, McIntosh LD, Shrivastava S, Nguyen TT, Payton JE, Watson

MA, Crosby SD, Head RD, Mitra RD, Nagarajan R, Kulkarni S, Seibert K, Virgin HWT, Milbrandt J, Pfeifer JD (2014) Validation of a next-generation sequencing assay for clinical molecular oncology. J Mol Diagn 16(1):89–105. doi:10.1016/j.jmoldx.2013.10.002

86. Lipson D, Capelletti M, Yelensky R, Otto G, Parker A, Jarosz M, Curran JA, Balasubramanian S, Bloom T, Brennan KW, Donahue A, Downing SR, Frampton GM, Garcia L, Juhn F, Mitchell KC, White E, White J, Zwirko Z, Peretz T, Nechushtan H, Soussan-Gutman L, Kim J, Sasaki H, Kim HR, Park SI, Ercan D, Sheehan CE, Ross JS, Cronin MT, Janne PA, Stephens PJ (2012) Identification of new ALK and RET gene fusions from colorectal and lung cancer biopsies. Nat Med 18(3):382–384. doi:10.1038/nm.2673

87. Grossmann V, Haferlach C, Weissmann S, Roller A, Schindela S, Poetzinger F, Stadler K, Bellos F, Kern W, Haferlach T, Schnittger S, Kohlmann A (2013) The molecular profile of adult T-cell acute lymphoblastic leukemia: mutations in RUNX1 and DNMT3A are associated with poor prognosis in T-ALL. Gene Chromosome Canc 52(4):410–422. doi:10.1002/gcc.22039

88. Murtaza M, Dawson SJ, Tsui DW, Gale D, Forshew T, Piskorz AM, Parkinson C, Chin SF, Kingsbury Z, Wong AS, Marass F, Humphray S, Hadfield J, Bentley D, Chin TM, Brenton JD, Caldas C, Rosenfeld N (2013) Non-invasive analysis of acquired resistance to cancer therapy by sequencing of plasma DNA. Nature 497(7447):108–112. doi:10.1038/nature12065

89. Bejar R, Stevenson K, Abdel-Wahab O, Galili N, Nilsson B, Garcia-Manero G, Kantarjian H, Raza A, Levine RL, Neuberg D, Ebert BL (2011) Clinical effect of point mutations in myelodysplastic syndromes. N Engl J Med 364(26):2496–2506. doi:10.1056/NEJMoa1013343

90. Jeck WR, Parker J, Carson CC, Shields JM, Sambade MJ, Peters EC, Burd CE, Thomas NE, Chiang DY, Liu W, Eberhard DA, Ollila D, Grilley-Olson J, Moschos S, Neil Hayes D, Sharpless NE (2014) Targeted next generation sequencing identifies clinically actionable mutations in patients with melanoma. Pigment Cell Melanoma Res. doi:10.1111/pcmr.12238

91. Mutz KO, Heilkenbrinker A, Lonne M, Walter JG, Stahl F (2013) Transcriptome analysis using next-generation sequencing. Curr Opin Biotechnol 24(1):22–30. doi:10.1016/j.copbio.2012.09.004

92. Lee EJ, Luo J, Wilson JM, Shi H (2013) Analyzing the cancer methylome through targeted bisulfite sequencing. Cancer Lett 340(2):171–178. doi:10.1016/j.canlet.2012.10.040

93. Chen K, Wallis JW, McLellan MD, Larson DE, Kalicki JM, Pohl CS, McGrath SD, Wendl MC, Zhang Q, Locke DP, Shi X, Fulton RS, Ley TJ, Wilson RK, Ding L, Mardis ER (2009) BreakDancer: an algorithm for high-resolution mapping of genomic structural variation. Nat Methods 6(9):677–681. doi:10.1038/nmeth.1363

94. Ye K, Schulz MH, Long Q, Apweiler R, Ning Z (2009) Pindel: a pattern growth approach to detect break points of large deletions and medium sized insertions from paired-end short reads. Bioinformatics 25(21):2865–2871. doi:10.1093/bioinformatics/btp394

95. Venkatraman ES, Olshen AB (2007) A faster circular binary segmentation algorithm for the analysis of array CGH data. Bioinformatics 23(6):657–663. doi:10.1093/bioinformatics/btl646

96. Chiang DY, Getz G, Jaffe DB, O'Kelly MJ, Zhao X, Carter SL, Russ C, Nusbaum C, Meyerson M, Lander ES (2009) High-resolution mapping of copy-number alterations with massively parallel sequencing. Nat Methods 6(1):99–103. doi:10.1038/nmeth.1276

97. Kumar P, Henikoff S, Ng PC (2009) Predicting the effects of coding non-synonymous variants on protein function using the SIFT algorithm. Nat Protoc 4(7):1073–1081. doi:10.1038/nprot.2009.86

98. Adzhubei IA, Schmidt S, Peshkin L, Ramensky VE, Gerasimova A, Bork P, Kondrashov AS, Sunyaev SR (2010) A method and server for predicting damaging missense mutations. Nat Methods 7(4):248–249. doi:10.1038/nmeth0410-248

99. Liu X, Jian X, Boerwinkle E (2013) dbNSFP v2.0: a database of human non-synonymous SNVs and their functional predictions and annotations. Hum Mutat 34(9):E2393–2402. doi:10.1002/humu.22376

100. Krzywinski M, Schein J, Birol I, Connors J, Gascoyne R, Horsman D, Jones SJ, Marra MA (2009) Circos: an information aesthetic for comparative genomics. Genome Res 19(9):1639–1645. doi:10.1101/gr.092759.109

101. Gargis AS, Kalman L, Berry MW, Bick DP, Dimmock DP, Hambuch T, Lu F, Lyon E, Voelkerding KV, Zehnbauer BA, Agarwala R, Bennett SF, Chen B, Chin EL,

Compton JG, Das S, Farkas DH, Ferber MJ, Funke BH, Furtado MR, Ganova-Raeva LM, Geigenmuller U, Gunselman SJ, Hegde MR, Johnson PL, Kasarskis A, Kulkarni S, Lenk T, Liu CS, Manion M, Manolio TA, Mardis ER, Merker JD, Rajeevan MS, Reese MG, Rehm HL, Simen BB, Yeakley JM, Zook JM, Lubin IM (2012) Assuring the quality of next-generation sequencing in clinical laboratory practice. Nat Biotechnol 30(11):1033–1036. doi:10.1038/nbt.2403

Methods in Pharmacology and Toxicology (2015): 259–270
DOI 10.1007/7653_2014_18

Published online: 16 September 2014

The Impact of Pre-analytic Variables on Tissue Quality from Clinical Samples Collected in a Routine Clinical Setting: Implications for Diagnostic Evaluation, Drug Discovery, and Translational Research

David G. Hicks

Abstract

The increasing application of biomarkers in both clinical practice and research has led to a greater emphasis on optimal sample preparation for such assays. However, in most institutions, current practices involved in tissue handling are highly variable and these variations can affect the quality of specimens used for analysis of macromolecules of potential clinical and research interest. In an attempt to address this important emerging issue, and to standardize tissue handling at our institution, the Surgical Pathology Laboratory has implemented an active tissue acquisition program in collaboration with the staff in the operating rooms and the Department of Surgery. Designated technical personnel from our laboratory assigned to the operating rooms are contacted as soon as a specimen is available. They are then responsible for collection of the specimen, checking accuracy on both the requisition and specimen label, and transportation of the specimen to the laboratory. With this implementation, we have seen significant improvements in the median time to fixation and the total number of specimens placed into formalin within 1 h of surgical removal from the patient. Rather than pushing an already busy OR staff to deliver specimens in a more timely fashion, we have found that a joint collaboration between departments, and proactively "pulling" specimens into the laboratory was more successful in achieving our goal of standardizing tissue handling and minimizing the time from surgical removal to the start of fixation. In addition, the type of tissue fixation and the time in fixative has the potential to impact the quality of tissue samples for molecular analysis of routinely collected clinical samples. These important issues will be reviewed in this chapter.

Key words Pre-analytic variables, Cold ischemic time, Biomarkers, Specimen integrity

1 Introduction

The introduction of targeted cancer therapies into routine clinical practice, in which novel treatment regimens are given based on companion molecular testing of tumor tissues, is rapidly ushering in a new era of individualized cancer care [1]. High-throughput molecular technologies have been used to define prognostic and predictive signatures that have been shown to outperform the standard clinical or pathologic variables currently available [2]. We are entering a time when cancer diagnosis and treatment will include molecular testing to guide the selection of the most appropriate therapies for individual patients on a case by case basis [3]. For hematologic malignancies and certain solid tumors such as breast, lung, gastric, and colon cancer, this new era has already arrived [4, 5]. These advancements in diagnosis and management are being driven by rapid technologic advancements in our ability to profile clinical samples at the molecular level, as well as by economic realities. The increasing number of new drugs, along with rising health care costs, necessitates the development and adoption of new diagnostic strategies and the implementation of new standards that will enable more effective therapeutic decisions and improved patient outcomes [6, 7]. As molecular analysis of tissue samples becomes increasingly clinically applicable, the accuracy, reliability, and relevance of this approach need to be addressed. These important issues are also significant for the discovery of new targeted cancer therapies and the development of companion diagnostic tests that will help guide the use of these agents.

The current standards for collecting and preserving clinical samples in most hospitals and clinics are decades old and involve 10 % neutral buffered formalin fixation to stabilize and preserve tissue for morphologic examination, and have paid little attention to the suitability of these tissues for further molecular analysis [8]. Significant variability in tissue handling and fixation in pathology laboratories has the potential to adversely affect the quality of clinical samples for diagnosis and research. Historically, the traditional diagnostic evaluation performed by anatomic pathologists involved the critical analysis and interpretation of morphologic features from routinely prepared, Hematoxylin & Eosin (H&E) stained tissue sections. For such cases, the goals of tissue handling and preparation were to ensure adequate tissue preservation and acceptable morphology, which will continue to be relevant in this new era of molecular profiling [9]. However, increasingly in current clinical practice, the evaluation of diagnostic samples requires further information beyond pure morphology, including an assessment of the level of expression of important target molecules within the diseased tissues. This is particularly true in breast cancer, where the evaluation of estrogen receptor (ER), progesterone receptor (PR), and the tyrosine kinase growth factor receptor (HER2) is now a part of the standard initial work up for any newly diagnosed tumor and is important for clinical decisions concerning appropriate adjuvant treatment. These critically important biomarkers help to identify subsets of patients who are appropriate candidates for specific treatments that target these major molecular drivers of disease progression [10–12]. These changes have led to a paradigm shift in surgical pathology practice related to cancer diagnosis, and the implementation of these biomarkers into clinical practice remains a challenge for these laboratories. The submitted tissue samples from patients with a suspected malignancy are no longer used only for morphologic interpretation. With the introduction of diagnostic assays for important tumor target molecules, the tissue sample also becomes in a very real sense an analyte, and specification of tissue quality becomes an important aspect of any validated assay performed on these samples [13].

Ligation of the blood supply to a living tissue that is resected during a surgical procedure will lead to hypoxia and metabolic stress, and will trigger progressive changes in gene expression and degradation of macromolecules that are of potential clinical interest [14]. The time interval between arterial ligation and tissue removal from the patient has been termed the warm ischemic time and can vary considerably depending on the experience of the surgeon and the complexity of the surgical procedure [14]. The cold ischemic time is the interval from removal of the sample from the surgical field until incision of the tissue in the laboratory and placement into a suitable fixative. While the warm ischemic time is difficult to control because it is dependent on the surgical procedure, the cold ischemic time is only dependent on the proximity of the operating rooms to the laboratory and how quickly the tissue is transported and prepared for tissue fixation [15]. Differing intervals of cold ischemic time due to variable tissue handling remains an important technical hurdle for the study of molecular targets in clinical samples. The current reality in most routine clinical practice is that specimen handling is diverse across different institutions and lacks strict standardization or well-defined standard operating procedures [8]. Furthermore, in many places both the time interval and the degree of variability are virtually unknown. With this in mind, recent national guidelines have recommended both prompt gross examination and the establishment of minimal and maximal fixation times for breast samples in an attempt to reduce the reported variability that can be seen in breast tumor predictive assays such as HER2 [16]. New ER and PR testing guidelines from the ASCO/CAP task force have taken this one step further and now require that breast biopsies and excised breast tissue samples be immediately assessed as rapidly as possible, sectioned and placed in formalin, ideally within 1 h from excision and removal from the patient [17]. These new guidelines have prompted a reevaluation of protocols and procedures involving tissue handling in a number of institutions. With the rising importance of being able to obtain molecular and genetic information from clinical samples, both surgeons and pathologists will need to reevaluate the traditional ways in which we have handled tissue samples coming from the operating rooms and their transportation to the pathology laboratory [18]. In light of these national guidelines, we have assessed the degree of variability in tissue handing from the operating rooms (OR) in our institution and, on the basis of our findings, took steps to try to standardize this potentially important pre-analytic variable for all tissue samples that are handled and processed in the surgical pathology unit at our medical center.

2 Establishing a Baseline for the Degree of Variability in Tissue Handling

The previous practice regarding tissue handling at our institution was for the OR nurse to transport all specimens to the pathology gross room at the completion of each operation via a pneumatic tube system. The times from collection to placement in formalin were markedly variable as a result of the various types and lengths of surgical procedures. In order to better understand how much variability existed and to establish a baseline demonstrating the effects of this practice, we began in 2008 to monitor the average length of time from specimen collection to receipt in the gross room, and from specimen collection to incision of the specimen and start of tissue fixation (cold ischemic time). Collection time was defined as the time that the specimen was removed from the operative field and fixation start time was defined as the time that the specimen was first placed in formalin after immediate gross assessment and incision.

3 Intervention: Pathology Take-Over of Specimen Collection and Transport

As a quality improvement initiative, for the second trial step in our analysis, designated staff from our surgical pathology accessioning team were stationed in the OR department ready to retrieve specimens as soon as they were collected and immediately transport them to the gross room via a pneumatic tube system. The length of time between collection and receipt in the gross room and between collection and fixation start time was recorded and monitored for each specimen. Our primary goals were to standardize specimen handling and reduce the time from collection to the start of fixation. We decided on a target of 60 min or less as our benchmark for all samples obtained from our operating rooms, given the ASCO/CAP task force recommendations. After a series of meetings with the OR leadership to discuss the plan, the technical staff members from pathology were provided with a cell phone and the OR staff were instructed to call them immediately upon collection of each specimen to communicate its availability for pickup. When called, the technical staff from the laboratory proceeded directly to the operating room to retrieve the specimen. A secondary aim of this study was to eliminate discrepancies and errors between specimens and their accompanying paperwork received in the surgical pathology laboratory from the OR. Upon retrieval, the pathology technician performed an immediate, "real-time," eight-point quality assurance and patient safety check of the requisition and the specimen for accuracy of labeling (see Table 1).

Table 1
Eight-point check—done at OR, at the time of pickup of tissue

1. Biopsy site on container and requisition correspond and are correctly highlighted
2. One requisition per specimen is provided
3. Specimen label matches requisition label
4. Ordering doctor is present
5. Clinical history is complete
6. Collection time is provided on requisition
7. OR number is listed on requisition
8. OR requisition has the Pathology watermark on the right-hand side at the top

4 Data Collection and Evaluation

On August 18th, 2008, pathology personnel assumed permanent responsibility for the collection and transport of tissue samples from the OR to the Surgical Pathology Laboratory after the hospital administration approved incremental personnel to support this initiative. Two goals of this rapid tissue acquisition program were to see what gains could be made in the standardization of tissue handling and the minimization of the cold ischemic time for clinical samples obtained from the OR at our institution. During the time interval from April 2008 through June of 2010, there were 21,887 specimens from the operating rooms with collection and fixation start times available for analysis. These data were subsequently reviewed to assess what improvements had been made.

5 Progress in Standardizing Tissue Handling

Figure 1 and its accompanying data demonstrate the percentage of collection to fixation start times that were within 1 h both prior to and after the implementation of our rapid tissue acquisition program. Specimens collected from April to August 17th 2008, prior to the implementation, met the 1 h to the start of fixation standard 37.48 % of the time. During the initial time period after the implementation of transport by pathology personnel on August 18th, 2008, specimens met the standard 85.82 % of the time, marking a 128.98 % increase in specimens meeting the 1 h standard from early 2008 to late 2008. In early 2009, specimens were meeting the standard 84.02 % of the time and specimens collected

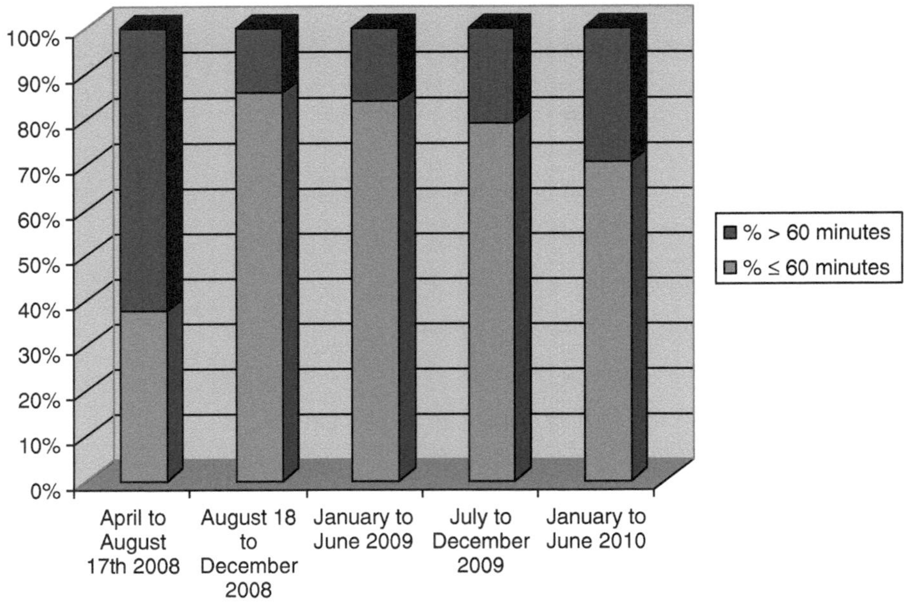

	Apr to Aug 17, 2008	Aug 18 to Dec 2008	Jan to Jun 2009	Jul to Dec 2009	Jan to Jun 2010
% ≤ 60 minutes	37.48	85.82	84.02	78.95	70.42
% > 60 minutes	62.52	14.18	15.98	21.05	29.58

Fig. 1 The percentage of clinical samples from the operating room that were within 1 h from collection to the start of tissue fixation both prior to and after the implementation of the rapid tissue acquisition program are shown here. During the initial time period after the implementation of transport of specimens by pathology personnel on August 18th, 2008, specimens met the 1 h standard 85.82 % of the time, marking a 128.98 % increase in specimens meeting the 1 h standard from early 2008 to late 2008

in late 2009 were meeting the standard 78.95 % of the time. Fixation of specimens occurred within an hour after collection 70.42 % of the time in early 2010. Overall, there has been an 87.89 % increase in occurrences of fixation within an hour from early 2008 to early 2010, but we saw slight percent decreases for each time period subsequent to late 2008. We found that periodically updating the operating room staff on our progress in standardizing tissue handling and, in particular pointing out this downward trend, would typically lead to improvements in times during the next period of analysis. To further illustrate these trends, Fig. 2 and its accompanying data display the percent of specimens meeting the 1 h goal from collection to the start of fixation for each month from January 2008 to June 2010.

Figure 3 represents the corresponding median time from tissue collection to the start of fixation for all specimens by month from early 2008 until June of 2010. For the time intervals examined, the

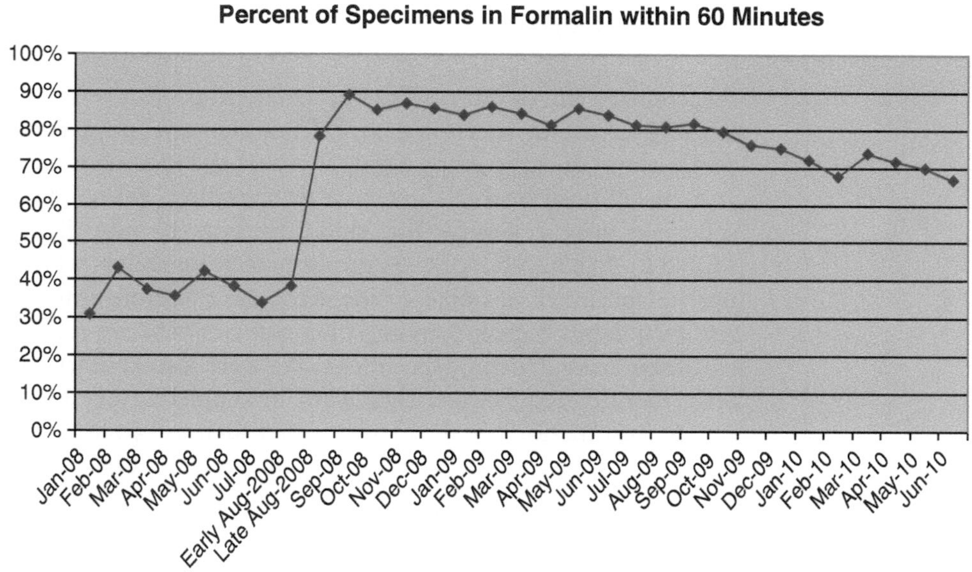

Fig. 2 The percent of specimens meeting the 1 h goal from collection to the start of tissue fixation for each month from January 2008 to June 2010 is shown here

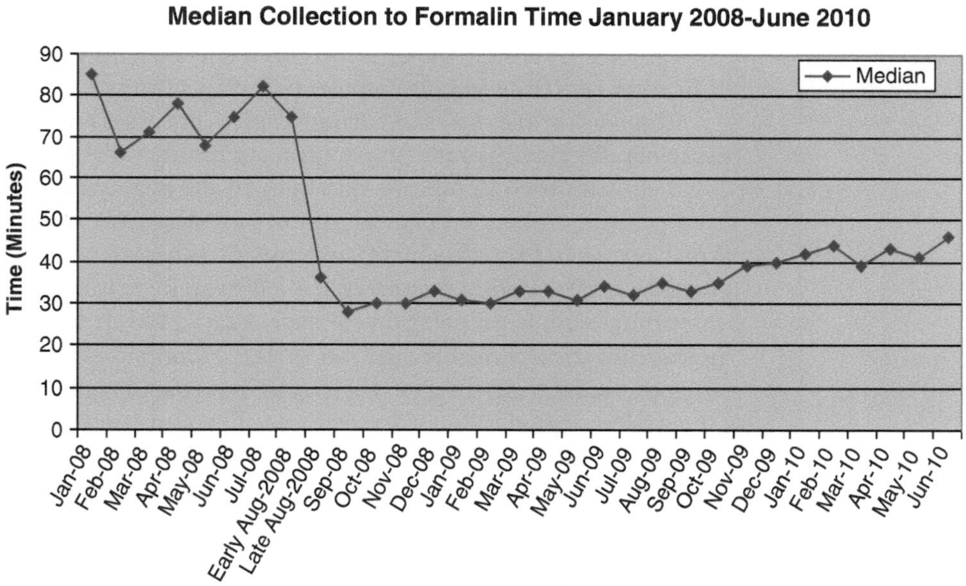

Fig. 3 The median time from tissue collection to the start of tissue fixation for all specimens by month from early 2008 until June of 2010 is shown here

median times were 30 min for late 2008, 32 min for early 2009, 35 min for late 2009, and 43 min for early 2010. A Chi Square Test for medians run with Analyse-It software produced a significant result ($P < 0.0001$, two-tailed), and subsequent pair-wise comparisons were performed and produced significant results comparing early 2008 with each subsequent time point.

6 Quality Assurance and Patient Safety

A secondary aim of our quality improvement project was to reduce labeling errors in specimens sent from our operating rooms by performing an eight-point safety check of the specimen container and the requisition real time in the operating room before the specimen was removed for transportation to the laboratory (Table 1). In a number of cases the pathology personnel who picked up the specimen was able to identify, and therefore immediately rectify, unlabeled specimens and discrepancies between requisitions and their accompanying specimens. We were also able to immediately identify requisitions that were incompletely filled out and obtain information for mandatory fields such as "collection time" and "clinical history" before the specimen left the OR.

7 How Important Is Standardizing Tissue Handling?

With this quality improvement project and the implementation of a rapid tissue acquisition program in our institution, we have shown that the introduction of a system for the proactive, rapid retrieval and delivery of tissue specimens from the OR to surgical pathology can significantly reduce the time interval from collection to the start of fixation for tissue samples removed during surgery.

In most institutions, the current clinical practices surrounding tissue handling and sample preparation are highly variable and thus could contribute to significant variability in the quality of formalin-fixed paraffin-embedded clinical samples for diagnosis and translational research [8]. The increasing use of biomarkers in clinical practice and translational research has led to an increasing emphasis on optimal sample preparation for such assays [19]. It has become increasingly clear that the ligation of the blood supply to tissues being excised during surgery will lead to hypoxia, ischemia, and the progressive degradation of macromolecules that are of potential clinical interest. Studies have shown that the level of expression of gene transcripts and proteins can change significantly during this ischemic interval [13, 14, 20, 21]. Of equal importance, delays to the start of formalin fixation may also affect tissue morphology and the mitotic index in tissue sections and negatively impact the accuracy of basic histological evaluation [22, 23]. The lability of different target analytes to ischemia will almost certainly confound research studies that have been performed on clinical samples with unknown or variable tissue handling [24], particularly in cases in which the ischemic interval is unknown. Additionally, for solid tumors such as breast cancer where targeted therapies are a possible treatment option, the potential impact of the quality of the tissue for diagnostic evaluation remains unclear and there are few studies available dealing with the impact of tissue handling on the accuracy of breast predictive factor assays [25]. Pinhel et al. [26] have shown that while ER, PR, and Ki67 immunohistochemical expression levels were similar between needle core biopsies (which are typically placed into fixative immediately after removal) and breast cancer excisions, the immunohistochemical reactivity for phospho-Akt and phospho-Erk1/2 was markedly reduced in the latter specimen type from the same patient. These differences are most likely attributable to variations in tissue handling of excisional samples. Given that protein kinases are targets for a significant number of new drugs under development for oncology, potentially important pharmacodynamic end points may be affected deleteriously by tissue handling practices that are routine, but suboptimal. These differences have major implications for future research, drug development, and patient management, and must be taken into consideration in the design of ongoing clinical trials of new therapeutic agents.

Neumeister et al. [27] have studied changes in antigenicity as a function of cold ischemic time in a series of 93 breast cancers with known time to fixation using the AQUA method of quantitative immunofluorescence. They found no evidence of loss of antigenicity with time-to-fixation in a 4 h time window. However, with a bootstrapping analysis, they observed a trend toward loss for ER and PgR, a statistically significant loss of antigenicity for phosphorylated tyrosine ($P = 0.0048$), and trends toward loss for other proteins. They reported evidence of increased antigenicity in acetylated lysine, AKAP13 ($P = 0.009$), and HIF1A ($P = 0.046$), which are proteins known to be expressed in conditions of hypoxia. The loss of antigenicity for phosphorylated tyrosine and increase in expression of AKAP13, and HIF1A were confirmed in a series of biopsies versus resection specimens from the same patients. These authors concluded that there is a need for further studies that extend the time range and normalize for intratumoral heterogeneity that can provide more comprehensive information on pre-analytic variation due to cold ischemic time and its potential impact on protein biomarker analysis.

Efforts to minimize and document the cold ischemic time will be important as we move forward with clinical/translational research and will allow investigators to determine which genes and proteins are potentially valid and reliable as biomarkers for clinical decision-making [13]. A greater emphasis needs to be placed on developing standardized methods of tissue procurement for diagnosis and molecular testing and this issue must be addressed by future bio-specimen research. Defining the pre-analytical requirement for the molecular analysis of clinical samples will be critical for moving next-generation molecular testing from the research laboratories into clinical practice. Such studies will in all likelihood lead to evidence-based guidelines for best practices in surgical pathology with regard to tissue handling and molecular testing.

8 Pre-analytical Variable and Breast Cancer Diagnosis

Accurate and reliable assessment of estrogen and progesterone receptors is important in adjuvant treatment planning for breast cancer patients [10, 17, 28]. Estrogen and progesterone receptors are thermolabile proteins whose levels of expression are altered by prolonged cold ischemic times [29, 30]. Recent studies have suggested that delays from tissue collection to the initiation of formalin fixation may adversely affect estrogen and progesterone receptor assessment [25, 29] as well as HER2 analysis [25]. In a study reported by Yildiz-Aktas et al. [30], breast resection specimens were subjected to variable cold ischemic times within the refrigerator and at room temperature. These samples were processed and stained for ER, PR, and HER2 and the results compared with the prior needle core biopsies from the same patient, which would have had a negligible cold ischemic time period before fixation. Significant reduction in IHC staining for hormone receptors and HER2 was not detected until 4 h for refrigerated samples and after 2 h for non-refrigerated samples. The authors concluded that the ASCO/CAP guideline of a cold ischemic time period of <1 h is a prudent guideline to follow and that refrigeration of specimens that may encounter delays until the start of fixation may be warranted.

Variable tissue handling leading to excessive cold ischemic times for excised breast tumor specimens could potentially result in patients being falsely classified as receptor negative. The consequences of invalid breast cancer testing for these important therapeutic targets have the potential to change the type of adjuvant therapeutic regimen offered, which in turn could adversely affect patient outcome.

New ER and PR testing guidelines from the ASCO/CAP task force now require that breast biopsies and excised breast tissue samples be incised and placed into formalin within 1 h from excision [17]. Accomplishing this goal will require that collection times and the fixation start times are recorded so that the cold ischemic time for each clinical sample can be calculated and monitored. The data from our quality improvement initiative suggests that the 1 h window proposed by the ASCO/CAP task force is challenging and will require an emphasis on standardization of tissue handling as well as a commitment of resources and personnel to accomplish this goal.

9 Conclusions

Taking ownership of specimen acquisition and transport on the part of our laboratory has led to improved standardization in tissue handling in our institution. Ensuring proper tissue handling of clinical samples is the joint responsibility of pathologists, surgeons, radiologists, and their respective staff [31] and will require good communication, collaboration, and partnership to help ensure the best possible clinical care. The standardization of tissue handling will, in all likelihood, lead to marked improvements in the overall quality of patient specimens, which will benefit both clinical and research efforts. We have concluded that a joint collaboration between departments to proactively "pull" specimens into the laboratory was more successful in achieving our goal of standardizing tissue handling than trying to push an already busy OR staff to deliver specimens in a more timely fashion.

Acknowledgments

The author would like to thank Ms. Mary Jackson for her assistance in manuscript preparation. The author would also like to thank and acknowledge the work of Kristin McCarthy, Robyn Sage, Laurie Baxter, and Michelle Fuller from the department of Pathology as well as the Operating Room staff for their help and support for this quality improvement project.

References

1. Hicks DG (2012) Targeted therapies in breast cancer: tailoring treatment to the molecular drivers of disease progression. MLO Med Lab Obs 44(36):38–39
2. Van de Vijver MJ, He YD, van't Veer LJ et al (2002) A gene-expression signature as a predictor of survival in breast cancer. N Engl J Med 347:1999–2009
3. Hicks DG, Kulkarni S, Hammond ME (2008) The role of the indispensable surgical pathologist in treatment planning for breast cancer. Arch Pathol Lab Med 132:1226–1227
4. Romond EH, Perez EA, Bryant J et al (2005) Trastuzumab plus adjuvant chemotherapy for operable HER2-positive breast cancer. N Engl J Med 353:1673–1684
5. Hicks D, Whitney-Miller C (2013) The evolving role of HER2 evaluation for diagnosis and clinical decision making for breast and gastric adenocarcinoma. Biotech Histochem 88:121–131
6. Parkinson B, Pearson SA, Viney R (2013) Economic evaluations of trastuzumab in HER2-positive metastatic breast cancer: a systematic review and critique. Eur J Health Econ Feb 24. [Epub ahead of print] PubMed PMID: 23436142
7. Barron JJ, Cziraky MJ, Weisman T, Hicks DG (2009) HER2 testing and subsequent trastuzumab treatment for breast cancer in a managed care environment. Oncologist 14:760–768
8. Hicks DG, Boyce BF (2012) The challenge and importance of standardizing pre-analytical variables in surgical pathology specimens for clinical care and translational research. Biotech Histochem 87:14–17
9. Hicks DG, McMahon L (2010) The pros and cons of automation for immunohistochemistry from the prospective of the pathology laboratory. In: Shi S, Taylor C (eds) Immunohistochemistry & antigen retrieval technique based research & diagnostics. Wiley, Hoboken, NJ, Chapter 7
10. Yaziji H, Taylor CR, Goldstein NS et al (2008) Consensus recommendations on estrogen receptor testing in breast cancer by immunohistochemistry. Appl Immunohistochem Mol Morphol 16:513–520
11. Hicks DG, Kulkarni S (2008) Trastuzumab as adjuvant therapy for early breast cancer: the importance of accurate human epidermal growth factor receptor 2 testing. Arch Pathol Lab Med 132:1008–1015

12. Hicks DG, Kulkarni S (2008) HER2+ breast cancer: review of biologic relevance and optimal use of diagnostic tools. Am J Clin Pathol 129:263–273
13. Hewitt SM, Lewis FA, Cao Y et al (2008) Tissue handling and specimen preparation in surgical pathology: issues concerning the recovery of nucleic acids from formalin-fixed, paraffin-embedded tissue. Arch Pathol Lab Med 132:1929–1935
14. Liu NW, Sanford T, Srinivasan R et al (2013) Impact of ischemia and procurement conditions on gene expression in renal cell carcinoma. Clin Cancer Res 19:42–49
15. Hicks DG, Kushner L, McCarthy K (2011) Breast cancer predictive factor testing: the challenges and importance of standardizing tissue handling. J Natl Cancer Inst Monogr 2011:43–45
16. Wolff AC, Hammond ME, Schwartz JN, American Society of Clinical Oncology/College of American Pathologists et al (2007) American Society of Clinical Oncology/College of American Pathologists guideline recommendations for human epidermal growth factor receptor 2 testing in breast cancer. Arch Pathol Lab Med 131:18–43
17. Hammond ME, Hayes DF, Dowsett M et al (2010) American Society of Clinical Oncology/College of American Pathologists guideline recommendations for immunohistochemical testing of estrogen and progesterone receptors in breast cancer. J Clin Oncol 28:2784–2795
18. Balch CM (2011) Reexamining our routines of handing surgical tissue in the operating room. J Natl Cancer Inst Monogr 42:39–40
19. Sherman ME, Howatt W, Blows FM et al (2010) Molecular pathology in epidemiologic studies: a primer on key considerations. Cancer Epidemiol Biomarkers Prev 19:966–972
20. Dash A, Maine IP, Varambally S et al (2002) Changes in differential gene expression because of warm ischemia time of radical prostatectomy specimens. Am J Pathol 161:1743–1748
21. Miyatake Y, Ikeda H, Michimata R et al (2004) Differential modulation of gene expression among rat tissues with warm ischemia. Exp Mol Pathol 77:222–230
22. Cross SS, Start RD, Smith JH (1990) Does delay in fixation affect the number of mitotic figures in processed tissue? J Clin Pathol 43:597–599
23. Start RD, Flynn MS, Cross SS et al (1991) Is the grading of breast carcinomas affected by a delay in fixation? Virchows Arch A Pathol Anat Histopathol 419:475–477
24. van Maldegem F, de Wit M, Morsink F et al (2008) Effects of processing delay, formalin fixation, and immunohistochemistry on RNA recovery from formalin-fixed paraffin-embedded tissue sections. Diagn Mol Pathol 17:51–58
25. Khoury T, Sait S, Hwang H et al (2009) Delay to formalin fixation effect on breast biomarkers. Mod Pathol 22:1457–1467
26. Pinhel IF, MacNeill FA, Hills MJ (2010) Extreme loss of immunoreactive p-Akt and p-Erk1/2 during routine fixation of primary breast cancer. Breast Cancer Res 12:R76
27. Neumeister VM, Anagnostou V, Siddiqui S et al (2012) Quantitative assessment of effect of preanalytic cold ischemic time on protein expression in breast cancer tissues. J Natl Cancer Inst 104:1815–1824
28. Goldstein NS (2010) Standardized predictive receptor immunohistochemical assays: the view forward from our past. Am J Clin Pathol 133:681–683
29. Nkoy FL, Hammond ME, Rees W et al (2010) Variable specimen handling affects hormone receptor test results in women with breast cancer: a large multihospital retrospective study. Arch Pathol Lab Med 134:606–612
30. Yildiz-Aktas IZ, Dabbs DJ, Bhargava R (2012) The effect of cold ischemic time on the immunohistochemical evaluation of estrogen receptor, progesterone receptor, and HER2 expression in invasive breast carcinoma. Mod Pathol 25:1098–1105
31. Kulkarni S, Hicks DG (2008) HER2-positive early breast cancer and trastuzumab: a surgeon's perspective. Ann Surg Oncol 15:1677–1688

Methods in Pharmacology and Toxicology (2015): 271–279
DOI 10.1007/7653_2014_23

Published online: 30 October 2014

Adapting HER2 Testing for a Different Organ: New Wine in Old Wineskins

Michael D. Lunt and Christa L. Whitney-Miller

Abstract

With the availability of increasing numbers of targeted monoclonal therapies, the determination of patients who are eligible for a given therapy is critical. One of the best studied examples is the use of trastuzumab, an anti-HER2 therapy, in breast cancer. With the publication of the results of the trial of Trastuzumab in Gastric Cancer (ToGA) and the subsequent FDA approval of trastuzumab for use in gastric and gastro-esophageal junction adenocarcinoma, assessment of HER2 status in these upper GI malignancies has become crucial. This chapter will review the current status of HER2 testing in both breast and gastric adenocarcinoma.

Key words Stomach, Adenocarcinoma, HER2, Trastuzumab, Testing, Gastroesophageal junction

1 Introduction

Human epidermal growth factor receptor 2, or HER2, is a transmembrane receptor tyrosine kinase whose activation is known to signal events necessary for cell growth, survival, and differentiation [1]. Originally discovered in rodents and named neu, this member of the ErbB receptor family has been found to be over-expressed in a variety of human malignancies. These tumors are generally more aggressive and have worse outcomes than those with normal HER2 expression [2, 3]. Because the HER2 protein is located on the cell surface, and because of its role in the biology of cancer cells and its relative overexpression in cancer cells compared to normal cells, HER2 is an important potential target for therapy.

2 Breast Cancer: The Old Wineskins

The first drug to target HER2 successfully was trastuzumab. Trastuzumab is a monoclonal antibody that blocks the dimerization of HER2, thereby preventing its downstream signaling. In 2001, researchers [4] showed that chemotherapy plus trastuzumab was more effective than chemotherapy alone in the treatment of

Table 1
IHC Scoring systems used to assess HER2 in breast and gastric carcinomas

Score	Breast (resection or biopsy)	Gastric (resection)	Gastric (biopsy)
0	≤10 % of tumor cells with faint incomplete staining	<10 % of tumor cells stained	<10 % of tumor cells stained
1+	>10 % of tumor cells with faint incomplete staining	>10 % of tumor cells with faint basolateral-to-circumferential staining (seen only at 40×)	>10 % of tumor cells with faint basolateral-to-circumferential staining (seen only at 40×)
2+	>10 % of tumor cells with weak/moderate circumferential staining, *or* ≤10 % of tumor cells with strong circumferential staining	>10 % of tumor cells with weak-to-moderate (evident at 10×), basolateral-to-circumferential staining	>10 % of tumor cells with weak-to-moderate (evident at 10×), basolateral-to-circumferential staining
3+	>10 % of tumor cells with strong circumferential staining	>10 % of tumor cells with moderate-to-strong (visible at low power), basolateral-to-circumferential staining	Tumor cell clusters ≥5 cells with moderate-to-strong (visible at low power), basolateral-to circumferential staining

Refs. [6–10]

HER2-positive metastatic breast cancer. Since that time, trastuzumab has been shown to be effective as adjuvant therapy for all HER2-positive breast cancer and is now being studied in the neoadjuvant setting.

The first commercial test to evaluate the status of HER2 in breast cancer was the HercepTest, marketed by Dako. This test used an immunohistochemical stain with a four-tiered scoring system from 0 to 3+, based on intensity and quality (circumferential) of staining and quantity of tumor cells stained. With proper specimen handling and control of pre-analytic variables and appropriate interpretation, there is good correlation between IHC and FISH results, such that IHC scores of 0 or 1+ can be considered negative and 3+ positive. 2+ is considered equivocal; in these cases, FISH may be used to determine if the HER2 gene is amplified.

In 2007, the American Society of Clinical Oncology and the College of American Pathologists first jointly released guideline recommendations to standardize the use of IHC and FISH in categorizing breast cancer as HER2 positive or HER2 negative [5]. The goal was to improve accuracy, reliability, and reproducibility. They provide specific recommendations for standardization of test performance, including interpretation and reporting criteria. There are also guidelines for proficiency testing and initial assay validation. These guidelines are part of the accreditation for laboratories that perform HER2 testing. The guidelines were updated in 2013 [6]. The current guidelines for interpreting HER2 IHC in breast cancer are summarized in Table 1.

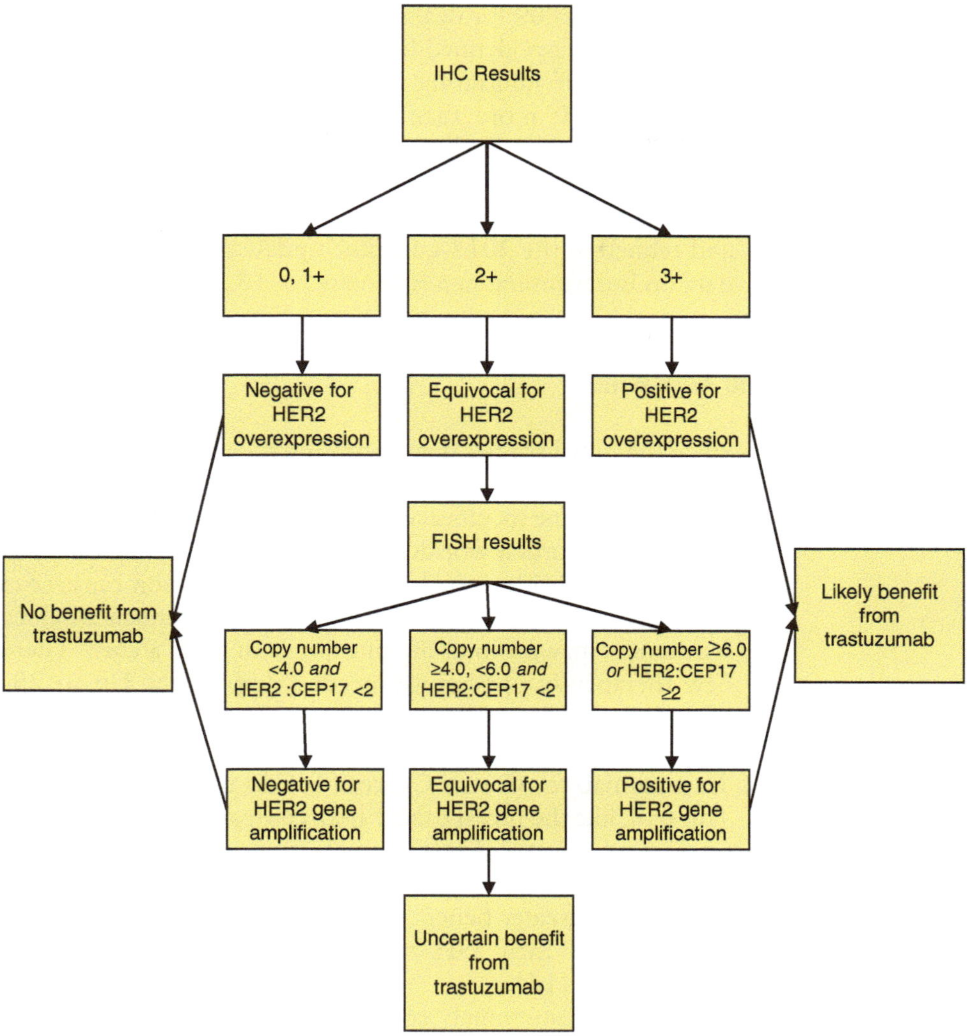

Flowchart 1 2013 ASCO/CAP guidelines for HER2 testing in breast cancer

The ASCO/CAP guidelines also provide guidance about in situ hybridization analysis for HER2 overexpression. According to the guidelines, both the ratio of HER2 signals to CEP17 signals (centromeric region of chromosome 17) and the average HER2 copy number should be assessed when using FISH to determine HER2 status. Cases having an average HER2 copy number <4.0 and a HER2: CEP17 ratio <2.0 are scored as negative/not amplified; cases having an average copy number ≥4.0 and <6.0 with a HER2:CEP17 ratio <2.0 are scored as equivocal; cases having an average copy number ≥6.0 OR a HER2:CEP17 ratio ≥2.0 are scored as positive/amplified. Flowchart 1 depicts the algorithm for determining HER2 status in breast cancer as described by ASCO/CAP.

Recognizing that a subpopulation of breast cancers exhibits intratumoral heterogeneous HER2 overexpression, CAP published guidelines in 2009 for dealing with genetic heterogeneity in breast cancer [12]. They defined heterogeneity as more than 5 % but less than 50 % of infiltrating tumor cells having a HER2:CEP17 ratio >2.2. (If more than 50 % of tumor cells have a HER2: CEP17 ratio >2.2 the tumor is considered amplified.) The document also included guidelines for reporting heterogeneity. This definition for intratumoral heterogeneity has been updated and refined in the 2013 guideline update to reflect new published data on intratumoral heterogeneity [6, 13].

3 Gastric Cancer: The New Wine

Since the initial studies demonstrating the utility of trastuzumab in HER2-positive breast cancer, a number of studies have been performed to define its efficacy in other body sites. Gastric cancer is a tremendous burden worldwide with almost one million diagnoses annually worldwide; it is the second most common cause of cancer mortality worldwide [14]. Gastric cancer is a surgical disease; chemotherapy may prolong life but does not offer a cure. Therefore, new therapeutic options are needed desperately. Up to 25 % of gastric cancers have been shown to overexpress HER2, making it a good candidate for a trastuzumab trial. This fact prompted the Trastuzumab for Gastric Cancer (ToGA) trial, published in 2010. This trial, like the breast cancer trial, compared chemotherapy alone to chemotherapy plus trastuzumab. The results showed an overall increase in survival from 11.1 versus 13.8 months in all patients, and an even greater benefit in patients who were IHC+ for HER2 (regardless of their ISH results) compared to those who were FISH + IHC [15].

Before trastuzumab could be used in patients with gastric cancer, it was necessary to determine which patients would be eligible for therapy; i.e., HER2 testing had to be validated in gastric cancer. These efforts were reported by Hofmann et al. They performed IHC and FISH on 168 [7] gastric cancer samples, applying the breast cancer scoring system and then correlating the IHC results with the FISH results. There was good concordance initially, but examination of the discrepant cases led to some observations about the differences in biology of gastric and breast cancer, and necessitated some modifications in the proposed scoring criteria and testing algorithm.

Complete circumferential staining is important in interpreting HER2 IHC in breast cancer; however, some gastric cancers exhibit basolateral rather than complete circumferential staining, yet still show HER2 amplification with FISH (see Fig. 1). As a result, it was recommended that either circumferential *or* basolateral staining be interpreted as positive. Gastric cancers exhibit more heterogeneous staining than breast cancer (see Fig. 2). Hofmann et al.

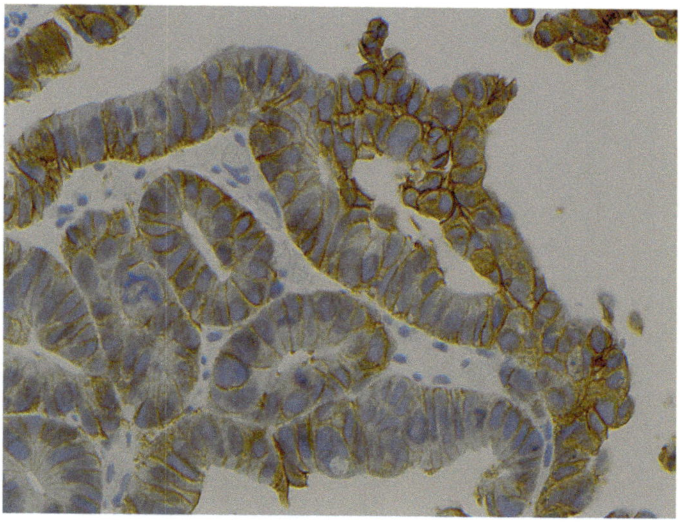

Fig. 1 Gastric adenocarcinoma with lateral staining on HER2 IHC. Gastric adenocarcinomas with basolateral/lateral staining are typically amplified with FISH and should be interpreted as IHC positive

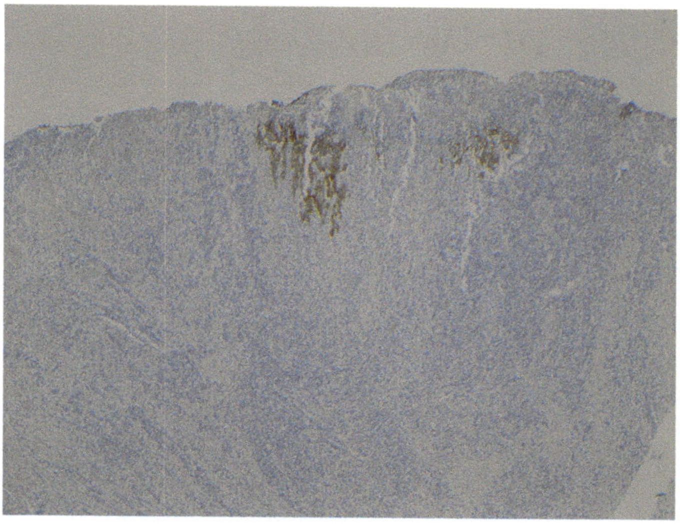

Fig. 2 Gastric adenocarcinoma with focal positivity with HER2 IHC. Such heterogeneous staining is more common in gastric cancer than breast. Targeted FISH shows these areas are amplified; therefore, they should be interpreted as positive on IHC

recommended some adjustment to the percentage of cells that must exhibit staining when rendering a positive interpretation in the biopsy specimens (see Fig. 3). These criteria are also summarized in Table 1.

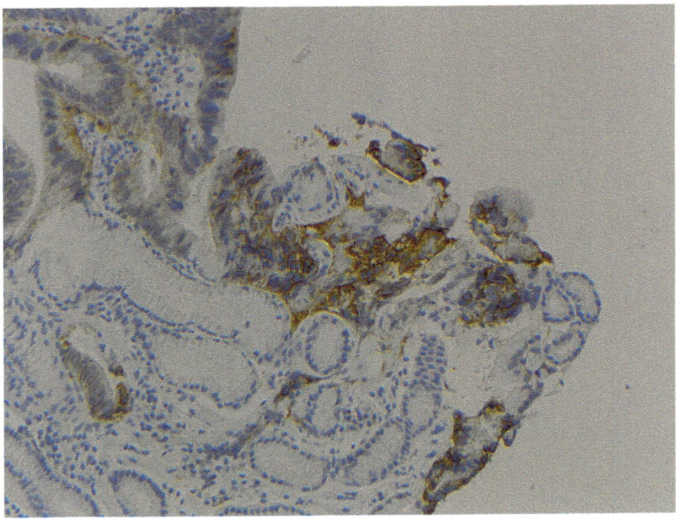

Fig. 3 Biopsy of gastric adenocarcinoma showing focal strong IHC positivity. Targeted FISH shows these areas are amplified; therefore, a cluster of five or more cells with otherwise strong IHC positivity can be interpreted as positive on IHC

ISH analysis for gene amplification of HER2 in gastric cancer is similar to breast, but as with IHC has some important differences (although these differences are fewer with the 2013 update to the breast criteria). The number of HER2 and CEP17 signals is compared; however, there are only two result categories: cases with a HER2:CEP17 ratio <2 are negative/not amplified and those with a HER2:CEP17 ratio ≥ 2 are positive/amplified. More recently, Ruschoff has recommended considering gastric cancer with an average HER2 copy number of >6 to be amplified, regardless of the ratio, similar to the updated recommendations in breast cancer [11, 13].

Lastly, the overall testing algorithm for gastric cancer is slightly different than for breast cancer. In breast cancer, if a case is negative by IHC, the case is considered HER2 negative/not amplified and no additional testing is needed; only IHC equivocal cases undergo ISH testing. In the ToGA trial, all gastric cancers underwent dual testing and if either result (IHC or FISH) was positive, the patient was eligible for randomization. The optimal HER2 testing algorithm in gastric cancer has yet to be determined, but to replicate the TOGA trial, all non-positive (i.e., equivocal and negative) IHC results should undergo ISH testing (see Flowchart 2).

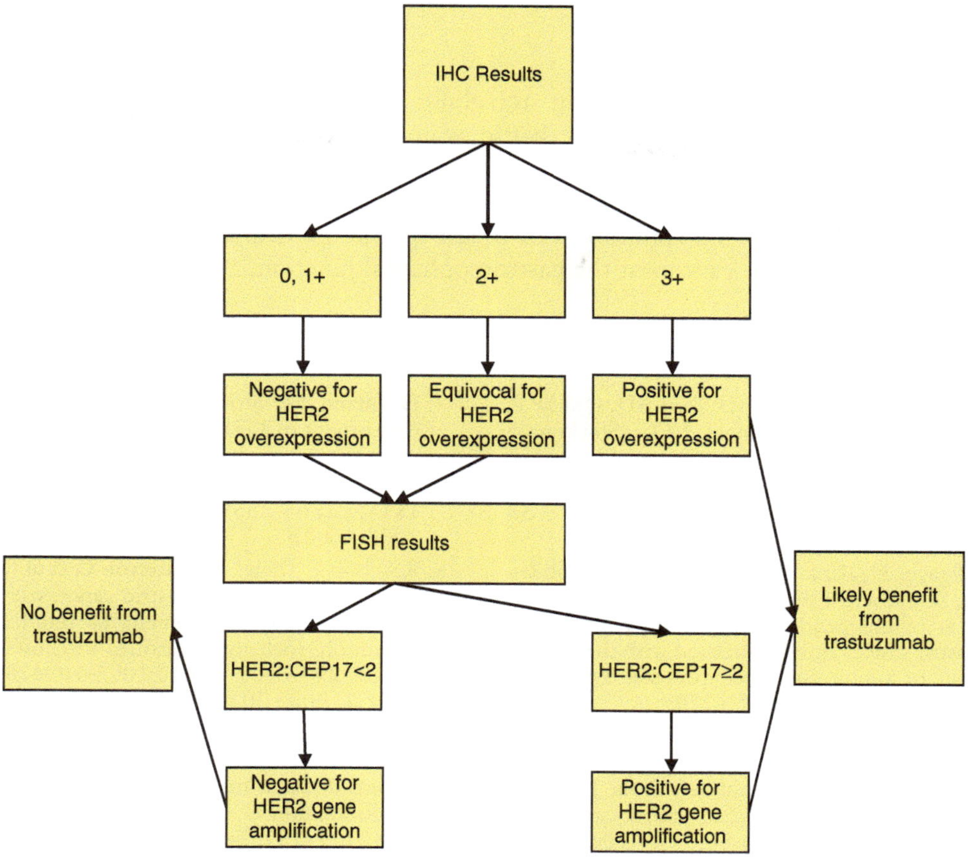

Flowchart 2 Proposed HER2 testing algorithm

4 Unresolved Issues

The publication of the ASCO/CAP guidelines for HER2 testing and interpretation in breast cancer have been very helpful in promoting uniformity and reliability of results, though some controversies still exist. In particular, the significance of aneusomy and intratumoral heterogeneity remains uncertain in terms of the potential for benefit from HER2-targeted therapy for patients whose tumors harbor these unusual HER2 genotypes. Currently, test performance, interpretation, and reporting for HER2 in gastric cancer are based on the scientific articles discussed in this chapter. In all likelihood, with increasing clinical experience and new publications on targeted therapies in upper GI-tract adenocarcinoma, refinements in testing recommendations will follow. A committee has been convened to draft guidelines for HER2 testing in gastric cancer, the report of which is much anticipated.

5 Summary

The increasing availability of targeted therapies is a welcome weapon in the battle against cancer. This chapter provides some insight into the testing methodologies necessary to determine eligibility for these targeted therapies, with particular emphasis on the application of breast-specific guidelines to gastric cancer and cancers at the gastroesophageal junction.

References

1. Koeppen HK, Wright BD, Burt AD, Quirke P, McNicol AM, Dybdal NO, Sliwkowski MX, Hillan KJ (2001) Overexpression of HER2/neu in solid tumours: an immunohistochemical survey. Histopathology 38 (2):96–104
2. Allgayer H, Babic R, Gruetzner KU, Tarabichi A, Schildberg FW, Heiss MM (2000) c-erbB-2 is of independent prognostic relevance in gastric cancer and is associated with the expression of tumor-associated protease systems. J Clin Oncol 18(11):2201–2209
3. Tanner M, Hollmén M, Junttila TT, Kapanen AI, Tommola S, Soini Y, Helin H, Salo J, Joensuu H, Sihvo E, Elenius K, Isola J (2005) Amplification of HER-2 in gastric carcinoma: association with Topoisomerase IIalpha gene amplification, intestinal type, poor prognosis and sensitivity to trastuzumab. Ann Oncol 16 (2):273–278
4. Slamon DJ, Leyland-Jones B, Shak S, Fuchs H, PatonV, Bajamonde A, Fleming T, Eiermann W, Wolter J, Pegram M, Baselga J, Norton L. Use of chemotherapy plus a monoclonal antibody against HER2 for metastatic breast cancer that overexpresses HER2. N Eng J Med. 344(11):783–792
5. Wolff AC, Hammond ME, Schwartz JN et al (2007) American Society of Clinical Oncology/College of American Pathologists guideline recommendations for human epidermal growth factor receptor 2 testing in breast cancer. J Clin Oncol 25:118–145, 344 (11):783–792
6. Wolff AC, Hammond ME, Hicks DG, American Society of Clinical Oncology/College of American Pathologists et al (2013) Recommendations for human epidermal growth factor receptor 2 testing in breast cancer: American Society of Clinical Oncology/College of American Pathologists clinical practice guideline update. J Clin Oncol 31(31):3997–4013
7. Hofmann M, Stoss O, Shi D et al (2008) Assessment of a HER2 scoring system for gastric cancer: results from a validation study. Histopathology 52(7):797–805. doi:10.1111/j.1365-2559.2008.03028.x, Epub 2008 Apr 18
8. Rüschoff J, Dietel M, Baretton G et al (2010) HER2 diagnostics in gastric cancer-guideline validation and development of standardized immunohistochemical testing. Virchows Arch 457(3):299–307. doi:10.1007/s00428-010-0952-2, Epub 2010 Jul 28
9. Hicks D, Whitney-Miller C (2013) The evolving role of HER2 evaluation for diagnosis and clinical decision making for breast and gastric adenocarcinoma. Biotech Histochem 88 (3–4):121–131, Epub 2013 Jan 7
10. Lunt M, Hicks D, Whitney-Miller C (2012) Gastric and gastro-esophageal adenocarcinoma: emerging therapeutic options and diagnostic testing challenges related to HER2. J Oncopathol 1(2):39–45
11. Rüschoff J, Hanna W, Bilous M, Hofmann M, Osamura RY, Penault-Llorca F, van de Vijver M, Viale G (2012) HER2 testing in gastric cancer: a practical approach. Mod Pathol 25 (5):637–650. doi:10.1038/modpathol.2011.198, Epub 2012 Jan 6, PubMed PMID: 22222640
12. Vance GH, Barry TS, Bloom KJ, Fitzgibbons PL, Hicks DG, Jenkins RB, Persons DL, Tubbs RR, Elizabeth M, Hammond H (2009) Genetic heterogeneity in *HER2* testing in breast cancer: panel summary and guidelines. Arch Pathol Lab Med 133 (4):611–612
13. Hanna WM, Rüschoff J, Bilous M, Coudry RA, Dowsett M, Osamura RY, Penault-Llorca F, van de Vijver M, Viale G (2014) HER2 in situ hybridization in breast cancer: clinical implications of polysomy 17 and genetic heterogeneity. Mod Pathol 27(1):4–18

14. Ferlay J, Shin HR, Bray F et al (2010) Estimates of worldwide burden of cancer in 2008: GLOBOCAN 2008. Int J Cancer 127 (12):2893–2917
15. Bang YJ, Van Cutsem E, Feyereislova A, ToGA Trial Investigators et al (2010) Trastuzumab in combination with chemotherapy versus chemotherapy alone for treatment of HER2-positive advanced gastric or gastro-oesophageal junction cancer (ToGA): a phase 3, open-label, randomised controlled trial. Lancet 376 (9742):687–697

Methods in Pharmacology and Toxicology (2015): 281–304
DOI 10.1007/7653_2014_26

Published online: 30 January 2015

Tissue-Based Companion Diagnostics: Development of IHC Assays from an Industry Perspective

Miu Chau and Jon Askaa

Abstract

In the last decade an increasing number of collaborations were established between pharma and diagnostic companies on the development of companion diagnostic assays to identify patients that will most likely respond to targeted therapies under development. This chapter focuses on the development of immunohistochemistry assays as companion diagnostics, which includes biomarker expression and specimen types, antibody selection and optimization, assay protocol development and optimization, assay verification and validation, and clinical validation. Past examples of companion diagnostic assay development, timeline coordination of drug-diagnostic co-development, and regulatory considerations are also briefly discussed.

Key words Immunohistochemistry assay development, Verification and validation, Drug-diagnostic co-development, Personalized medicine, Companion diagnostics

1 Introduction

In the last decade, the number of targeted therapies developed or in development for cancer treatment is increasing [1]. The goal is to increase and improve treatment effectiveness in selected patient populations that will most likely respond to therapy based on certain selection criteria, e.g., overexpression, amplification, or genetic changes of biomarkers. Recent contemporaneous approvals of crizotinib and vemurafenib, and their respective companion diagnostic (CDx) assays, ALK translocation and BRAF V600E mutation, are success stories of personalized medicine.

However, despite the increase in knowledge of drug targets, molecular understanding of cancer, disease biology, and signaling pathways, which could facilitate the discovery of new predictive biomarkers, only few targeted therapies are co-approved with companion diagnostic assays. Why hasn't the tremendous effort on biomarker discoveries translated into identification and validation of new predictive biomarkers? The challenges are many. From a biomarker identification and assay development perspective the challenges include (1) the lack of standardized assays used for biomarker discovery leading to conflicting results published and the difficulty to establish the clinical utility of a biomarker, (2) the presumption that a single biomarker would be sufficient in identifying the patient population that would most likely benefit from a targeted treatment, and (3) the uncertainty of whether the detection of protein expression, mRNA expression, or genetic changes of the biomarker would be most predictive of treatment effectiveness. From a clinical perspective the challenges include (1) the retrospective nature of biomarker hypothesis testing in early clinical studies that are usually small and not powered for subgroup analysis, and (2) samples may not be available from all enrolled patients thus negatively impacting the ability to fully test the biomarker hypothesis. Recent publications also described the challenges associated with companion diagnostics and drug co-development from regulatory and diagnostic perspectives [2, 3].

This chapter focuses on the development of immunohistochemistry (IHC) assays as companion diagnostics from a technical and clinical perspective. As such, challenges related to intellectual property, diagnostic partner/platform selection, regulatory approval, and commercialization/launch will not be discussed.

2 Personalized Medicine and Companion Diagnostic Assay Co-development

2.1 Past Examples

The discovery of HER2 amplification/overexpression in association with breast cancer development ushered in the era of development of targeted therapy [4]. During the development of trastuzumab, a humanized anti-HER2 antibody, in metastatic breast cancer setting it became evident that patients with HER2 protein overexpression were the most likely to respond to trastuzumab therapy. For the clinical studies conducted, a clinical trial assay (CTA) was developed by researchers and the testing was performed at a commercial laboratory. However, the CTA would not be suitable for commercial launch and a commercial diagnostic assay would be needed for worldwide testing at the time of the approval of the drug. A diagnostic (Dx) partner was identified and the commercial assay was developed. This technical validated assay was then compared with the CTA to establish concordance using commercially available samples, as the clinical study samples were not available/suitable for re-testing. The first companion diagnostic assay, HER2 IHC, was approved in the USA by FDA in September of 1998 simultaneously with the drug trastuzumab. At the time of approval HER2 positivity was defined as IHC 2+ and 3+. Later on HER2 gene amplification, HER2:CEP17 ratio $\geq$2.0 by fluorescence in situ hybridization (FISH), was also found to be important in identifying patients that would benefit from trastuzumab therapy. A HER2 FISH assay was approved by FDA in 2001 for the assessment of patients for trastuzumab therapy. After the approval of both HER2 IHC and FISH assays there were publications that revealed testing issues and the need for proficiency testing programs to ensure commercial labs performing HER2 testing meet certain performance criteria. In 2007, ASCO/CAP guideline on HER2 testing was published and HER2 testing algorithm was recommended for both IHC and FISH assays [5]. This speaks to the importance of developing not only a robust commercial companion diagnostic assay but also proficiency programs, and continual education/training of pathologists to ensure testing accuracy worldwide in both large and small institutions. The importance of accurate testing has far-reaching effects as patients wrongly diagnosed could either be deprived of a life-saving treatment, or be subjected to treatments of no benefit and having an adverse side effect.

With the success of trastuzumab and the companion diagnostic HER2 assays it is believed that targeted therapies would be most efficacious in patients bearing the drug target in their tumor cells. The EGFR-targeted therapy cetuximab and panitumumab followed the path of trastuzumab and prospectively enrolled only colorectal cancer (CRC) patients with EGFR expressing tumor cells as defined by an EGFR IHC assay, i.e., the EGFR negative patients were not enrolled in the pivotal clinical studies. The drugs and the companion diagnostic EGFR IHC assay were approved simultaneously for patients with metastatic colorectal cancer. Shortly after the approval of cetuximab, publications revealed patients that were EGFR IHC negative (low or no detectable level of EGFR protein by IHC) also seemed to benefit from EGFR-targeted therapies [6]. Retrospective biomarker subgroup analysis later identified that CRC patients with KRAS mutations in codon 12 or 13 do not seem to benefit from cetuximab or panitumumab treatment [7, 8]. Eventually, combined retrospective subset analyses of seven clinical studies with either cetuximab or panitumumab showed that patients with KRAS mutations should not receive monoclonal EGFR-targeted therapy, as they do not benefit from such treatments [9, 10]. The drug labels for cetuximab and panitumumab were updated in 2009 to include a statement that the drugs are not to be recommended for patients with KRAS mutations in codon 12 or 13 [9, 10]. Due to the retrospective nature of the analyses and the delay in CDx assay development, the FDA first approved the companion diagnostic KRAS mutation assay in 2012. This speaks to the hurdles a Dx manufacturer faces when the assay was used in a retrospective manner.

There are several past examples of companion Dx development, summarized in Table 1. In 2010, the FDA did not approve the drug omacetaxine mepesuccinate (Omapro™) for the treatment of chronic myeloid leukemia, as the agency believed that a standardized commercial assay for the detection of Bcr-Abl T315I mutation was needed at the time of drug approval. In the clinical study two lab-developed assays were used for patient enrollment and

Table 1
Examples of targeted therapies and companion diagnostics co-development

Drug	CDx	Indication	CDx testing	Regulatory path
Trastuzumab	HER2 IHC (HER2 FISH)	mBC	Prospective CTA[a], retrospective (CDx)	Contemporaneous—method comparison to CTA[a]
Cetuximab/ Panitumumab	EGFR IHC	mCRC	Prospective	Contemporaneous (wrong assumption)
Cetuximab/ Panitumumab	KRAS mutations	mCRC	Retrospective	Approval of CDx several years after drug approval
Crizotinib	ALK translocation (FISH)	NSCLC	Prospective	Contemporaneous
Vemurafenib	BRAF V600E mutation	Melanoma	Prospective	Contemporaneous
Pertuzumab and Ado-trastuzumab emtansine	HER2 IHC and FISH	mBC	Prospective	Contemporaneous
Dabrafenib and Trametinib	BRAF V600E and V600K mutations	Melanoma	Prospective CTA[a], retrospective (CDx)	Contemporaneous—bridging study compared to CTA[a]
Erlotinib and Afatinib	EGFR mutations	NSCLC	Prospective CTA[a], retrospective (CDx)	Contemporaneous—bridging study compared to CTA[a]

[a]Clinical trial assay

these assays used different technologies. This highlights the importance of using a standardized assay in clinical studies.

2.2 Timeline Coordination

The most critical parameter for successful drug-companion diagnostics co-development is timeline alignment. During early drug discovery phase (research and early development) preclinical models and drug mechanisms of actions or signaling pathways are usually employed for biomarker discovery. Many different technologies are often used and in-house research assays are developed using commercially available reagents. Sometimes proprietary reagents, e.g., antibodies raised for IHC or immunoassays, are developed if these are not available commercially. Once potential biomarkers are identified scientists have to decide which technology or assay platform is to be used for development of the companion diagnostic assay. Sometimes the reagents used for the research assay may not be compatible with the selected Dx platform. This may necessitate the development of new reagents, thus adding significant time to the CDx assay development timeline. Also, it is important to retain the samples used, from commercial vendors or clinical studies, for development of research assay to be re-tested by the

CDx assay to ensure the performance of CDx assay matches that of the research assay. This is critical to ensure that the biomarker hypothesis generated or verified using the research assay is still valid with the CDx assay.

Ideally, the final companion diagnostic assay, including assay reagents, software and hardware for the testing platform, and sample type (e.g., tumor tissue, blood, or serum/plasma) should be developed, and the design locked, for the intended use disease indication before the start of pivotal study. To meet this timeline, the development of CDx assay must begin at or shortly after the initiation of phase 2 clinical studies, as it usually takes at least 18–24 months, if not longer, to develop a CDx assay. This in turn requires identification of the potential predictive biomarker(s) during research or early phase 1 stage, which may or may not be feasible depending on the mechanism of action of the drug or the clinical program. The timing of initiation of CDx assay development is even more critical for accelerated clinical programs where phase 2 is skipped or when a phase 2 study is converted into a registrational study in areas of unmet medical need and when the drug efficacy is evident in early clinical development.

In many occasions several potential biomarkers are identified during early development. However, it may not be until the availability of outcome data at the end of phase 2 that the biomarker with the highest predictive potential is identified. In these situations, it is usually unrealistic to initiate in vitro diagnostic (IVD) development of several biomarkers as companion diagnostics due to cost concern and uncertainty of the drug program when little is known about the efficacy and safety of the drug. Thus, it is very unlikely that the final CDx assay will be ready for the registrational study in these circumstances. To mitigate the risk of not having the CDx assay ready for the registrational study one may be able to gate the CDx investment by adopting a stepwise approach—start with a prototype assay in early phase and then move on to an IVD CDx assay when there is more certainty on the biomarker and drug program (see Fig. 1 in reference [11]).

In situations where the CDx development timeline lags behind the drug development timeline, there is no avoidance of conducting the registrational study with a preliminary version of CDx assay and then performing a bridging study at the end to validate the final CDx assay. The FDA has very clear guidance on how a bridging study should be conducted which includes the banking of ALL screened patient samples and re-testing with the final CDx assay. To successfully conduct the bridging study one must be sure that the samples banked are stable over the course of the clinical study, which can be 2–3 years. This might not be a problem for blood/serum/plasma samples but could be a challenge for tumor tissue samples for IHC testing as epitopes in cut slides are usually not stable for several years. In these situations one may have to mandate storage of tissue blocks so that fresh sections can be made for re-testing at the end of the study. However, due to different practices in different countries or institutions tissue blocks may not be available for long-term storage by the study sponsor. When samples are missing then one must address the risk of bias in the clinical trial's statistical analysis plan. Early discussion with health authorities, e.g., Pre-IDE meeting(s) with the FDA, is (are) recommended when the final assay is not available for the registrational study in order to obtain feedback and buy-in on the design and conduct of the bridging study.

In summary, it is critical that pharma companies initiate timely discussions and establish collaborations with Dx partners early on to align the drug and CDx co-development timelines.

2.3 Regulatory Considerations

In the USA, the FDA considers companion diagnostic and drug co-development critical when the diagnostic is essential for the safe and effective use of the corresponding drug. This means that drug usage is dependent on the IVD assay results, i.e., the drug could be improperly administered if the assay does not work or gives erroneous results. As such, the companion diagnostic assay must be approved by the FDA (OIR branch, Office of In Vitro Diagnostics and Radiological Health; formerly known as Office of In Vitro Diagnostic Device Evaluation and Safety, OIVD) concurrently with the drug approval. There is also co-labeling of the drug and the device to ensure that an FDA-approved test is used to identify the right patients for the approved drug. FDA published a draft concept paper on drug-diagnostic co-development in 2005 [12], as well as a draft and final guidance documents on in vitro companion diagnostic devices in 2011 and 2014, respectively [13], that details the process for CDx development and approval process. OIR encourages drug and Dx sponsors to have early and frequent dialogue through the pre-IDE (Investigational Device Exemption) process to discuss CDx development. Whether contemporaneous approval of drug and companion diagnostics is required for breakthrough therapies is yet to be determined. The two recently published FDA draft guidance in 2014 about expedited access and balancing pre- and post-market data collection for premarket approval of medical devices could also have an impact on CDx development [14, 15].

In Europe and Asia there is not a requirement for concurrent drug and diagnostic approval. In EU, an IVD assay is only required to be CE-mark labeled—a self-certification process by the Dx manufacturer, and this does not require regulatory approval. However, the situation in EU is evolving as there are major changes in EU IVD medical device legislation and a new IVD regulation has been proposed. If the proposal is adopted then the regulatory landscape in EU will be very different from its current state as CDx will be required to be examined by a notified body [16]. Regardless of the changes, it would be prudent to obtain feedback on biomarker validation and companion diagnostic development from the European Medicines Agency's Pharmacogenomics Working Party (PgWP) or Scientific Advice Working Party (SAWP). In Japan and China the registration of diagnostics requires additional testing of regional patient samples at local institutions as part of approval requirements. In Japan, the Pharmaceuticals and Medical Devices Agency (PMDA) published a technical guidance on development of in vitro companion diagnostics and corresponding therapeutic products in 2013 [17]. Whether other countries will follow the footsteps of the FDA and require contemporaneous approval of drug and CDx remains to be determined.

3 Companion Diagnostic IHC Assay Development

3.1 Early Research and Biomarker Discovery Phase

During preclinical and early drug development phases researchers routinely focus on biomarker research related to disease biology, molecular target and mechanism of action of the drug, pharmacodynamics and pharmacokinetics, etc. These studies help inform the clinical development program in the assessment and selection of potential drug candidates, combination of drug treatments, disease indications, etc. It is very rare that potential predictive biomarkers are identified during preclinical or phase I studies except for drugs targeting certain genetic changes, e.g., mutations or translocations, or antibody–drug conjugates that require the presence of the target to deliver the chemo payload.

Early biomarker research is usually performed with assays developed in-house using commercially available reagents, as these are easily accessible and relatively inexpensive for studying a wide array of targets. However, one should be aware that these reagents are usually "research use" grade and their quality could vary, thus impacting the validity and reproducibility of the results generated. Lot-to-lot variation could lead to variable results when the assay was performed at different times, thus making it difficult to reproduce early findings. It is not uncommon to find in the literatures conflicting biomarker data and one of the reasons could be due to the use of different reagents/assays or assay conditions in various studies.

There are many factors that could affect the performance of an IHC assay, including the reagents used, pre-analytical conditions, and assay parameters. A recent publication on recommendations for improved standardization of immunohistochemistry succinctly summarized the importance of these factors [18]. These recommendations should be followed in the biomarker assay development process even for research assays. When developing an in-house IHC assay, antibody selection is most critical as it dictates the sensitivity and specificity of the assay. For a given target, there can be several different antibody clones on the market and the same antibody clone could be available from several manufacturers, which underscores the possibility of differences in performance. The available antibody could be a polyclonal, mouse monoclonal, or rabbit monoclonal antibody. There are pros and cons of these classes of antibodies and deciding which one to use could be based on multiple factors, e.g., general availability, specificity, and sensitivity. In general, researchers will test most, if not all, commercially available antibodies and select the one that performs the best with their assay system. Detection system used can also impact the sensitivity and background level of the IHC assay. It is not uncommon that an in-house IHC assay developed for early research and Phase 1 or 2 clinical studies uses antibody, detection system, and pretreatment reagents from different manufacturers. This might be acceptable when the assay is only used in-house but could pose a challenge when it needs to be transferred to a Dx partner's platform, as these reagents might not be compatible with the selected Dx platform. Thus, it is important to consider the ease of transfer of an IHC assay to a potential Dx partner's platform when developing an in-house assay that could become a CDx.

3.2 Clinical Trial Considerations for CDx Assay Development

Phase 1 study of a new molecular entity (NME) focuses on the safety of the NME. The study will start with a very low dose on a few patients and then the dose is gradually increased to investigate the safety and tolerability of the NME. As usually only a few doses (four to six) are studied and only a few patients are studied per dose, the total number of patients studied in phase 1 could be just a couple of dozens. In oncology it is common to enroll patients with different diseases in phase 1 clinical studies as the main indication has not been determined. As phase 1 studies are small and enroll patients with different diseases it is not uncommon that tumor samples are either optional or not collected. If tumor tissues are not collected or only collected from a few patients one would lose the opportunity of early assessment of potential predictive biomarker(s), especially if the NME is active and elicits significant responses in phase 1. Thus, collecting mandatory tumor samples is strongly recommended even for phase 1 studies.

For phase 2 studies (proof of concept, POC) it is critical to have mandatory tumor sample requirements in order to fully evaluate the potential predictive biomarker candidate(s). If sample collection is optional in a study the percent of evaluable samples received could be 50 % or less due to low consent rate and samples not received or, not evaluable, even if patients consented to provide tumor samples. Even with mandatory sample collection it is rare that evaluable samples from all patients are available. So it is important to educate the investigators on the importance of providing the "right" samples to facilitate evaluation of the potential predictive biomarker.

As POC studies are also used to establish the cutoff of a predictive biomarker they should be powered to allow proper assessment of the predictive nature of the biomarker(s) and the cutoff value of the assay. The total number of patients to be enrolled will depend on the prevalence of the biomarker in the disease indication studied. In addition, it is best to use a prototype assay developed by the identified Dx partner in these studies. This will ensure that the final companion Dx assay developed will have similar performance as the prototype assay and the results obtained in the POC study could be validated in the registrational study. If an in-house IHC assay is used in the POC study it is very important to retain tumor samples from the study for re-testing with the technical validated Dx assay to confirm that the performance of the two assays is the same and the cutoff value determined with the in-house assay is still applicable with the final Dx assay. As antigen stability on cut slides is often not very long, ranging from weeks to several months, it is important to retain tumor tissue blocks, with proper informed consent, for re-testing as POC studies usually last more than a year.

3.3 Biomarker Expression

As IHC is an assessment of the protein expression of a biomarker it is important to understand the expression level in the disease indication(s) for which the NME is intended. Prevalence studies should be conducted to evaluate the expression range of the biomarker in different disease indications and also the tumor subtypes for a certain indication to better inform the POC and phase 3 study designs. For example, EGFR expression is more prevalent in head and neck cancer compared to lung cancer [19], and HER2 expression is more prevalent in intestinal vs. diffuse subtypes and in gastroesophageal junction vs. gastric cancer [20]. Also, it is not uncommon that different tumor types may have different expression pattern of the biomarker of interest. Thus, it is important not only to collect information on the expression level but also expression pattern (homogeneous vs. heterogeneous, staining pattern—membrane, nuclear, or cytoplasmic) in the prevalence studies. One example is HER2 expression in breast cancer versus gastric cancer. For breast cancer HER2 expression is more homogeneous and complete membrane staining is important to define HER2 positivity. However, based on a validation study performed in gastric cancer, HER2 expression in gastric cancer is more heterogeneous and incomplete membrane staining can be frequent [21]. This led to different positivity definitions for HER2 in breast vs. gastric cancer, and biopsy vs. resection specimens for gastric cancer due to tumor heterogeneity [22]. In addition, biomarker expression may also be different in different ethnicity groups, e.g., EGFR mutation rate is much higher in some Asian subpopulations compared to Caucasians in patients diagnosed with NSCLC [23]. Thus it is also important to conduct biomarker prevalence studies in different ethnicity groups.

In oncology, NMEs are often evaluated in the metastatic setting first and if efficacy is demonstrated then earlier disease settings, e.g., adjuvant or neo-adjuvant, might be investigated. It is common practice in clinical studies that archival tumor samples are used for IHC testing. These samples could be from the primary or metastatic lesions (lymph nodes or other organs). It is thus important to understand if biomarker expression is the same for primary vs. metastatic lesions. Prior to conducting clinical studies paired primary vs. metastatic samples should be studied to evaluate the expression level and pattern of the biomarker of interest. Additionally, it is also important to understand whether biomarker expression changes with disease progression or prior treatments. If significant changes are noticed then new biopsy samples as an alternative to the diagnostic tumor specimen might be required for clinical studies in the metastatic setting to ensure an accurate assessment of the biomarker level at the time a patient is being considered for the study.

4 IHC Assay Development

4.1 Design Control

As companion Dx assays are class III in vitro diagnostics (IVD), and these are classified with the highest risk and require premarket approval by the FDA in the USA, the development process must follow the Design Control process as described in the guidance document issued by the FDA in 1997 (Fig. 1 in ref. [24]). The purpose of design control is to ensure that products meet the need of end users. As depicted in Fig. 1 in the FDA guidance document design control includes many steps, including design input, design output, design review, design verification and validation, design transfer, design change, and design history file. Design input defines product requirements such as detailed specifications and well-defined acceptance criteria that will meet the needs of patients and clinicians for the intended use. The product developed following quality system requirements of the manufacturer will be verified against the original design (design verification) by objective methods of analysis and testing to meet the design input requirements. The design output consists of the device, its packaging and labeling, and the device master record. Formal design review must be conducted at appropriate stages of the device's design development and documented in the design history file. Design transfer is a process to document transfer of the device from R&D to manufacturing with specific production specifications to ensure the device can be repeatedly and reliably produced. The initial production lots should be validated under user conditions (design validation) to ensure the device conforms to defined user needs and intended uses. The design control processes must be documented, reviewed, and approved and any changes to the design of the product must be documented in the design change control process. All the design control documents must be maintained in the design history file.

4.2 Antibody Selection: Sensitivity and Specificity

As mentioned earlier the choice of polyclonal vs. monoclonal antibodies or rabbit vs. mouse monoclonal antibodies is the manufacturer's decision when developing an IHC assay. This section will focus on the assessment of specificity and sensitivity of the antibody regardless of origin.

It is well known that different antibody clones that target the same protein often bind to different epitopes with varying specificity and binding affinity, leading to different staining patterns in some samples [25, 26]. This is further complicated by the fact that two antibody clones which show the same staining pattern in cancer cell lines and xenografts could give different staining patterns on certain human tissue samples. It is thus advisable to follow a standard process when evaluating antibodies for IHC assay development [27].

It is routine practice to first evaluate the sensitivity and specificity of antibodies on cancer cell lines with known levels of expression of the biomarker of interest. If the antibodies are shown to be specific in cell lines then they are further tested on xenografts as these provide a matrix to assess background staining, if any, of the antibody. Once specificity and sensitivity of the antibody are established on cancer cell lines and xenografts the next step is to test the antibodies on a wide variety of human tissue samples, e.g., tissues from different organs, normal and/or malignant tissues, and tissues from various sources to include diverse fixation parameters. Specificity of an antibody can also be demonstrated in IHC using recombinant proteins or peptides (if the epitope is known) in inhibitory experiments. If the antibody is specific for the target the IHC signal should be reduced/eliminated using recombinant protein or peptides for the target of interest but not when using an irrelevant protein/peptide. In addition to these analyses, the specificity of the antibody selected should also be confirmed by other methods, e.g., Western blot can demonstrate that the antibody does not cross-react with other proteins. Epitope mapping of the antibody is also important as one can assess if there might be potential cross-reactivity with other proteins by performing blast searches. If hypothetical cross-reactivity is identified then one can further evaluate the specificity of the antibody with tissues or cell lines that express the potential cross-reactive protein. Even if an antibody cross-reacts with another protein it might still be used for IHC assay development as long as the cross-reacting protein is not expressed in the tumor type of interest. However, caution is warranted when choosing an antibody that cross-reacts with other proteins as the assay might be used for multiple disease indications and it will be important to make sure that the cross-reacting protein is not expressed in all disease indications studied.

4.3 Assay Development

Once the antibody is selected for IHC assay development the next step is to establish the assay protocol. This involves selection of pretreatment reagents/conditions, primary antibody dilution/concentration and incubation time/temperature, detection reagents, incubation time/temperature, and counterstain. Another important parameter is whether the assay will be performed manually or on an automated instrument as the assay protocol/condition may be different. Detailed discussion of IHC assay development and the other aspects can be found in the CLSI guidance document [27] and this section only aims at providing a general description of the process.

Pretreatment reagents or conditions (deparaffinization, rehydration, and antigen retrieval) to be selected are dependent on the antibody, target, tissue type, and fixation condition (fixative and fixation time). Formalin-fixed paraffin-embedded (FFPE) specimens require deparaffinization and rehydration and in general require antigen retrieval for most targets. In contrast, cytology specimens that are lightly fixed in general require only rehydration without deparaffinization or antigen retrieval. The antigen retrieval condition (pH, proteolytic digestion, time and temperature) is highly dependent on the target and fixation condition—nuclear targets may require more stringent antigen retrieval whereas cytoplasmic/membrane proteins may require less. Over-fixed specimens will require more stringent antigen retrieval compared to under-fixed specimens. Also, the type of fixatives used, e.g., formalin, Bouin's, and new molecular fixatives, will have an impact on antigen retrieval condition required. In recent years "three-in-one" pretreatment reagents become more common offering one-step approach for deparaffinization, rehydration, and antigen retrieval. The suitability of such pretreatment reagent will be assay-dependent. Some targets require not only a mild antigen retrieval step but also proteolytic digestion (trypsin or proteinase K is commonly used) for effective detection. Extra caution should be exercised when selecting proteolytical digestion condition as slight variation of this step could have huge impact on the performance of the assay—destruction of the epitope of interest or poor tissue morphology. Thus, in most IHC protocols proteolytic digestion step is generally omitted unless absolutely necessary due to the difficulty of having precise control on the extent of digestion. The final selection of optimal pretreatment reagent and condition is accomplished through the process of trial and error and cannot be determined a priori.

The optimal primary antibody dilution and incubation time/temperature for best signal-to-noise ratio is highly dependent on the antibody chosen. Antibodies with higher affinity to the target usually require shorter incubation time and lower dilution

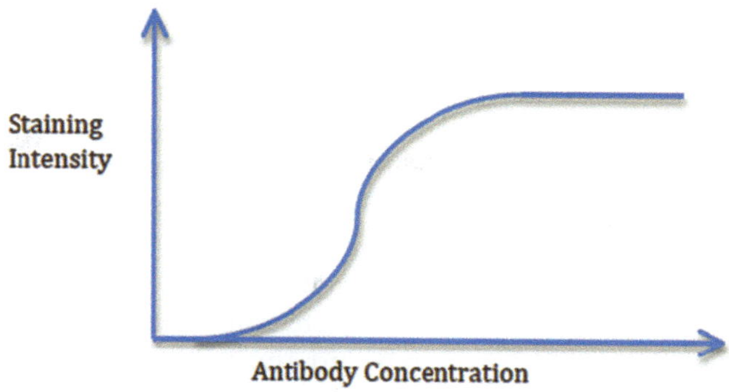

Fig. 1 Antibody titration curve

compared to those with lower affinity. Antibody incubation time and temperature in general have less impact on assay performance compared to antibody dilution. In general most antibodies are incubated at room temperature for 30 min to 2 h, or at 37 °C for 15–60 min. If an antibody requires prolonged incubation time due to low affinity then it should not be chosen for companion diagnostic assay development as it will likely not meet robustness requirements. Antibody diluent is usually manufacturer-specific as each manufacturer has their preferred diluent and usually there are few to choose from for ease of manufacturing. When performing antibody titration experiments to ascertain the optimal concentration to be used it is important to test a wide concentration range to determine the linear range and plateau (see Fig. 1). For commercialization purpose it is advisable to choose a titer that is on the plateau for robustness reason that still demonstrates sufficient dynamic range of staining intensity required. Choosing an antibody concentration that is too high (saturation) may make it hard to differentiate tissues with weaker expression from stronger ones, whereas choosing an antibody concentration within the linear range ("slippery slope") could have an impact on assay reproducibility. In addition, it is important to use tissues covering the expected expression range of the target protein and the disease types when performing antibody titration experiments so that the final antibody concentration chosen could be applicable for all tissue types of interest. When performing antibody titration experiments one may start with a few tissues and testing antibody concentrations in multiple fivefold dilutions. Once a narrower range of antibody concentration is identified then it will be important to include tissues with very low and also very high expression levels in the titration experiment with multiple twofold dilutions to determine the optimal dilution. For robustness of the commercial IHC assay the antibody might be tittered at a slightly higher concentration than the optimal concentration to accommodate user variations to ensure consistent performance on a daily basis. The final antibody dilution chosen is the most critical step of assay development as it has the most impact on assay performance and robustness.

Detection of antigen–antibody complex is usually performed via indirect methods as direct methods in general do not provide the required level of sensitivity. There are several types of indirect detection methods using different technologies, e.g., enzyme-conjugated secondary antibody, avidin-biotin, polymer-based, or catalyzed signal amplification (CSA) detection systems [27]. The choice of detection method depends on the level of sensitivity required and the target tissues. For higher sensitivity polymer-based or CSA detection systems may be used. For tissues with high endogenous level of biotin it might be advisable to avoid the use of avidin-biotin detection system even though a blocking step is employed. The detection system to be used should be the one that produces the best contrast (crisp staining) with minimal background. The most common enzyme label used for signal visualization is horseradish peroxidase (HRP) or alkaline phosphatase (AP) with the precipitation of a chromogenic substrate, i.e., 3,3′-diaminobenzidine (DAB) or fast blue/fast red, respectively, generating a brown or blue/red staining, respectively. Endogenous HRP, or AP, activities should be quenched by treating the tissue with hydrogen peroxide or levamisole, respectively, prior to the application of primary antibody. Counterstain and mounting media used will depend on the chromogen as some are soluble in alcohol and thus not compatible with alcohol dehydration step.

Each IVD manufacturer has its own "preferred" pretreatment and detection reagents/conditions that are applicable to most of their products for ease of use by customers. For pharma researchers developing an in-house IHC assay to be transferred to a Dx partner it will be prudent to understand what the "preferred" assay reagents/conditions are to facilitate a smooth transfer, especially when timelines are tight. Otherwise there could be situations where the performance established with the in-house assay might not be attainable on the Dx partner's platform. This could lead to the need for extensive assay development by the Dx partner or even result in a change of the antibody clone. All of these will have a negative impact on the timeline of availability of the CDx assay for clinical studies.

Once the assay protocol is finalized or design-locked, including all reagents and assay conditions, the following areas should be addressed prior to performing assay robustness testing:

(a) *Control Cell Lines*

Companion diagnostic IHC assays usually include development of cell line control slides as reagent and run controls to monitor staining performance. The control slides in general will have several cell pellets covering no staining, weak, moderate, and strong staining to ensure the staining run is performed correctly and the run is valid. The selection of proper control cell lines is not a trivial task as these cell lines must represent different levels of marker expression desired, and have similar staining pattern and sensitivity to assay condition changes as the target tissue to serve as a proper run control. In addition, the expression level of these cell lines should be stable upon many cell passages and under the conditions used during the manufacturing process for lot-to-lot consistency. As the control cell lines are provided as cut slides they must have a minimal shelf life of 12 months or more for ease of manufacturing and commercial viability.

As IHC interpretation is subjective it would be ideal to have reference standards to assist pathologists in gauging staining intensity levels. Health authorities in the past have inquired about the use of internal reference controls, e.g., normal cell types with a consistent staining intensity, but this has not been feasible so far. Another possibility is to use the control slides as reference standards for staining intensity. As IHC staining intensity is usually a continuum and the control cell lines only represent several discrete staining intensities it might be misleading to use the control cell lines as reference standards. Also, the manufacturing process must be extremely tight to ensure that each lot of cell line controls will have identical staining intensities in order for these to serve as reference standards.

(b) *Pre-analytical variables*

As most IHC assay development rely on the use of commercially available samples it is important to understand the pre-analytic conditions used, e.g., time to fixation, fixative used and fixation time, and storage condition, when these samples were collected and processed as these will have an impact on the performance of the assay.

The most commonly used fixative is neutral buffered formalin, but new fixatives, especially those developed to facilitate molecular analysis, are being used more frequently. An IVD developer must define the fixative and fixation conditions that the assay will function properly under. It is also important to identify those fixative or fixation conditions known to have significant impact on staining performance to avoid false positives or false negatives. Users must be warned that any deviation from the recommendations could have an impact on assay performance. Precautions and recommendations on pre-analytical variables can be found in CLSI guidance document [27].

(c) *Specimen types*

The type of specimens used for design verification must be carefully chosen to reflect that of the target indication (intended use population). For example, lung cancer samples could be from different sites, e.g., primary or metastatic, and collected using different methods, e.g. resection, biopsy, or fine needle aspiration (FNA). One must confirm if the assay will perform properly with the anticipated specimen types and it is thus important to include the anticipated sample types in the verification and validation studies. As most IHC assays are developed for formalin-fixed paraffin-embedded specimens, FNA samples that are usually fixed in alcohol or prepared as a smear would not work well unless the cells are pelleted and fixed in formalin and embedded in paraffin.

(d) *Staining pattern and scoring system*
The staining pattern and scoring system must be defined prior to the initiation of design verification. Staining pattern could be staining localization, i.e., membranous, cytoplasmic, and/or nuclear, or cell types stained, i.e., tumor cells, stromal components, and/or inflammatory cells. Sometimes certain staining patterns could be the result of nonspecific staining, e.g., nuclear staining could be nonspecific if the targeted staining pattern is membranous. For IHC assays it might be important to define the minimum number of tumor cells present in the specimen to be used for scoring.

The scoring system could be a combination of staining intensities, staining pattern, and percent of cells stained. For example, HER2 IHC assays require the assessment of staining pattern (complete or incomplete membrane staining), membrane staining intensity, and percent of tumor cells stained. Usually the scoring system is not defined until the end of phase 2 studies as the cutoff for positivity can only be defined with outcome data.

(e) *Software and hardware systems*
In the era of automation IHC assays are generally developed for use on automated instruments. It is thus very important to identify the platform(s) and the associated software (SW) version(s) to be used for assay verification/validation. The market situation at the time of launch must be taken into consideration, i.e., if more than one platform will be used or if a new platform will be introduced. US FDA considers the companion diagnostic a "system," i.e., this include the reagents, staining protocol, scoring system, SW and HW, and all the components will be specified in the approval.

4.4 Assay Verification and Validation

Assay verification, transfer to manufacturing, and validation processes should follow the quality system regulations established by each IVD manufacturer. For USA, one can also consult "Guidance Document for Submission of Immunohistochemistry Applications to the FDA" published by the FDA [28] and also "Quality Assurance for Design Control and Implementation of Immunohistochemistry Assays, Approved Guideline" published by Clinical and Laboratory Standards Institute (CLSI) [27].

In general, assay verification ensures that the IVD device meets its specifications and could include the following components: assay sensitivity and specificity, assay robustness/precision (see Appendix A1 in ref. 27 for details), reagent stability, antigen stability on cut slides, section thickness, stability of reagents and antigen, and packaging. It is extremely important to demonstrate that performance of the CDx assay is robust around the cutoff to ensure correct classification of Dx positive and negative status. If the assay is not very robust around the cutoff, i.e., samples near the cutoff could be misclassified as false positive or false negative, it could have a negative impact on the demonstration of efficacy of the targeted therapy. The extent of the impact will be dependent on the magnitude of benefit of the targeted therapy.

The FDA also requires testing of the specificity of the primary antibody on a panel of normal tissues from different anatomic sites from three different individuals (see Appendix C for the list of anatomic sites in ref. 28). In addition, IVD manufacturers also routinely perform testing on a panel of malignant tissues to evaluate expression patterns.

The assay can be transferred to manufacturing for commercial production after successful verification testing. The FDA requires manufacturing of three lots of reagents and all three lots must meet the manufacturing specifications. This is to ensure that the companion diagnostic assay can be reproducibly manufactured for uninterrupted supply after commercialization.

Prior to commercialization the CDx assay should be validated against the intended use under user conditions to ensure that it meets user needs (design validation). US FDA also requires external validation of the companion diagnostic to be performed at three different sites to assess day-to-day, site-to-site, and observer-to-observer agreements to ensure that CDx testing could be performed accurately by various users after commercialization.

5 Clinical Validation and Clinical Utility

5.1 Study Design

The clinical utility and validation of companion diagnostic assays are demonstrated in pivotal studies that provide evidence for the safety and effectiveness of the targeted therapy and associated CDx used to randomize patients in the clinical study. For FDA approval, the final technical validated CDx assay, i.e., commercialized version, must be used to prospectively randomize patients in the pivotal study. All components of the final CDx assay, from sample preparation to test results, must not change from the pivotal study to commercialization. If there are changes to any components then a bridging study must be performed with the final commercial version to demonstrate clinical utility and validity of the CDx test.

As discussed earlier, if the final version is not available then one must bank ALL samples, including screen-failed and randomized, to be re-tested with the final assay. The testing of screen-failed samples is extremely important if the pivotal study is a selection trial, i.e., only enrolling biomarker positive or negative patients. Health authorities want to ensure that the final assay will select the same patient population for drug treatment after approval as that of the initial device in the clinical study. If there is significant discordance between the final assay and the assay used in the study, additional analyses must be performed to address any bias introduced due to discordance. This is especially important in the situation of "false negatives" by the initial assay but otherwise positive by the final assay as the false negative patients will not be included in the study. In this situation, it is unclear whether the "false negative" patients would derive any benefit from the targeted treatment. It is thus important to determine the false positive and false negative rate by the initial assay used in the study compared to the final assay; this will not be possible if the screen-failed samples are not retained in selection studies for re-testing.

For IHC CDx assay, bridging study is particularly challenging as this necessitates the storage of patient sample blocks since cut sections are not stable for the duration of the pivotal study, which can last for 2–3 years. For certain countries the storage of patient samples, especially screen-failed samples, is not permitted per local ethics regulation. In the case of missing samples one must develop a statistical analysis plan to address potential bias introduced by missing samples. Even if all samples are available there could be potential imbalance between the two arms of the randomized pivotal study if the performance of the final assay and the assay used in the study exhibits significant differences owing to changes in classification of biomarker status. If the final assay will not be ready prior to phase 3 study start it will be prudent to have early dialogues with the FDA to seek input on the bridging study design and also the statistical analysis plan to address any potential bias/ imbalances.

The most important question when designing a clinical study that includes a companion diagnostic is whether to conduct a selection study, i.e., only including biomarker positive or negative patients, or a stratification study, i.e., enrolling both biomarker positive and negative patients and stratifying them among the two arms—active drug vs. placebo. Many past examples of CDx approvals are based on selection studies, e.g., HER2, EGFR, BRAF, and ALK, enrolling only biomarker positive patients. Selection studies are attractive as they require fewer patients and thus lower cost and potentially shorter recruitment timelines. However, selection studies suffer from the shortfall of not providing concrete evidence that the biomarker is predictive of drug benefits, as the biomarker negative population was not assessed to definitively

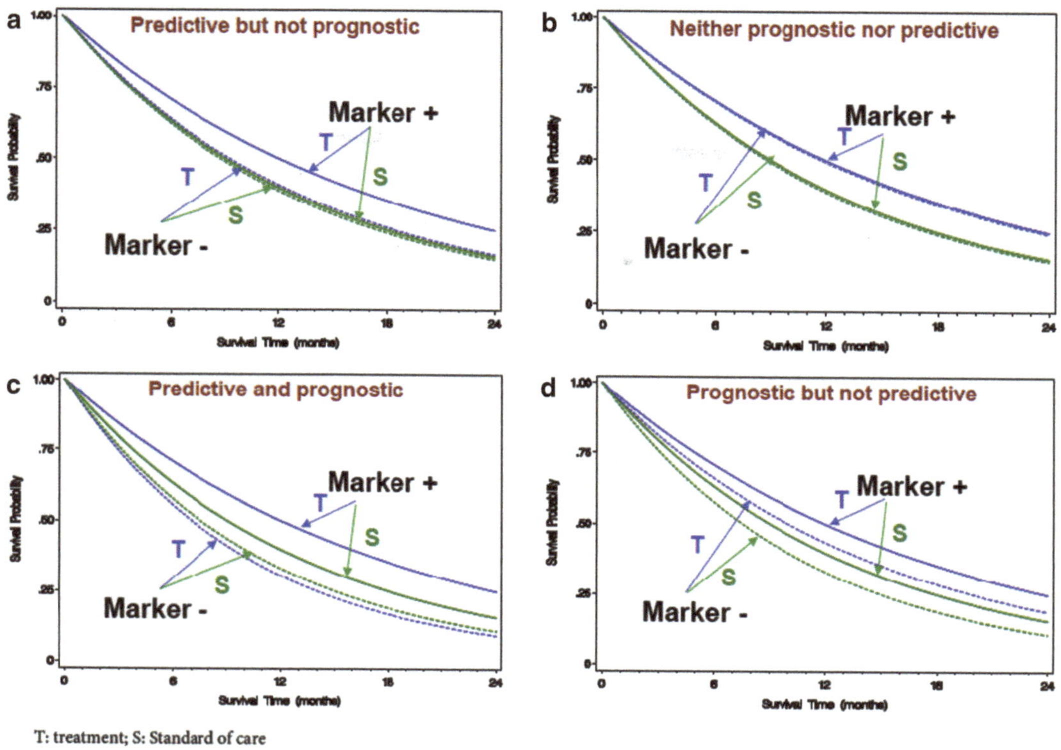

Fig. 2 Biomarker Scenarios (courtesy of Gary Clark)

conclude that the drug has no benefit in that population. The clear examples are EGFR inhibitors, cetuximab and panitumumab, and the CDx EGFR IHC assay. If the original studies had included both EGFR IHC positive and negative patients the results would have demonstrated that EGFR IHC assay is not predictive of drug benefit as later studies showed that EGFR IHC negative patients also derived benefits [6].

When conducting selection studies it is important to avoid pre-screening by sites, i.e., sites only subject biomarker positive or negative patients, as determined by a local method, to be screened for enrollment into the pivotal study. Local pre-screening has a detrimental effect on the effective demonstration of the clinical utility of the companion diagnostic as not all potential patients are tested by the CDx assay. This is because false negative patients per local pre-screening will not be subjected to testing by the CDx assay and hence excluded from the study. At public presentations the FDA has strongly advised sponsors to prevent pre-screening or be prepared to address pre-screening issues in order to establish the clinical utility of the CDx device.

A stratification study, i.e., studying all patients, is the best way to thoroughly demonstrate whether a biomarker is predictive or not as shown in Fig. 2. If the biomarker negative population is not studied, as in a selection trial, one would not be able to differentiate whether a biomarker is predictive, prognostic, both predictive and prognostic, or neither predictive nor prognostic. Stratification study is the preferred study design by health authorities and selection studies are only accepted if there is detrimental effect of the study drug on the biomarker negative patient population, or if there is very strong biological rationale to only study a subgroup of patients identified by the biomarker, e.g., ALK translocation in NSCLC and BRAF mutations in melanoma. Stratification studies also have the added benefit of allowing retrospective analysis of additional potential predictive biomarkers as no patients are excluded by the prospective biomarker studied in case the prospective biomarker studied does not pan out. Overall, stratification study design is better even though it means a larger study, resulting in potentially higher cost and longer timeline.

5.2 Study Setup and Conduct

As the pivotal study is usually conducted and managed by the pharma (Rx) partner while the companion diagnostic is managed by the Dx partner both must come to an agreement on how the companion diagnostic testing would be managed during the entire study prior to study start. It is also important to determine if additional data need to be collected for CDx filing that are usually not routinely collected in a drug study. If these aspects are not thoroughly addressed prior to study start there could be a negative impact on the acceptability of the CDx submission by health authorities. As the FDA has repeatedly emphasized at various conference presentations—no test, no drug—both Rx and Dx partners must work together to ensure the pivotal study conducted meets the requirements for both drug and CDx approvals. The study is used to demonstrate safety and effectiveness of not only the drug but also the CDx.

When choosing test laboratories to perform prospective CDx IHC testing the following aspects should be considered: experience with prospective testing and result reporting to sites, turnaround time from sample receipt to results, accreditation, competency of technical staff and pathologists, familiarity with the Dx partner's technology and instrument platform, sample management system, quality system, experience with FDA inspection, etc. The test lab(s) is considered an investigative site for the CDx and it must meet the requirements of the Dx partner. In addition, as the test lab(s) could be inspected by the FDA prior to approval of the CDx it is important that the one chosen has a good quality system in place which meets FDA requirements. Both Rx and Dx partners should agree upon the total number of test lab(s) to be used for the study. As most pivotal studies are global involving many countries in different regions, e.g., US, EU, Asia, and China, it is natural that several test labs be selected encompassing different geographic areas for quick turnaround time. If the study will have clinical sites in China then it might be more convenient to have a test lab set up in China to avoid lengthy and uncertain export license application process.

Once the test lab(s) is chosen it is the responsibility of the Dx partner to train the test lab(s) on the CDx assay, including sample handling, staining, scoring, and result reporting. This approval is separate from the drug investigator sites' IRB/EC approvals. Test lab personnel must be qualified by the Dx partner prior to conducting testing on clinical samples. If multiple test labs are chosen then it is very important to demonstrate that these labs are significantly concordant with each other to ensure consistency of assay performance across the test labs. As IHC scoring could be subjective it can be challenging to demonstrate concordance among many pathologists. Thus one might consider limiting the number of pathologists evaluating clinical study samples at each test lab.

As an investigative site the test lab(s) must have IRB/EC approval prior to conducting clinical sample testing. It is the responsibility of the Dx partner and the test lab(s) to obtain IRB/EC approval. For test labs in the USA a central IRB approval will be sufficient while test labs outside of the USA will have to follow local regulations. In some countries local ethics committee approval must be obtained, which could be a lengthy process. As the CDx is an investigative device the reagents must be labeled as "Investigation Use Only" (IUO) in the USA or "Performance Evaluation Only" (PEO) in the EU, and the usage must be tracked during the duration of the study.

The Dx partner is also responsible for monitoring the test lab(s) for the duration of the study, which includes site initiation, routine monitoring, and study close out visits, to ensure that CDx testing is performed according to the testing plan. Any deviations from the testing protocol or non-conformances must be recorded. It is also important to monitor the performance of the assay by tracking the positivity rate in each lab, and also for each reader, so as to identify potential testing issues early on.

More detailed description of conducting clinical testing at test lab(s) can be found in Chapter 24 (Mark Kockx).

For clinical studies that do not have a companion diagnostics component it is routine practice of the Rx partner not to collect information on screen-failed patients as these patients are not part of the study. However, for studies with a companion diagnostics ALL patients tested are considered part of the study for evaluating the performance of the CDx assay. It is thus important that patient demographics and certain clinical characteristics are collected for ALL patients tested whether they are enrolled in the clinical study or not. This is particularly important for selection studies as health authorities want to ensure that there is no bias in the patient population enrolled, vs. not enrolled, with respect to demographics and clinical characteristics. It is equally important that the demographic and clinical characteristics data collected on screen-failed patients are monitored in the same manner as enrolled patients to ensure data integrity, as these data would have to be submitted to the health authorities in the CDx submission. It is thus important that both Rx and Dx partners reach an agreement early on regarding how (database to be used) and what (demographics and clinical characteristics parameters) data should be collected and eventually shared for regulatory submission.

6 Conclusion

After successful development of a companion diagnostic assay and demonstration of its effectiveness in identifying the right patient population that would benefit from a targeted therapy there are still challenges at the time of commercialization. For patients to receive targeted therapies they need easy access to the CDx tests and the testing quality is also very important. In the past, proficiency testing programs have been developed for some companion diagnostics by professional societies, e.g., CAP, UKNEQAS, and NordicQC, but these programs were not available immediately at the time of drug and CDx launches. Moving forward with the increase in companion diagnostic approvals there is a need to implement proficiency testing programs closer to time of drug and CDx launch to ensure that test laboratories have the means to monitor their performance. False negative or false positive results for a companion diagnostics have serious implications on patient welfare, as a patient could miss the opportunity to receive an effective treatment or could develop adverse side effects from an unnecessary treatment, respectively. It is thus very important that healthcare providers, including clinicians and lab personnel, understand that the effectiveness of a targeted therapy is highly dependent on an accurate testing result. Traditionally, pharma companies focus on educating clinicians on the drug while Dx companies focus on educating pathologists on the IHC assay. However, a paradigm shift is needed as both Rx and Dx partners now need to educate their "customers" on both the drug and the companion diagnostics to highlight the importance of the use of the right test and the importance of obtaining an accurate test result due to the serious implications on patient health.

In summary, companion diagnostics development is a long and complex process and Rx partners are advised to engage Dx partners early on to ensure that the Dx development timeline is aligned with the drug development timeline. In recent years the pharmaceutical industry is looking for innovative ways to accelerate drug development timelines, e.g., applying for breakthrough therapy designations in the USA, jumping from phase 1 studies to pivotal P2/3 studies, and adaptive design. These innovative ways of shortening drug development timeline add to the challenges in developing a companion diagnostic as they require the identification of potential predictive biomarker(s) very early on. In addition, the skipping of clinical trial phases may not allow the thorough evaluation of the optimal cutoff for the predictive biomarker, as this usually requires a randomized all-comers study with enough statistical power. There may be a need for a more creative approach in the development and validation of companion diagnostics in the future in order to cope with the ever-changing drug development landscape. Health authorities also play a critical role in the future success of drug and companion diagnostics co-development, as new regulations are established to hasten drug development similar changes in the regulations on companion diagnostics need to be implemented. Close collaboration between the Rx and Dx partners and the early and frequent engagement of health authorities for input are key for successful drug and Dx co-development.

References

1. Aggarwal S (2010) Targeted cancer therapies. Nat Rev Drug Discov 3(9):427–428
2. Phillip R, Carrington L, Chan M (2011) US FDA perspective on challenges in co-developing in vitro companion diagnostics and targeted cancer therapeutics. Bioanalysis 3(4):383–389
3. Beeler J (2013) Integrating companion diagnostic assays into drug development: addressing the challenges from the diagnostic perspective. Drug Dev Res 74:148–154
4. Slamon D, Godolphin W, Jones L et al (1989) Studies of the HER-2/neu proto-oncogene in human breast and ovarian cancer. Science 244:707–712
5. Wolff AC, Hammond EH, Schwartz JN et al (2007) American Society of Clinical Oncology/College of American Pathologists guideline recommendations for human epidermal growth factor receptor 2 testing in breast cancer. Arch Pathol Lab Med 131:18–43
6. Chung KY, Shia J, Kemeny NE et al (2005) Cetuximab shows activity in colorectal cancer patients with tumors that do not express the epidermal growth factor receptor by Immunohistochemistry. J Clin Oncol 23:1803–1810
7. Karapetis CS, Khambata-Ford S, Jonker DJ et al (2008) K-ras mutations and benefit from cetuximab in advanced colorectal cancer. N Engl J Med 359:1757–1765
8. Amando RG, Wolf M, Peeters M et al (2008) Wild type KRAS is required for panitumumab efficacy in patients with metastatic colorectal cancer. J Clin Oncol 26:1626–1634
9. Cetuximab USPI. http://packageinserts.bms.com/pi/pi_erbitux.pdf
10. Panitumumab USPI. http://pi.amgen.com/united_states/vectibix/vectibix_pi.pdf
11. Jergensen JT (2012) Companion diagnostics and the drug-diagnostic codevelopment model. Drug Dev Res 73:390–397
12. (2005) FDA draft concept paper "Drug-Diagnostic Co-Development". www.fda.gov/downloads/Drugs/ScienceResearch/ResearchAreas/Pharmacogenetics/UCM116689.pdf
13. (2014) FDA guidance "In Vitro Companion Diagnostic Devices". www.fda.gov/downloads/MedicalDevices/DeviceRegulationGuidance/GuidanceDocuments/UCM262327.pdf
14. (2014) FDA draft guidance "Expedited Access for Premarket Approval Medical Devices Intended for Unmet Medical Need for Life Threatening or Irreversibly Debilitating Diseases or Conditions". www.fda.gov/downloads/MedicalDevices/DeviceRegulationandGuidance/Guidancedocuments/ucm393878.pdf
15. (2014) FDA draft guidance "Balancing Premarket and Postmarket Data Collection for Devices Subject to Premarket Approval". www.fda.gov/downloads/MedicalDevices/DeviceRegulationandGuidance/Guidancedocuments/ucm393994.pdf
16. Olsen D, Jorgensen JT (2014) Companion diagnostics for targeted cancer drugs—clinical and regulatory aspects. Front Oncol 4:105
17. (2013) "Technical Guidance on Development of In vitro Companion Diagnostics and Corresponding Therapeutic Products" PMDA. www.pmda.go.jp/english/service/pdf/in_vitro/PFSB-ELD20131226.pdf
18. Goldstein NS, Hewitt SM, Taylor CR et al (2007) Recommendations for improved standardization of immunohistochemistry. Appl Immunohistochem Mol Morphol 15:124–133
19. Gullick WJ (1991) Prevalence of aberrant expression of the epidermal growth factor

receptor in human cancers. Br Med Bull 47(1):87–98

20. Gravalos C, Jimeno A (2008) HER2 in gastric cancer: a new prognostic factor and a novel therapeutic target. Ann Oncol 19 (9):1523–1529
21. Hofmann M, Stoss O, Shi D et al (2008) Assessment of a HER2 scoring system for gastric cancer: results from a validation study. Histopathology 52:797–805
22. Ruschoff J, Dietel M, Baretton G et al (2010) HER2 diagnostics in gastric cancer—guideline validation and development of standardized immunohistochemical testing. Virchows Arch 457(3):299–307
23. Paz Ares L, Soulieres D, Melezinek I et al (2010) Clinical outcomes in non-small-cell lung cancer patients with EGFR mutations: pooled analysis. J Cell Mol Med 14:51–69
24. (1997) Design control guidance for medical device manufactures published by the FDA on March 11, 1997. www.fda.gov/downloads/MedicalDevices/DeviceRegulationandGuidance.Guidancedocuments/ucm070642.pdf
25. Pallares J, Bussaglia E, Martinez-Guitarte JL et al (2005) Immunohistochemical analysis of PTEN in endometrial carcinoma: a tissue microarray study with a comparison of four commercial antibodies in correlation with molecular abnormalities. Mod Pathol 18:719–727
26. Schrohl AS, Holten-Andersen M, Sweep F et al (2003) Tumor markers: from laboratory to clinical utility. Mol Cell Proteomics 2:378–387
27. CLSI (2011) Quality assurance for design control and implementation of immunohistochemistry assays; approved guideline, 2nd edn. CLSI document I/LA28-A2. Clinical and Laboratory Standards Institute, Wayne, PA
28. Guidance Document for Submission of Immunohistochemistry Application to the FDA. Accessed 3 June 1998. www.fda.gov/downloads/ Medical Devices/DeviceRegulationandGuidance/GuidanceDocuments/ucm094015.pdf

Methods in Pharmacology and Toxicology (2015): 305–324
DOI 10.1007/7653_2014_24

Published online: 28 October 2014

Development of Tissue-Based Companion Diagnostics: The Relationship Between the Pharmaceutical Company, Diagnostic Partner, and the Biomarker Laboratory

Mark Kockx, Stefanie de Schepper, and Christopher Ung

Abstract

Companion diagnostics has fully gained its footing in the development cycle of targeted therapeutics. There is universal acknowledgement for the need of companion diagnostics from regulatory bodies, drug development organizations, physicians, and insurance payers. The companion diagnostic plays an integral role in selecting or excluding patients for a specific therapy and its development is a significant endeavor. The drug developer and the diagnostic manufacturer have visible roles in this undertaking. The role of a third entity—the biomarker laboratory—is vital to the successful regulatory approval; however its responsibilities are less known. When utilized well, the biomarker laboratory can serve as a potent development partner for both the drug developer and the diagnostic manufacturer from the origin of the assay to its fully realized function as a companion diagnostic. In this chapter we describe the roles of the biomarker laboratory, examine the required infrastructure and skills sets, distinguish the differences between itself and the traditional diagnostic laboratory, and provide guidelines of operations based on our experiences with multiple companion diagnostics programs.

Key words Companion diagnostic, Biomarkers, Personalized oncology, IHC, Immunohistochemistry, Assay development, PMA, 510(k), Targeted therapy

1 Introduction

Historically, pharmaceutical companies unilaterally controlled the discovery, development, marketing, and post-marketing surveillance of their oncologic therapeutic drug products. Pathology laboratories and diagnostic entities were minimally involved, if at all, in the drug development cycle.

The landscape changed dramatically in 1998 with the introduction of HercepTest™, ushering in the era of personalized oncology with its inherent need for companion diagnostics. Two stakeholder groups entered the development arena. Pathologists became essential for the morphological interpretation of tissue-based companion diagnostics and providing guidance in the selection of treatment. Diagnostic assay manufacturer companies, with its role as the developer and distributor of these tissue biomarker assays, emerged as vital contributors. These new roles and the new rules imposed new interplays among the diagnostic constituents and the companion diagnostics industry birthed, often untidily, its own set of guidelines and requirements. Downstream, the regulatory agencies, such as the Food and Drug Administration (FDA) shifted its own interface, coordinating its drug arm (CBER) and its medical device arm (CDRH) to review submissions. At the far end of this development and distribution chain, insurance companies, clinics, and patients all grasped to understand the dynamics of co-pays and reimbursement of tests that appeared to increase in costs overnight.

Today while there is broad acknowledgement of this sea change and the multi-priorities that need to come together, the work that is required to develop a companion diagnostic, obtaining regulatory approval in global markets and commercializing it in tandem with pharmaceutical entities remains daunting. It is an orchestra that requires dedication, foresight and a commitment among those involved to ensure the harmony that can lead to real benefit for the cancer patient.

In this chapter the authors describe the activities that lead to the development and commercialization of a companion diagnostic. This arena has seats with different vantage points—e.g. development, validation, service, regulatory, and commercialization. The authors have been able to participate from these various perspectives having collectively worked on all the approved tissue-based companion diagnostics in the USA, Europe, and China. Much has been written on this subject from the perspective of pharmaceutical companies and diagnostic manufacturers but perhaps less so, from the biomarker laboratory's perspective. This chapter emphasizes the biomarker laboratory's participation. The use of the term "biomarker laboratory" is intentional. The chapter illustrates the differences in skills set, daily operations, and mindset between a routine diagnostic laboratory with a biomarker laboratory that are necessary to provide development services for companion diagnostics.

2 New Models of Oncologic Drug Development

Worldwide, regulatory agencies expect the inclusion of a companion diagnostic as part of a new targeted therapeutic application [1]. This expectation necessitates the inclusion of diagnostic companies and biomarker laboratories, thus creating the tripartite—the drug company, the diagnostic manufacturer, and the biomarker laboratory. Pharmaceutical companies typically trigger the need for a companion diagnostic and hold responsibility for their therapeutic agent, which includes the companion diagnostic. To that end, the pharmaceutical companies rely on one of two approaches. The first is the vertical integration of the diagnostic test where the pharmaceutical company owns the diagnostic entity. This is most vividly seen with Roche Pharmaceuticals and its deployment of diagnostic tests through wholly owned entities such as Ventana and Roche Molecular Systems. The second method relies on partnerships held together by agreements based on common interests. A recent example is seen with the approval of GlaxoSmithKline's dabrafenib and trematinib based on BioMerieux's development of the THxID®-BRAF real-time PCR test [2]. The former method offers a greater degree of control but is costly and requires management of diverse businesses, each with its own profit goals. The latter is attractive since it is much more economical yet the penalties are often surprises that are hatched from the differing nuances of stakeholder priorities. Both methods have successfully brought companion diagnostics to the market. Both methods have also suffered setbacks.

Regardless of the selected model, clear communication and co-operation between all stakeholders, the pharmaceutical company, the assay developer, and the biomarker lab, is essential in planning and execution of studies and clinical trials. The mixture of complex content, deep financial investments and consequences, tight and often inflexible timelines, and a highly regulated environment amplify the tension that naturally exists between multiple parties. Thus, the often-elusive elements of recognition, respect and trust between the partners are vital to ensure that the companion diagnostics project is effectively implemented in the drug development cycle.

Finally, even as the industry is adapting to the evolved model of companion diagnostics, at least two new forces are clamoring for additional adaptation. The economical sequencing of the human genome along with powerful technological innovations lends greater confirmation that multiplatform biomarker analyses can create greater targeting. The immunotherapy revolution drives this conclusion powerfully as it demands assessments of immune entities together with the tumor itself. It also forces us to question the "best biomarker" for predicting patient benefit especially in combination with standard targeted therapies.

3 The Companion Diagnostic Device

A companion diagnostic device is an in vitro diagnostic (IVD) device that provides information that is essential for the safe and effective use of a corresponding therapeutic product [3]. An IVD companion diagnostic device would be employed in the following scenarios:

- Identify patients who are most likely to benefit from a particular therapeutic product.

- Identify patients likely to be at increased risk for serious adverse reactions as a result of treatment with a particular therapeutic product.
- Monitor response to treatment for the purpose of adjusting treatment (schedule, dose, discontinuation) to achieve improved safety or effectiveness.

The use of a companion IVD device with a particular therapeutic product is stipulated in the instructions for use in the labeling of both the diagnostic device and the corresponding therapeutic product, as well as in the labeling of any generic equivalents of the therapeutic product.

4 In Vitro Diagnostic (IVD) Device Versus the Laboratory-Developed Test (LDT) as Companion Diagnostics

A companion diagnostic test may be developed in one of the ways: as an in vitro diagnostic (IVD) device or as a laboratory-developed test (LDT). The difference between an IVD and an LDT companion diagnostic test is its path to market and the official body authorized to regulate it. Under the current US regulatory framework, molecular diagnostic tests (including genetic tests) may enter the market via compliance with the rules that govern either IVD or LDT development.

An IVD is developed and manufactured by device manufacturers under the Quality System Regulation (QSR) and normally require pre-market approval (PMA) or 510(k) notification prior to sale and distribution. The FDA has the authority to regulate devices that are distributed as IVDs as well as those that are distributed for "Investigational Use Only" (IUO) and for "Research Use Only" (RUO).

An LDT is "an IVD that is intended for clinical use and designed, manufactured and used within a single laboratory [4]." Therefore, a diagnostic test may be developed and utilized by a Clinical Laboratory Improvement Act (CLIA)—certified laboratory under the auspices of their Quality Management System. A CLIA laboratory presumably would develop these tests especially for unmet diagnostic needs and for their own exclusive use. Under CLIA guidelines, laboratories assay the performance characteristics and perform the analytical validation for these LDTs in order to obtain licenses to offer them as diagnostic services. The Centers for Medicare & Medicaid Services (CMS) regulates clinical laboratories and LDTs according to guidelines in the Clinical Laboratory Improvement Act of 1988 (CLIA'88). In addition to the adherence to these regulations, laboratories must be certified by the CLIA'88 and are subjected to regular inspections.

The role of the FDA with LDTs is a subject of intense discussion and impassioned discourse as the mapping of the human genome and the parallel availability of innovative platforms have dramatically increased the use of LDTs. In addition, the "single laboratory" clause becomes increasingly clouded as laboratories acquire multiple entities and declare in-licensed diagnostic tests from another entity as LDTs. Additionally, the complexity of LDTs have increased substantially and often contain software driven, "black-box" algorithms. Despite two citizen petitions [5] and years of "enforcement discretion" of LDTs, the FDA announced its intention [6] to regulate LDTs in July 2014. The proposed framework would phase in FDA regulation over 9 years, with the level of premarket and postmarket oversight varying according to risk. The emphasis on "highest-risk" tests naturally includes companion diagnostics—the crucial element to personalized medicine for oncology and other conditions. The implementation of FDA regulation over LDTs will introduce another dynamic with far-reaching impact across all the constituents of the companion diagnostics industry.

5 The Regulatory Phases of an IVD as a Medical Device

In the eyes of the FDA, an IVD companion diagnostic is viewed as a medical device and subject to medical device classification and regulation. The FDA defines a medical device as an instrument, apparatus, implement, machine, contrivance, implant, in vitro reagent, or other similar or related article, including a component part, or accessory, which is:

- Recognized in the official National Formulary, or the United States Pharmacopoeia, or any supplement to them.
- Intended for use in the diagnosis of disease or other conditions, or in the cure, mitigation, treatment, or prevention of disease, in man or other animals.
- Intended to affect the structure or any function of the body of man or other animals, and which does not achieve its primary intended purposes through chemical action within or on the body of man or other animals and which is not dependent upon being metabolized for the achievement of any of its primary intended purposes.

Ideally, a therapeutic product and its corresponding IVD companion diagnostic device would be developed contemporaneously, commonly referred to as co-development. In this endeavor, the clinical performance and clinical significance of the IVD companion diagnostic device are established using data from the clinical

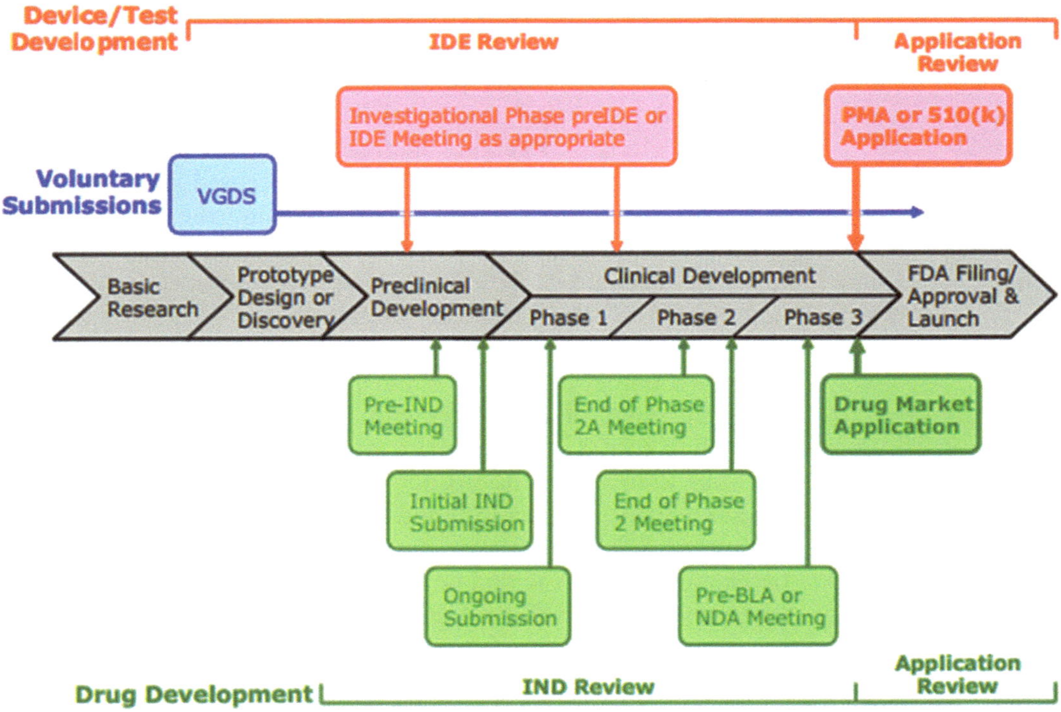

Fig. 1 Prototype of an idealized approach to developing and regulating combined diagnostic tests and drugs. Drug-Diagnostic Co-Development/Concept Paper. http:/www.fda.gov/downloads/drugs/scienceresearch/researchareas/pharmacogenetics/ucm116689.pdf

development program of its corresponding therapeutic product. However, the FDA recognizes that there may be cases when parallel development may not be possible (*see* Fig. 1).

6 IVD/Medical Device Classification

Device classification depends on the intended use and indications for use of the device. A major component of device classification is the risk the device poses to the patient and/or the user where class I includes devices of lowest risk and Class III includes those with the greatest risk.

Pre-market approval (PMA) is the FDA process of scientific and regulatory review to evaluate the safety and effectiveness of Class III medical devices. Class III devices are those that support or sustain human life, are of substantial importance in preventing impairment of human health, or are which present a potential, unreasonable risk of illness or injury. Due to the level of risk associated with Class III devices, the FDA has determined that general and special controls alone are insufficient to assure the safety and effectiveness of class III devices. PMA is the most stringent type of device marketing application required by FDA. The applicant must receive FDA approval of its PMA application prior to marketing the device. PMA approval is based on a determination by FDA that the PMA contains sufficient valid scientific evidence to assure that the device is safe and effective for its intended use(s).

In the event that PMA assessment is not required for a Class I, II, and III device intended for human use, a 510(k) application must be submitted to FDA. A 510(k) is a pre-market submission made to FDA to demonstrate that the device to be marketed is at least as safe and effective and substantially equivalent to a legally marketed device (21 CFR 807.92(a)(3)) that is not subject to PMA. Applicants must compare their device to one or more similar, legally marketed devices and make and support their substantial equivalency claims. In the July 31 guidance, the FDA states in a footnote that "most IVD companion diagnostics will be Class III devices."

7 The "Tri-Party" of Companion Diagnostic Assay Development

The development phase of a companion diagnostic assay brings three players in close proximity: the drug company developing the Investigational New Drug (IND), the manufacturer of the companion diagnostic assay that develops the medical device and, finally, the biomarker laboratory which evaluates the companion diagnostic during a clinical trial.

These three partners have different goals and interests.

The pharmaceutical partner must demonstrate that its applicant therapeutic is safe and effective through submission of a New Drug Application (NDA) via clinical trials designed. The labeling of the drug must comprise the intended use of the IND in association with the outcome of the companion diagnostic test. The final goal is the regulatory approval of the NDA.

The diagnostic partner needs to prove the diagnostic utility of the assay to support the therapeutic effect of the IND. Specifically, the diagnostic sponsor must show that the product is analytically and clinically validated and has clinical utility. The labeling of the companion diagnostic test must comprise its intended use in association with the use of the IND. The final goal is the regulatory approval of the companion diagnostic test.

The biomarker laboratory has to demonstrate the analytical utility of the companion diagnostic test. At the behest of either the pharmaceutical and/or the diagnostic partner, the biomarker laboratory performs the assay and collects evidence (reliability, robustness and accessibility) to support the intended use of the test. The final goal is to use the companion diagnostic test according to its detailed Instructions for Use (IFU) and obtain data with rigorous documentation that becomes part of the regulatory application package.

In spite of different priorities, the partners are tethered by a shared focus on the tissue biomarker's potential diagnostic application or therapeutic support. Each partner is essential to the process and brings a unique set of skills, experience and capabilities to the development pipeline. Meanwhile, as the constituents adjust to the optimal engagement practices, the biomarker evolution continues to demand dynamic adjustments.

8 The Role of the Biomarker Laboratory in Companion Diagnostics Development

The pharmaceutical or diagnostic partner collaborates with a biomarker laboratory in the development of a companion diagnostic assay. The biomarker laboratory may serve several functions: (1) prototype development of the assay, (2) independent validation of the companion diagnostic, (3) clinical site for the companion diagnostic registration trial, and (4) as an early launch site for the newly approved companion diagnostic.

Immunohistochemistry (IHC), for instance, is a common companion diagnostic platform but the validation of an IHC biomarker is nontrivial. An experienced biomarker lab will have superior insight and experience into preliminary assessment, selection, optimization, and parameters that govern the eventual utility of an IHC assay. The success of any IHC assay is dependent upon the quality of the specimen [7], which is governed by the tissue source and how it was handled, fixed, and processed prior to testing. It is important that the effect of inadequate processing on the IHC performance is checked and that guidelines, with possible exclusion criteria, are provided to guarantee consistent, sufficient sample quality, and reproducible results.

A well-designed IHC companion diagnostic assay will be robust enough to thrive in the complicated, time-scarce, and high-volume environment of a diagnostic clinical laboratory. The biomarker laboratory's workflow and logistics input on the IHC assay can greatly improve its scalability, stability and ease of use. To this end, development and optimization are best performed on an automated staining platform using ready-to-use reagents. The pathologists of an experienced biomarker laboratory will have indispensable input on the type of controls, positive and negative—preferably tissue slides with predefined staining—which will become important components of the assay.

The final critical aspect of IHC assay development is the interpretation of the staining outcome. A clear-cut scoring system must be implemented, preferably by on-site training or via interactive live meetings. The biomarker laboratory can advise on the feasibility and effectiveness of the scoring system, and assist in designing proficiency assessments program that can be used upon regulatory approval of the companion diagnostic.

These capabilities can be incorporated into the prototype development and subsequent analytical validation. Then as the companion diagnostic takes form as an Investigational Use Only device, ideally these early investments pave the way for the biomarker laboratory to serve as the registration clinical trial site. As a final step, the biomarker laboratory can contribute significantly, offering itself as a knowledgeable and experienced diagnostic laboratory for the newly approved companion diagnostic. In this post-approval step, the biomarker laboratory transforms into a diagnostic laboratory with the intent to ensure the companion diagnostic is performed effectively and patients are appropriately included in treatment. It also serves as a reference center for additional testing and an educational resource for routine diagnostic laboratories. The authors have engaged in collaborations as described above with a high degree of success.

9 Implementing Companion Device Clinical Trials in the Biomarker Laboratory

The assays that are developed with the aim to be used as medical device require medical device clinical trials (510k or PMA) as part of the application package to the FDA. Typically, the diagnostic partner collaborates with a pharmaceutical partner(s) in clinical trials where the assay is used to screen participants or assess the efficacy of the therapeutic. The device manufacturer transfers this companion diagnostic candidate assay to one or more external testing laboratories where the testing would be performed. Thus, these labs represent the third or even fourth party in this collaboration.

The biomarker laboratory then performs the test as described in the study protocol and according to Good Clinical Laboratory Practice (GCLP) guidelines [8]. All deviations from the protocol are documented and reported to the diagnostic partner. These deviations take multiple forms given the numerous compliance requirements. Examples include improper fixation, insufficient specimens, improper labeling of specimens and non-matching patient information. The frequency and variety of deviation often accelerate and accentuate the complexity of this exercise. The probability of success increases if the diagnostic manufacturer and the biomarker laboratory have clearly defined processes for routine operations in addition to thorough contingency planning. The biomarker laboratory depends on the diagnostic manufacturer for the regular and timely supply of lot-controlled and consistent reagents on well-supported instrumentation. The diagnostic manufacturer relies on the biomarker laboratory for high quality testing practices and its experience with real-world highly variable patient specimens. Both the diagnostic manufacturer and the biomarker laboratory work in concert to meet the regulatory requirements, time lines, and financial obligations imposed by the pharmaceutical partner. The establishment of regular tripartite communications with a direct and open agenda is necessary to rapidly formulate solutions to the inevitable logistics and technical issues that emerge.

10 Surviving the PMA Clinical Trial: The Biomarker Laboratory's Perspective

This section highlights the additional responsibilities and substantial challenges that the biomarker laboratory faces compared to a routine pathology laboratory when performing a PMA clinical trial. The following steps are crucial to successful survival of a PMA (or 510(k)).

Step 1—Ethical Approval
The pathologist of the biomarker laboratory usually assumes the role of a principle investigator for the medical device registration. This requires obtaining approval of the local ethics committee to perform the trial.

Step 2—Analytical Plan
The study parameters are described in two seminal documents: the study protocol outlined by the diagnostic partner and the analytical plan, which provides a detailed description of the laboratory's activities. These documents are the basis to define responsibilities of the biomarker laboratory. The workflow and all procedures must be defined and described in detail in the analytical plan, which is written collaboratively by the diagnostic partner and the biomarker laboratory's Quality Assurance (QA) department. For example, the analytical plan may contain the following sections: study identification and design, the sample and data flow, laboratory requirements, the analytical processes, maintenance of records, reporting requirements, QA, and the communication plan.

Step 3—Instrument Set-up
The candidate companion diagnostic assay commonly includes an instrument platform with analysis software, such as an automated staining platform or a PCR cycler. The diagnostic partner may elect to use a dedicated instrument and provide the requisite units to the biomarker laboratory or it may also use an existing platform at the biomarker laboratory site. In either case, the diagnostic partner manufacturer would assume the responsibility of installing the instrument (if not already on-site), performing installation qualification (IQ), operational qualification (OQ), and performance qualification (PQ) of the instrument. Furthermore, the diagnostic partner is responsible for the provision of guidelines and training for correct use and maintenance of the instrument and to repair or replace the instrument if needed. Finally, an arrangement must be

made as to actions required in case of an emergency, an adverse event or a technical breakdown that may compromise the turnaround time of the test results.

Step 4—Training and Assay Implementation
After installing and qualifying the instrument, the diagnostic partner will train laboratory personnel on the instrument, the correct application of the candidate medical device kit and assay results interpretation. The training tracks a qualification plan containing clear success criteria for the trainees. When the criteria are met, the trainee obtains a certificate or a validation report declaring the qualification of the individual to perform the assay and/or read the assay. Some device manufacturers do not subscribe to the "train the trainer" approach and in such cases, the laboratory will need to allocate sufficient personnel for the training session. To meet the tight turnaround times required by companion diagnostics device trials, the laboratory will also need to ensure that sufficient staff is trained to guarantee continuity both for performance and interpretation of the assay. For qualitative or semi-quantitative tests such as IHC, subjectivity is an inevitable contributor to results variability. Proficiency programs based on a known test set are useful as a standardization assessment tool for both the diagnostic laboratory and the biomarker laboratory. These can be administered at predefined intervals across the biomarker laboratory's pathology team.

Step 5—Assay Translation
Before initiating the training programs, the biomarker laboratory may choose to translate all the companion diagnostic clinical trial technical procedures including standard operating procedures (SOPs), work instructions, and accompanying forms to its documentation system. This exercise surfaces important details that may have been overlooked or neglected. While seemingly insignificant, clarifying issues such as storage conditions or the policy for using different lot numbers, may allow the biomarker laboratory to initiate the clinical trial and maintain a smooth process.

All documents must be compliant with the quality standards of both the diagnostic partner and the biomarker laboratory. The laboratory sends these procedure documents to the diagnostic partner for final review and approval before implementation. This step crucially identifies small nonconformities that may have been introduced during translation of the procedure to the laboratory's workflow and information system. Again, direct communication between the diagnostic partner and the biomarker laboratory is a key component to a successful assay transfer.

Step 6—Reporting
In most studies, the results are reported in a case report form (CRF), which can be paper based or automated electronically. A paper-based reporting system requires the pathologist to

complete and to sign the report form. The paper-based system is often subjected to double data entry procedures to ensure transcription accuracy. Finally, the completed form must be sent to both the diagnostic and the pharmaceutical partner. The use of automated reporting reduces labor and inaccuracies and the diagnostic partner may offer this feature. Alternatively, the laboratory can leverage its 21CFR part 11 compliant Laboratory Information Management System (LIMS) to customize the design of the report form to match the required format of the reporting as described in the study protocol.

Step 7—Monitoring
The diagnostic partner is required to monitor the progress of the clinical trial at the laboratory. This typically begins with the site initiation visit upon completion of all preparations including the protocol transfer, instrument and assay installation and qualification, training, and SOP clearance. During the course of the trial, the diagnostic partner has to monitor the testing laboratories on a regular basis and at least twice a year. During a monitoring visit, the diagnostic partner reviews all paperwork including sample reception documents, staining run reports, report forms and records of protocol deviations or other irregularities. The biomarker laboratory requires frequent contact with the diagnostic partner, clarifying emerging issues as the specimens arrive at the laboratory. This active communication allows the biomarker laboratory to process specimens effectively, maintaining crucial results delivery times while coordinating the required documentation. This vital communication dynamic between the test manufacturer and the biomarker laboratory will enable a "no-surprises" monitoring audit.

11 Impact on the Workload and Laboratory Workflow

Although the testing processes are usually very straightforward, the impact on the workload and laboratory workflow should not be underestimated. The increase in workload is mainly caused by the burden of administration that accompanies a PMA/510(k) trial. In addition, the biomarker laboratory frequently forfeits the opportunity to gain efficiencies through batch testing due to the requirement of short turnaround time between sample receipt and reporting.

We have repeatedly emphasized communication as an element of success. We espouse the following laboratory team makeup to ensure a successful outcome. The project manager and the laboratory scientist—two dedicated project individuals—lead this team. The former coordinates communication with the pharmaceutical and diagnostic partner while the latter oversees all technical aspects of the assay with the diagnostic partner. These two individuals guide the efforts of at least 3–4 laboratory technologists who have been trained by the diagnostic partner. Together, this team is responsible for continuous testing and results delivery and to ensure that the trained lab technicians are current on their skills and knowledge. Finally, a qualified pathologist closely monitors the quality control slides of each staining run.

As described earlier, all deviations from the protocol must be reported and documented. These deviations are quite varied and the testing laboratory should consult with the diagnostic manufacturer to adopt the appropriate action(s) and document it accordingly for each case. For example, since PMA/510(k) trials contain criteria to include/exclude paraffin slides for testing, such as glass type and section thickness. When non-compliant slides arrive (e.g., the glass type differs from the study protocol requirement), the laboratory would inform the diagnostic partner and await feedback before further testing can be started, or terminated. The diligent adherence to these details specified in the Instructions for Use is an important hallmark of an effective testing laboratory.

12 Final Recommendations

We provide some final thoughts for a successful PMA/510(k) venture:

12.1 Invest in Pre-trial Communication

The start-up of the trial is the most intense period of communication with the diagnostic and pharmaceutical partners. In this first phase, organize weekly calls with each partner and monthly calls with the entire team. During this period there will be many issues to discuss, including the details of the analytical plan and communication map, how data and results will be transferred and shared, the progress of training, assay transfer, and ethical approval and the method of assay quality control.

12.2 Form Mutually Supportive Working Relationships

The tripartite of pharmaceutical partner, diagnostic partner and biomarker laboratory must form an extended team with varying but compatible goals. The working relationships with partners, which are forged in the pre-trial period, will facilitate the communication and support required during the trial. Regular, direct contact with both partners will be necessary. Issues regarding sample quality and accession will require contact between the sample management team and the pharmaceutical partner. The technical issues with regard to the assay will be the domain of the diagnostic partner.

12.3 Validate Documentation

Besides communication, documentation is a crucial component of a successful PMA trial. All procedures must be written, and approved by the diagnostic partner and internal QA and medical director before the trial starts. In addition, all documents related to the trial must be version controlled, valid and always up to date. All deviations from the protocol or the procedures must be documented, either as a protocol deviation or a note to study file. Management of this documentation system and preexisting templates need to be in place. The specimen and data custody process are tested and revalidated to avoid mislabeling or incorrect reporting.

12.4 Define Responsibilities

Individual responsibilities as outlined in the analytical plan must be very clearly communicated. All staff involved must list their role in the study and complete the delegation of functions form indicating their study-related tasks. The biomarker laboratory management should hold an internal meeting with all participants before a study commences to clarify final processes and activities.

13 FDA Audits and Inspections

These processes and documentation, when well described and implemented, will incline a positive outcome for the on site FDA inspections and data audits conducted by Bioresearch Monitoring (BIMO) Program. These audits are designed to monitor all aspects of the conduct and reporting of FDA regulated research. The BIMO Program was established to assure the quality and integrity of data submitted to the agency in support of new product approvals, as well as, to protect the rights and welfare of the thousands of human subjects involved in FDA regulated research

If no major issues are identified, the FDA representative(s) will list observations made during the inspection of your facility. They are inspectional observations, and do not represent a final agency determination regarding compliance.

14 Conclusion

Elizabeth Mansfield, Deputy Office Director for Personalized Medicine at FDA, recently stated that companion diagnostics are "here to stay" and furthermore, the PMA pathway is the most likely journey to approval for a candidate companion diagnostic [9]. From concept to commercialization, the journey requires commitments from multiple parties based on the promise of effective patient selection. The biomarker laboratory plays a significant support role to obtain the validation, analytical and clinical data required for regulatory approval. This endeavor is usually deployed under tight timelines for the overall project, very short test turnaround times for patient enrollment, and a rigorous document administration infrastructure. With each experience of supporting a companion diagnostic clinical trial, a biomarker laboratory can improve its personnel and processes to effectively manage emerging issues. Importantly, it does so by anticipating the problems that may emerge as patients are recruited and is able to administer solutions using the critical experience it has gained in addition to optimizing its internal processes prior to trial initiation. Its ability to fluently converse in tripartite language between the parties is also an important contributor to success.

15 Examples of Existing Tissue-Based Companion Diagnostics

The table below shows examples of cleared or approved companion diagnostic devices/for an updated list consult the FDA website:

http://www.fda.gov/MedicalDevices/ProductsandMedicalProcedures/InVitroDiagnostics/ucm301431.htm

	Trade name	PMA	Manufacturer	Intended use (IU)/indications for use (IFU)
1	INFORM HER-2/ NEU	P940004 S001	Ventana Medical Systems, Inc.	The Inform Her-2/Neu gene detection system is a fluorescence in situ hybridization (FISH) DNA probe assay that determines the qualitative presence of Her-2/Neu gene amplification on formalin-fixed, paraffin-embedded human breast tissue as an aid to stratify breast cancer patients according to risk for recurrence or disease-related death. It is indicated for use as an adjunct to existing clinical and pathologic information currently used as prognostic indicators in the risk stratification of breast cancer in patients who have had a priori invasive, localized breast carcinoma and who are lymph node negative
2	PATHVYSION HER-2 DNA Probe Kit	P980024 S001-S008	Abbott Molecular Inc.	The PathVysion HER-2 DNA Probe Kit (PathVysion Kit) is designed to detect amplification of the HER-2/neu gene via fluorescence in situ hybridization (FISH) in formalin-fixed, paraffin-embedded human breast cancer tissue specimens. Results from the PathVysion Kit are intended for use as an adjunct to existing clinical and pathologic information currently used as prognostic factors in stage II, node-positive breast cancer patients. The PathVysion Kit is further indicated as an aid to predict disease-free and overall survival in patients with stage II, node-positive breast cancer treated with adjuvant cyclophosphamide, doxorubicin, and 5-fluorouracil (CAF) chemotherapy. The PathVysion Kit is indicated as an aid in the assessment of patients for whom Herceptin (trastuzumab) treatment is being considered (see Herceptin package insert)

(continued)

(continued)

	Trade name	PMA	Manufacturer	Intended use (IU)/indications for use (IFU)
3	PATHWAY ANTI-HER-2/NEU (4B5) Rabbit Monoclonal Primary Antibody	P990081 S001-S013	Ventana Medical Systems, Inc.	Ventana Medical Systems' PATHWAY Her2 (clone CB11) is a mouse monoclonal antibody intended for laboratory use for the semiquantitative detection of c-erbB-2 antigen in sections of formalin-fixed, paraffin-embedded normal and neoplastic tissue on a Ventana automated immunohistochemistry slide staining device. It is indicated as an aid in the assessment of breast cancer patients for whom Herceptin treatment is being considered
4	DAKO EGFR PharmDx Kit	P030044 S001-S002	Dako North America, Inc.	The EGFR pharmDx™ assay is a qualitative immunohistochemical (IHC) kit system to identify epidermal growth factor receptor (EGFR) expression in normal and neoplastic tissues routinely fixed for histological evaluation EGFR pharmDx specifically detects the EGFR (HER1) protein in EGFR-expressing cells EGFR pharmDx is indicated as an aid in identifying colorectal cancer patients eligible for treatment with Erbitux (cetuximab) or Vectibix (panitumumab)
5	DAKO C-KIT PharmDx	P040011 S001-S002	Dako North America, Inc.	The c-Kit pharmDX assay is a qualitative immunohistochemical (IHC) kit system used on the Dako Autostainer, for the identification of c-kit protein/CD 117 antigen (c-kit protein) expression in normal and neoplastic formalin-fixed paraffin-embedded tissues for histological evaluation. The c-Kit pharmDX rabbit polyclonal antibodies specifically detect the c-kit protein in CD 117 antigen-expressing cells The c-Kit pharmDx is indicated as an aid in the differential diagnosis of gastrointestinal stromal tumors (GIST). After diagnosis of GIST, results from c-Kit pharmDx may be used as an aid in identifying those patients eligible for treatment with Gleevec/Glivec (imatinib mesylate) Results from hematoxylin and eosin (H&E) stains and a panel of antibodies can aid in the differential diagnosis of GIST. Interpretation must be made by a qualified pathologist, within the context of a patient's clinical history, proper controls, and other diagnostic tests

(continued)

(continued)

	Trade name	PMA	Manufacturer	Intended use (IU)/indications for use (IFU)
6	INSITE HER-2/ NEU KIT	P040030	Biogenex Laboratories, Inc.	InSite Her-2/neu Mouse Monoclonal Antibody (Clone C1B11) kit is intended for In Vitro Diagnostic use in Immunohistochemistry (IHC) assays to semiquantitatively localize by light microscopy the over-expression of Her-2/neu (i.e., c-erbB-2) in formalin-fixed, paraffin-embedded normal and neoplastic tissue sections. InSite Her-2/neu is indicated as an aid in the assessment of breast cancer patients for whom Herceptin (trastuzumab) therapy is being considered. Clinical interpretation of InSite Her-2/neu immunostaining results (absence or presence; semi-quantitative intensity score) should be complemented by appropriate controls and morphological tissue analysis and be evaluated by a qualified pathologist within the context of patient clinical history and other diagnostic results
7	SPOT-LIGHT HER2 CISH Kit	P050040 S001-S003	Life Technologies, Inc.	For In Vitro Diagnostic Use: The SPOT-Light HER2 CISH Kit is intended to quantitatively determine HER2 gene amplification in formalin-fixed, paraffin-embedded (FFPE) breast carcinoma tissue sections using chromogenic in situ hybridization (CISH) and bright-field microscopy This test should be performed in a histopathology laboratory The SPOT-Light HER2 CISH Kit is indicated as an aid in the assessment of patients for whom Herceptin (trastuzumab) treatment is being considered. The assay results are intended for use as an adjunct to the clinicopathological information currently being used as part of the management of breast cancer patients. Interpretation of test results must be made within the context of the patient's clinical history by a qualified pathologist
8	Bond Oracle Her2 IHC System	P090015	Leica Biosystems	The Bond Oracle Her2 IHC system is a semi-quantitative immunohistochemical (IHC) assay to determine Her2 (human epidermal growth factor receptor 2) oncoprotein status in formalin-fixed, paraffin-embedded breast cancer tissue processed for histological evaluation following automated staining on the bond-max slide staining instrument. The Bond Oracle Her2 IHC system is indicated as an aid in the assessment of patients for whom Herceptin (trastuzumab) treatment is being considered

(continued)

(continued)

	Trade name	PMA	Manufacturer	Intended use (IU)/indications for use (IFU)
9	HER2 CISH PharmDx Kit	P100024 S001-S002	Dako Denmark A/S	HER2 CISH PharmDx kit is intended for dual-color chromogenic visualization of signals achieved with directly labeled in situ hybridization probes targeting the HER2 gene and centromeric region of chromosome 17. The kit is designed to quantitatively determine HER2 gene status in formalin-fixed, paraffin-embedded breast cancer tissue specimens. Red and blue chromogenic signals are generated on the same tissue section for evaluation under bright-field microscopy. The CISH procedure is automated using Dako Autostainer instruments HER2 CISH pharmDx Kit is indicated as an aid in the assessment of patients for whom Herceptin (trastuzumab) treatment is being considered. Results from the HER2 CISH pharmDx Kit are intended for use as an adjunct to the clinicopathologic information currently used for estimating prognosis in stage II, node-positive breast cancer patients This kit is for in vitro diagnostic (IVD) use only
10	INFORM HER2 DUAL ISH DNA Probe Cocktail	P100027 S001-S002	Ventana Medical Systems, Inc.	The INFORM HER2 Dual ISH DNA Probe Cocktail is intended for use in determining HER2 gene status by enumeration of the ratio of the HER2 gene to chromosome 17. The HER2 and chromosome 17 probes are detected using two color chromogenic in situ hybridization (ISH) in formalin-fixed, paraffin-embedded human breast cancer tissue specimens following staining on Ventana BenchMark XT automated slide stainers (using NexES software), by light microscopy. The INFORM HER2 Dual ISH DNA Probe Cocktail is indicated as an aid in the assessment of patients for whom Herceptin (trastuzumab) treatment is being considered This product should be interpreted by a qualified reader in conjunction with histological examination, relevant clinical information, and proper controls This reagent is intended for in vitro diagnostic (IVD) use

(continued)

(continued)

	Trade name	PMA	Manufacturer	Intended use (IU)/indications for use (IFU)
11	VYSIS ALK Break Apart FISH Probe Kit	P110012 S001	Abbott Molecular Inc.	The Vysis ALK Break Apart FISH Probe Kit is a qualitative test to detect rearrangements involving the ALK gene via fluorescence in situ hybridization (FISH) in formalin-fixed, paraffin-embedded (FFPE) non-small-cell lung cancer (NSCLC) tissue specimens to aid in identifying patients eligible for treatment with Xalkori (crizotinib). This is for prescription use only
12	COBAS 4800 BRAF V600 Mutation Test	P110020 S001-S005	Roche Molecular Systems, Inc.	The Cobas 4800 BRAF V600 Mutation Test is an in vitro diagnostic device intended for the qualitative detection of the BRAF V600E mutation in DNA extracted from formalin-fixed, paraffin-embedded human melanoma tissue. The Cobas 4800 BRAF V600 Mutation Test is a real-time PCR test on the Cobas 4800 system, and is intended to be used as an aid in selecting melanoma patients whose tumors carry the BRAF V600E mutation for treatment with vemurafenib
13	HERCEPTEST	P980018 S001-S015	Dako Denmark A/S	For in vitro diagnostic use HercepTest is a semiquantitative immunocytochemical assay to determine HER2 protein overexpression in breast cancer tissues routinely processed for histological evaluation and formalin-fixed, paraffin-embedded cancer tissue from patients with metastatic gastric or gastroesophageal junction adenocarcinoma HercepTest is indicated as an aid in the assessment of patients for whom Herceptin (trastuzumab) treatment is being considered and for breast cancer patients for whom Perjeta (pertuzumab) treatment is being considered (see Herceptin and Perjeta package inserts)
14	HER2 FISH PharmDx Kit	P040005 S001-S008	Dako Denmark A/S	For in vitro diagnostic use HER2 FISH PharmDx kit is a direct fluorescence in situ hybridization (FISH) assay designed to quantitatively determine HER2 gene amplification in formalin-fixed, paraffin-embedded (FFPE) breast cancer tissue specimens and FFPE specimens from patients with metastatic gastric or gastroesophageal junction adenocarcinoma

(continued)

(continued)

	Trade name	PMA	Manufacturer	Intended use (IU)/indications for use (IFU)
				HER2 FISH PharmDx Kit is indicated as an aid in the assessment of patients for whom Herceptin (trastuzumab) treatment is being considered and for breast cancer patients for whom Perjeta (pertuzumab) treatment is being considered (see Herceptin and Perjeta package inserts) For breast cancer patients, results from the HER2 FISH pharmDx Kit are intended for use as an adjunct to the clinicopathologic information currently used for estimating prognosis in stage II, node-positive breast cancer patients
15	*therascreen* KRAS RGQ PCR Kit	P110030	Qiagen Manchester, Ltd.	The *therascreen* KRAS RGQ PCR Kit is a real-time qualitative PCR assay used on the Rotor-Gene Q MDx instrument for the detection of seven somatic mutations in the human KRAS oncogene, using DNA extracted from formalin-fixed paraffin-embedded (FFPE) colorectal cancer (CRC) tissue. The *therascreen* KRAS RGQ PCR Kit is intended to aid in the identification of CRC patients for treatment with Erbitux (cetuximab) based on a KRAS no mutation detected test result

References

1. Mansfield E. Companion diagnostics and targeted therapy for cancer [FDA PowerPoint document]. Targeted Anticancer Therapies 2014 Congress Online Web site: http://tatcongress.org/wp-content/uploads/2014/05/140307-mansfield.pdf
2. FDA (2014) Tramateib and Dabrafenib. [Press release]. http://www.fda.gov/Drugs/InformationOnDrugs/ApprovedDrugs/ucm381451.htm
3. Aug 6, 2014. In vitro companion diagnostic devices. guidance for industry and food and drug administration staff. p. 7. http://www.fda.gov/downloads/MedicalDevices/DeviceRegulationandGuidance/GuidanceDocuments/UCM262327.pdf
4. 31 July 2014. Laboratory developed tests. http://www.fda.gov/MedicalDevices/ProductsandMedicalProcedures/InVitroDiagnostics/ucm407296.htm
5. Petition of the American Clinical Laboratory Association, Docket No. FDA-20130P-0667 (Jun. 4, 2013); Petition of Genentech, Inc., Docket No. FDA-2008-P-0638 (Dec. 5, 2008); Petition of the Washington Legal Foundation (WLF), Docket No. FDA-2006-P-0149 (Sept. 28, 2006)
6. Howard S (2014, July 31) Anticipated Details of the Draft Guidance for Industry, Food and Drug Administration Staff, and Clinical Laboratories: Framework for Regulatory Oversight of Laboratory Developed Tests (LDTs). http://www.fda.gov/downloads/MedicalDevices/ProductsandMedicalProcedures/InVitroDiagnostics/UCM 407409.pdf
7. For some epitopes, staining of intrinsic control structures, such as endothelial cells for PTEN and perineurium for EGFR, can be included as an aid to determine the quality of a specimen
8. April 1996. http://www.fda.gov/downloads/Drugs/Guidances/ucm073122.pdf
9. Mansfield E. Companion diagnostics and targeted therapy for cancer [FDA PowerPoint document]. Targeted Anticancer Therapies 2014 Congress Online Web site: http://tatcongress.org/wp-content/uploads/2014/05/140307-mansfield.pdf. Slide 15

Methods in Pharmacology and Toxicology (2015): 325–343
DOI 10.1007/7653_2015_40

Published online: 18 February 2015

Navigating Regulatory Approval for Tissue-Based Companion Diagnostics

Joseph S. Krueger, Holger Lange, G. David Young, and Steven J. Potts

Abstract

Companion diagnostics requires successful partnership between a company supporting commercialization of a drug and in vitro diagnostics (IVD) manufacturers. This partnership may also include other participants such as a pharmaceutical company sponsor, antibody manufacturers, development firms with diagnostics expertise, reference laboratories, and developers/manufacturers who may create specific components of the device. A previous chapter outlined how such entities work together to deliver an FDA-cleared companion diagnostic, especially emphasizing the work required at the reference laboratory.

This chapter looks in more detail at specific roles and processes involved in delivering a tissue-based FDA premarket cleared (510(k)) or premarket approved (PMA) diagnostic, either as part of a companion diagnostic or separately. The chapter draws on the authors' past experience with seven successful tissue-based breast cancer medical device diagnostic clearances and on-going companion diagnostics (CDx) development programs with pharmaceutical partners. An efficient approach to earning a clearance is to develop a plan of action, working backwards, from what is specifically required in a PMA or 510(k) submission. Each critical part of the process will be discussed, to facilitate the planning of a successful clearance. Since there is not a one-size-fits-all approach to gaining successful registration of the diagnostic, the goal of this chapter is not to form a template of actions and processes, but rather to educate pharmaceutical decision makers about best practices as companion diagnostic programs are planned and executed.

Key words Companion diagnostics, Premarket approval, Tissue biomarkers, Immunohistochemistry, Digital pathology, 510 k

1 Introduction

More than one executive has commented on the complexity of sponsoring and planning both a therapeutic approval and a diagnostic companion. Pharmaceutical companies with decades of experience in successful NDA filings may have few or no personnel in-house with direct diagnostics experience. Additionally, while diagnostic manufacturers may have significant experience in the components of kit or device manufacturing and device regulatory requirements, they may not have significant expertise in how the drug component performs in a manner dependent on the diagnostic. While pharmaceutical and diagnostics can share common application in clinical areas, the objectives and business approaches of the device manufacturer and pharma sponsor are strikingly different. Like personal marriages, the marriage of the drug and the device requires a substantial mutual understanding and means of communication between the two parties. While there is a large field of professional experts devoted to sustaining and improving personal marriages, there does not appear to be sufficient expertise or support in sustaining or improving the marriage between the stakeholders from the drug and device side. Furthermore, there seems to be a general underestimation of the amount of diligent effort which is required to make this marriage successful. As these partnerships continue to be formed and developed across the industry, critical experience lessons are earned at substantive time, effort, and dollar cost. Thus, it is important that each side develop the necessary internal expertise to facilitate this marriage and ensure success of the drug/device effort.

2 Three Companion Diagnostic Development Paths

Apart from the substantial equivalence (510(k)) route, three potential paths can be taken related to the regulatory approval timing of novel companion diagnostic and the therapeutic if PMA is required. The drug and CDx can be approved together (co-development); the diagnostic can be approved after the drug (post-approval); or a new marketing indication can be sought for an approved drug with a new diagnostic (post-marketing). Much of pharmaceutical drug development is managing risk, and while one path may be chosen by a team, it is good practice to have an alternative option for a second approach, should the first face delays. Co-development is the most risky due to interdependency of the drug and diagnostic for simultaneous approval, but often a necessary strategy. The best-laid plans for a co-development approach will drift toward post-approval to eliminate the co-development approach, should the diagnostic hit regulatory roadblocks and the therapeutic show sufficient efficacy in an indication without the diagnostic selecting patients. While post-marketing is the least risky approach, it is only applicable to the drugs with existing indications and proven efficacy in a new indication. For these reasons, most drugs presumptively start with a co-development approach as the primary strategy, with the alternative strategies being implemented only when the circumstances facilitate.

2.1 Co-development

In a co-development strategy, the diagnostic and the therapeutic are developed side-by-side and submitted with simultaneous and interdependent approval. The current examples of co-developed drug/diagnostic combos include vemurafenib (Zelboraf®, Roche/Genentech) and the Roche Cobas BRAF V600E mutation test (Roche/Genentech); and crizotinib (Xalkori®, Pfizer), and ceritinib (Zykadia®, Novartis), and the Abbott Vysis EML4–ALK translocation test. Co-development is the default for a novel targeted therapy where the efficacy hypothesis is naturally based on the presence of the drug target in the disease. For example, with many biologics and particularly antibody-drug conjugates, efficacy is intuitively linked to patients who have a positive test for the therapeutic target. However, even within this example, there are considerable nuances pertinent to the class of drug or its mechanism of action (MOA). For example, with a naked antibody whose MOA is based on functional blocking of protein which is endogenous in healthy tissue but is a disease driver in the context of amplification or overexpression, the diagnostic may serve to select only the highest expressing patients (inclusive diagnostic). In contrast, if the drug target is unique to the disease, the diagnostic may serve only to screen out target negative patients (exclusive diagnostic).

In the case of co-approval the diagnostic is considered critical for improving the likelihood of therapeutic approval (e.g., improving efficacy or minimizing toxicity) and can allow higher flexibility in designing the trials and patient selection cutpoints to support the best clinical trials outcome. There are potential disadvantages to this approach: the higher risk that a failed diagnostic clearance will adversely impact the drug approval; there is higher planning complexity and coordination; and potentially higher cost for the CDx development. Furthermore, the pharmaceutical company will likely bear the brunt of the cost of the diagnostic development, as few diagnostic companies will be interested in internal R&D funding for a diagnostic that requires a yet-to-be approved therapeutic product to establish a clinical need.

The cost of the diagnostic is not limited to the cost of the device manufacturer for development of the diagnostic (if relevant); the cost is also related to the complexity of trials, number of patients needing enrollment, and any risk resulting from trials improperly designed for the performance of the diagnostic. For example, co-development strategies begin early in the discovery phase of a drug, where biomarkers for efficacy are vetted out. Furthermore, instead of early phase clinical trials focusing on toxicity, dosing, and scheduling; they are equally focused on evaluating patient selection strategies. These strategies may require the incorporation of special clinical protocols for biopsy collection or implementation of multiple prototype diagnostic assays prospectively or retrospectively. Finally, for a given patient selection target with low prevalence (as is typically seen), two to three times the number of patients typically needed for a trial are needed to evaluate the diagnostic/efficacy hypothesis in biomarker positive and negative patients. For these reasons, the amount of effort, complexity, cost, and risk is at least double that of a trial not evaluating a patient selection strategy.

2.2 Post-approval

In a post-approval strategy, the diagnostic is submitted independently following drug approval. The CDx is thus dependent on retrospective drug trial results, but the drug must be approved without any dependence on the diagnostic. Examples include EGFR IHC and K-ras mutations in colorectal cancer (Vectibix, Erbitux), EGFR mutations in NSCLC (Tarceva), and HER2 FISH in breast (Herceptin). The advantages to this approach are primarily logistics and cost saving for the diagnostic development. The CDx program does not have to be tied to complex milestones on the NDA path, diagnostic vendors may be more interested in sharing the development cost, and the launch and development timelines can be planned according to diagnostic rather than therapeutic needs. The disadvantage is that the CDx is not available to assist in the drug approval—executives should be certain that the drug can be approved without further patient stratification.

The above examples of early CDx cases demonstrate where drug efficacy for a given indication was sufficient for approval, or existing clinical tests available which serve to stratify patients by diagnosis. Vectibix, Erbitux, and Tarceva all had demonstrated efficacy in an indication, but were shown to be specifically beneficial or ineffective in particular patients with a mutation based on clinical trials ran after approval. In the case of HER2 FISH, it was added to breast cancer testing for Herceptin as part of "reflex testing" required for equivocal cases using IHC.

Outside of these cases, this approach is typically reserved for drugs which have been prioritized based on an unmet medical need, such as Orphan Drugs, drugs which have a fast-track approval status, and drugs which have a Special Protocol Assessment (SPA) mechanism to facilitate drug approval with the accomplishment of specific criteria. As the breadth of targeted therapies grows, we can expect to see many more examples of this approach employed.

2.3 Post-marketing

In a post-marketing approval, a new label is filed for an existing drug with an existing diagnostic. This is typically part of drug lifecycle management efforts, where the market for a drug can be increased with new indications or exclusivity prolonged through the gain of a new indication with a patient selection rationale. An example in IHC would include the use of Herceptin in gastric cancer with a modified HER2 test. This approach is common when a new tissue or cancer type might be targeted, as the Herceptin example indicates. This approach is the least risky and most straightforward to execute and manage, but the disadvantage is that the number of situations where such an approach may be utilized is limited.

Not all targeted therapies necessarily require CDx approval like the examples above. There are two examples of targeted therapies which were approved without a CDx requirement, despite their indication being classified by a molecular marker, and their MOA being dependent on expression of a particular biomarker: Biogen's anti-CD20 Rituximab (Rituxan, MabThera, and Zytux) in B-cell non-Hodgkin lymphomas and Seattle Genetics anti-CD30 Brentuximab Vedotin (Adcetris®) Antibody Drug Conjugate (ADC) in Hodgkin lymphoma (HL) and Systemic Anaplastic Large Cell Lymphoma (sALCL). These two examples demonstrate the difference in strategy between a diagnostic-guided decision to treat a patient and a medically guided decision to treat a patient, despite both relying on a test to identify the patient population. Normal clinical practice for lymphoma is to perform a series of molecular (flow cytometry, histology) based tests to identify the subtype. This is the key discriminating factor: in both cases of Rituximab and Brentuximab Vedotin, the patient selection was based on the indication, which is defined by a clinical test at the time of diagnosis to define the subtype of disease; not a test which is used to select patients for therapy (although the intent is very similar).

3 PMA Approval Versus 510 k Clearance

There are three levels of review for IDEs—Premarket Approval (PMA), De Novo, and Classic 510 k classifications. PMAs require a new demonstration of safety and effectiveness and include all Class III devices, including novel devices. The burden of clinical validation is very high, as the risk/benefit profile must be well proven. In the CDx setting, since PMA approval is tied to drug approval, the risk to drug approval is equally as high as the risk to approval of the device itself. Statutory timeline for review is 180 days, and the 2013 fees range from waived to $259,000, depending on the size of the company and various other factors.[1] The submission must include analytical and clinical validation, CMC information, design controls, and labeling information. Evaluation of the device as part of an Investigational Device Exemption (IDE) will require a site inspection, advisory panel, and annual reporting.

A De Novo submission is a reclassification and used for low or moderate risk devices with no predicates. The FDA has stated that "Class III devices can be down-classified to Class II when sufficient information becomes available to establish special controls that reasonably assure safety and effectiveness." This is a special case, where there is not existing approval of a substantially equivalent device to the proposed medical device, but there is sufficient evidence of similar utility in established clinical practice and/or minimal risk to the patient. Formally, the device may be down-classified using the 513G pathway (official request for classification of a

[1] See updated fees on the FDA website under PMA Review Fees: http://www.fda.gov/MedicalDevices/DeviceRegulationandGuidance/HowtoMarketYourDevice/PremarketSubmissions/PremarketApprovalPMA/ucm048161.htm

currently unclassified device) or through the pre-IDE submission process. The pre-IDE submission process is preferred as this informal interactive process allows early assessment of device class and facilitates the "least burdensome" regulatory route to approved product. Review time for a PMA is 90 days plus a possible additional 60 days and fees are $4 k. Submission will include all that is required of a PMA, but also a risk/benefit petition to establish the argument for reduced risk. Do Novo approvals are only done by recommendation of the FDA.

A 510(k) is a premarket clearance and involves claiming that the device is substantially equivalent to a predicate device (21 CFR 807.92(a)(3)). They are typically used only for Class I or II devices, where a predicate exists. They have a review time of 90 days and a $4 k fee. The submission includes analytical and clinical validation of the device as well as design controls which demonstrate clinical performance of the device similar to the predicate.

4 Development of an Image Analysis Based CDx

The FDA has made it clear that all oncology CDx require PMA process due to the inherent risk profile of oncology drugs and the critical life benefit imparted. However, in the image analysis field there has been much energy expended debating when a PMA or a 510(k) submission is required in all oncology settings. In particular, image analysis based companion diagnostic tests are designed to show equivalence or improvement to pathological approaches. In this case, the computer algorithm is to augment or improve the ability to quantify a biomarker from a microscope slide in the manner consistent with a pathologist's trained approach. However, the image analysis strategy provides the opportunity for better precision and reproducibility over manual pathology, and in some cases may be the only adequate level of performance required to stratify patients. It is likely that in such a use case, a PMA will be required the first time to demonstrate the required burden of proof. It is possible, however, follow-up approaches which employ this tactic would rely on 510(k) clearance.

In our experience, it is dangerous to speculate whether a PMA or a 510(k) will be required—a better approach is to discuss this with the FDA during the pre-IDE meeting with a well-prepared team. We have been surprised in the past at what was accepted under 510(k). For example, in previous clearances directed by one of the authors (1, 2), two of the 510(k) clearances successfully received had no direct precedents, the use of Tunable Image Analysis (K080564) and HER2/neu Reading on Monitor (K071671).

Since companion diagnostics when submitted in parallel with a therapeutic will require a PMA, it is likely that the first application of image analysis technology will be applied to a de novo oncology CDx. Over a dozen 510(k) image analysis clearances exist for the standard HER2, ER, PR, and Ki67 testing which is utilized for breast cancer, but these are truly substantial equivalency tools. With substantial precedent in the 510(k) space, the hurdle for the adoption of an image analysis approach which supplants manual pathologist interpretation appears viable. The first adopters will certainly be need-based, where a complex interpretation of a tissue cannot be reliably performed by a pathologist. The emerging classes of oncology drugs, particularly antibody-drug conjugates (ADCs) and Immuno-Oncology Modulators (IOM), stand to have the level of complexity which exceeds pathologists' ability due to the complicated biology, mechanism of action, and tissue complexities which may force the reliance of the application of image analysis.

As with any IHC-based CDx performed on an autostainer platform, there are three main components to the medical devices: (1) The Reagents; (2) The Autostainer; and (3) The Interpretation/scoring. Each of these components must have the appropriate design controls to ensure performance of the test. In the typical case of a pathologist-read IHC test, the reagent kits and autostainers are manufactured under GMP, quality systems are built into the test procedure, and performance of the pathologist relies on an established scoring paradigm and training. In the case of an image analysis based approach, the only substitutions occur in the interpretation component, where the pathologist is removed and a whole slide scanner and image analysis software are substituted (a manual pathology read may already rely on a whole slide scan). Since these components (the scanner and software) have already been proven and cleared, the creation of an image analysis based PMA is only waiting for the right opportunity.

Leadership to enable movement into this new space remains unclear, yet strong needs for such devices are emerging. Ideally, a reagent manufacturer with existing platforms for staining and scanning (Dako, Leica, and Ventana all have these components) will be the first to create such a medical device. However, the matching of disruptive technology with needs can occur anywhere, and existing device manufacturers are just as likely to rely on sources of opportunity with groups who have expertise outside of their domain. The recently announced proposed acquisition of the image analysis software manufacturer Definiens by the antibody-based drug manufacturer MedImmune is the first transparent movement in this direction, with others sure to follow.

5 Guiding a Novel IHC/Image Analysis Combination Through PMA

In the USA, an automated analysis algorithm which is designed to quantify an IHC test as a companion diagnostic test is regulated by the Center for Devices and Radiological Health (CDER) of the Food and Drug Administration (FDA). To date, the CDER has only approved whole slide imaging systems for viewing IHC within the clinical context, as well as automated algorithms for IHC analysis, under the 510(k) pathway as Class II devices which demonstrate substantial equivalence to a microscope or manual pathology approach for which an IHC-based test has already been approved.

Since medical device manufacturers are regulated by the FDA and can promote only FDA-cleared/approved systems and applications in the clinical market, they must follow guidance or established process for creating a device. However, the FDA has not yet provided any guidance on whole slide imaging and image analysis algorithms which would be used to create a device incorporating these approaches. For this reason, there is significant uncertainty in how such a device would be created and how CDER would view the device. This uncertainty may create perceived risk on behalf of the drug sponsors, yet all of the pieces required to create a device (IHC test, scanner, image analysis software) have already undergone FDA clearance or approval. From this viewpoint, the task in supporting an IHC test/image analysis combination device through PMA is defined (though without precedent).

A new companion diagnostic which would directly combine an IHC test with an image analysis approach as a de novo approach for quantification of the biomarker would require the PMA pathway, as there is currently no predicate for the test. However, if this test were an oncology companion diagnostic, PMA would be required regardless. There are no limiting hurdles to developing a combined IHC-image analysis device, but this path has not been forged as of yet. In absence of direct FDA guidance, drug companies and clinical laboratories wishing to support trials for such as device cannot expect much guidance from medical device manufacturers regarding the validation of digital pathology systems.

The FDA clearances for image analysis based quantifications are provided for specific tissue types, stains, and reagents. The gap between the current FDA clearance approach and incorporation of the image analysis based quantification into a new PMA is the tripartite cooperation between the device manufacturers of the image analysis components and the IHC antibody kit manufacturers. Two of the three major IHC antibody kit manufacturers (Dako, Leica, Ventana) have incorporated whole slide scanning and image analysis capability in-house through acquisition, indicating their intent to be capable of creating such a device. Ultimately, as drivers of innovation and adoption, the pharma drug sponsor will create the need and provide the support bringing such a device to market through the PMA pathway.

6 Components of a PMA Filing for an IHC Test

This description is meant to be a guide in consideration of the typical components requiring evaluation and validation for an IHC-based test to be filed as a PMA. These descriptions are for utility of the diagnostic developer and are apart from the manufacturing, marketing, and commercialization considerations. Typically, the pharma sponsor will directly support the efforts of the device manufacturer towards development of the device once the drug has reached a critical proof-of-concept milestone. Often, diagnostic strategies are explored in parallel with the normal drug development efforts, at the hands of the pharma company prior to hand-off to a manufacturer once a commercialization path is established. Here, there is rarely a completely internal-driven effort, but rather a collaborative effort between the pharma sponsor, clinical trial support providers, and device manufacturers. Many different paradigms for roles and responsibilities exist, but typically the pharma sponsor chooses to retain control of the prototype diagnostic vetting process until sufficient proof-of-concept for the diagnostic approach is shown, before handing commercialization efforts over to a diagnostic kit manufacturer.

The general elements of the PMA or 510(k) include:

- Intended use
- Device description
- PMA-specific sections (internal/external controls; preanalytical variables)
- Analytical performance
- Clinical performance
- Instrument and software, if applicable
- If multiple platforms, assay performance on each
- Labeling (package insert)

As always, direct and frequent communication with CDER about the intent of the device and validation approaches to be used during the evaluation and registration trials will ensure acceptance of the approach by the FDA before submission. This guide is meant to only discuss the general components of a PMA submission.

6.1 Intended Use Description

The Intended Use description should include a description of the Intended Population, the Analyte Tested, and the Indication for Use. The Intended Use Statement should be concise and include the name and identity of the analyte and purpose for use in the clinical setting. The Indication for use describes the context of use (the disease application, assay format, and nature of use), the nature of the test (prognostic/ predictive), and the drug the test is tied to. The exact components depend on the Intended Use claims and may also depend on the technology or assay format utilized.

6.2 Linearity/Range Testing

Linearity refers to the ability of the assay to demonstrate varying staining intensities in the same specimen with differing concentrations of the antibody. Range is closely related to linearity and is defined as the ability of the antibody to detect varying amounts of analyte using the same assay conditions and antibody concentration. We recommend serial sectioning of the same specimens stained with varying concentrations of the antibody. Image analysis should play a critical role here, and it is vital that the algorithm used be able to identify the right cells and the right components within those cells (e.g., nuclear/membrane/cytoplasm). In our programs we utilize a whole tissue alignment program (3) so that serial sections can be effectively compared. A biostatistician should be included on the team to determine the most appropriate number of replicates as well as a minimum number of samples. A sigmoidal curve of antibody concentration vs. staining intensity should be generated, with replicates and outlier analysis as appropriate. The descriptions below pertain to the activities the kit manufacturer will perform, from which the pharma sponsor can consider during prototype diagnostic efforts.

6.3 Precision Studies

Advance planning in the pre-IDE meeting can be useful for determining the scope and size of the required precision and accuracy studies. Assay precision refers to the technical reproducibility of the test and should be expressed as a coefficient of variation. A set of consecutive sections (more than three in order to evaluate any outlier is recommended) for a specimen is run on the same instrument to determine inter-run precision. Multiple specimens should be utilized upon consultation with the statistician. A set of consecutive sections run on at least three different instruments can determine the inter-instrument precision. In our experience, the FDA requires CVs less than 15 % to qualify as a precise assay.

Assay Specificity. Specificity refers to the ability of the assay to detect a specific epitope (analyte) over other epitopes. Control models should be developed along with the assay that have previously characterized epitopes, validated in some other way than with the companion antibody (e.g., other IHC antibodies, western blots, etc. RT-PCR can be used, but correlation between mRNA and protein can be problematic). These control models, which can be human tissues, xenografts, and/or cell lines, are processed with the finalized assay. In addition, a set of other normal human tissues should also be evaluated, potentially on a tissue microarray (TMA). The ability of the assay to detect expected levels of epitope is examined, along with the staining patterns in the appropriate cells and subcellular compartments. Image analysis can be extremely helpful in this step.

Assay Sensitivity. Sensitivity refers to the ability of the antibody to bind consistently to the epitope of interest. A large cohort of positive samples in both xenografts and clinical tissues is required to evaluate assay sensitivity. For rare epitopes, close collaboration with academic centers that can dig through a large number of specimens to find those positive for the antigen of interest is recommended. A cut-point analysis should have already been generated prior to this step, and image analysis is preferred over manual subjective scoring.

Preanalytic Variability. The two preanalytic variables that most impact antibody performance are time to fixation (cold ischemia time) and time in fixation. We recommend the use of specially prepared TMAs that have modeled both cold ischemia time and time in fixation. These can be created in both xenografts and clinical sections, and are highly informative about the antibody and epitope of interest. While not required for FDA submissions, preanalytic factors seem highly antibody specific, and understanding their impact as part of antibody development can be extremely advantageous.

Antibody Stability Testing. Antibody stability refers to the structural and functional stability of the antibody as it relates to shelf life. Multiple consecutive sections (more than three for outlier analysis) of a validation specimen are stained with the final assay using two different antibody formulations. The two formulations include a freshly prepared antibody and a 7-day-old antibody incubated at either 37 or 45 °C. The incubated antibody must produce comparable staining to the freshly prepared antibody. If the 45 °C antibody produces comparable staining, a shelf life of 12 months is assigned. If the 37 °C antibody produces comparable staining, but not the 45 °C antibody, a shelf life of 6 months is assigned. If neither incubated antibody produces comparable staining, the antibody is deemed to have failed.

6.4 Analytical Performance Measures

Other studies may be included and vary depending on the type of technology the device encompasses, the end user of the technology, the sample type/matrix, the potential for carry-over/contamination, and necessary sample preparation conditions. The requirement for inclusion of these studies should be determined in IDE discussions.

6.5 Manufacturing Section

The manufacturing components included in the PMA filing begin in the IDE process and are reported in the submission. These include pre-approval inspection (GMP compliance), BIMO (bioresearch monitoring visit to clinical and/or sponsor sites), and a recommendation for the possible Panel-Track for review. Additionally, post-approval activities should be described, including annual reports, and PMA supplements for well-defined modifications (http://www.fda.gov/cdrh/ode/guidance/1584.pdf).

7 Components of a PMA Filing Utilizing Image Analysis

While most diagnostic players, particular autostainer companies, tend to emphasize the antibody and decouple and delay the image analysis measurement, we recommend using image analysis as a de-risking mechanism. The analogy would be like insurance; it is there when it may be needed. The cost of an image analysis development clearance when done by an experienced team is far lower than the cost of the antibody development, and the image analysis will be there if needed for approval. Image analysis can assist with antibody optimization, evaluation, cut-point determination, and pathologist training. Actually, in each of the PMA steps outlined in the next section, it is difficult to identify a step where custom image analysis cannot play an important role. Executives who assume quantitative readouts in tissue are optional will likely be surprised when these approaches become critical, central, and even mandatory for FDA approvals.

The FDA is now investigating how to best address digital pathology from a regulatory perspective but is mainly focusing on the use of whole slide images for the replacement of H&E glass slides and conventional microscopy, specifically for purposes of making routine surgical pathology diagnosis. The FDA has not yet commented on the use of whole slide image analysis systems for patient selection strategies, but given the large number of FDA-cleared digital pathology systems and applications, there is not necessarily a direct need for additional comment. Device manufacturers wishing to create IHC test-image analysis combination devices need only to focus on the well-established individual paths for validation of an IHC kit and an image analysis algorithm, and combine them. To accomplish this, the activities should surround the key question: "What kind of performance validation studies should manufacturers provide?"

In the absence of specific FDA guidance, alternative regulatory proposals have been brought forward by stakeholders to address the FDA's role in regulating medical devices. The goal of the stakeholders is to generate dialog on the development and implementation of a regulatory framework for PMA approval of IHC/image analysis devices which fulfills the agency's requirement to regulate devices and IVDs in the "least burdensome" manner. The College of American Pathologists have made recommendations for performing performance validation studies of image analysis algorithms which address the key components of analytical validity, proficiency testing, and clinical validity. The CAP guidelines emphasized the importance of proper validation, and how the use of the tool matches the context of how a diagnosis is rendered in the clinical practice. The guidelines also focused on diagnostic concordance among many types of specimens and diagnoses, and a sample size large enough to provide adequate statistical power and provide a challenge set which reflects the spectrum of diagnostic labels and secondary measures. The CAP Laboratory General checklist (GEN) and Anatomic Pathology Checklist (ANP) directly address recommended approaches for validating whole slide scanners and image analysis algorithms, which are consistent with the general principles of validation required in any medical device submission.

Manual Versus *Automated Performance.* There is a major misconception in the industry regarding the necessity of use of an automated platform in tissue-based companion diagnostics development. In our studies, we prefer to focus our energies on the eventual GMP manufacturer of the IVD antibody, which may or may not also be the supplier of an autostainer platform. When the PMA is submitted, the data can be submitted with both a manual approach and comparisons using one or more autostainer platforms. This allows comparison of more than one platform in clinical usage and gives the pharmaceutical partner a much broader worldwide access to more than one install base.

Quality System Requirements and GMP. The quality system for all steps of the antibody production should follow 21 CFR Part 820 and manufactured under Good Manufacturing Process. The steps undertaken below should ideally follow GLP practices as outlined in both 21 CFR 11 and 21 CFR 58. Total traceability and documentation is required at every step of the diagnostic development process. All Class II and Class III IVDs must be developed under Design Control.

8 Validation of Clinical Utility

Ultimately, the performance of the device is measured against a comparison to a clinical outcome. The structure of the studies showing clinical utility may vary depending on the technology used and the end user of the device. Performance criteria will vary if the endpoint is a quantitative or qualitative assay. Validation of the scoring approach will depend on what is reported (individual analytes vs. composite score) and the performance around the cut-off.

The PMA filing should include tabulations of data from all individual subjects, any data analysis, and results of statistical analyses. Additionally, the filing should include any other information from the clinical investigations and supportive literature.

9 The PMA Submission

While the exact format of a PMA submission can vary, the submission should contain the following items:

General Information. Introduction to the epitope of interest, the relationship between the diagnostic and the therapeutic, and the disease area.

Indications and Contraindications. The target of interest, how it is measured in a clinical setting, and any contraindications.

Background and Study Objectives. A longer section on the disease area, the rationale for the diagnostic, the approach taken in quantitation, the cutpoints, etc.

Complete Investigational Device System Overview. This section describes the complete device, either the manual usage or the usage with automation. It should include detailed information on the kit involved, the staining procedure, any hardware and software utilized in the system, and the assay interpretation.

Materials of Construction and Biocompatibility. This section plays a more prominent role in other medical devices (e.g., implants); it can usually be very short in an in vitro diagnostic.

Investigational Plan. Describes the complete plan for how the study is carried out. Should include information on the study protocol, the study configuration, sample size justification, patient acquisition and enrollment criteria, validation and acceptance criteria for samples, instrument and reagent handling criteria, complete procedure, and data management and analysis processes.

Nonclinical Studies. Will include information on tests performed, as described in the previous section. Should include assay sensitivity and specificity, precision studies, antibody stability testing, scoring algorithm development, etc.

Image Analysis Quantification Plan and Validation. This is not a required section, but we have expanded this area to provide detailed information on how a customized image analysis algorithm is developed and utilized in the other data sections. It should include information on validation and test dataset, concordance with manual pathologist scoring, algorithm performance on tissue selection (e.g., tumor/stroma pattern recognition, if applicable), performance on cell-compartment analysis (nuclear/membrane/cytoplasm), quality control on scanning and image analysis results, and validation results.

Clinical Validation Section. The PMA filing should also contain a description of the studies performed to demonstrate clinical utility. The studies performed during the clinical validation will depend on the intended use/indications for use/end user. The studies should use real clinical samples where feasible and have used both prospective and retrospective evaluation of samples to demonstrate performance. When possible, there should be a comparison to a reference method or predicate device.

Patient-specific information should be included in the form of case report forms detailing any patient information, patient complaints for any of the studies done under IDE. The patient information should include:

- Diagnosis, residual disease, etc. (current state)
- Monitoring, recurrence (change in state)
- Risk of disease, prognosis, prediction (future state)
- Study protocols including IRB approval letters/informed consent

Safety and effectiveness data should also be included, listing any adverse reactions and complications, and any device failures and replacements. Full analytical and clinical validation data for each patient must be included to support the claims. The information should include:

- Each specimen type
- All matrices
- All specimen collection devices
- All transport media
- All transport and storage conditions
- All collection methods

Risk Management and Risk Analysis Plan. This section is much larger in an implanted device than in an in intro diagnostic, depending on the risk class of the device. For an IVD Class II or III submission, it should include a review of sources for risk assessment, including scientific literature, manufacturing experience (including nonconforming material reports, corrective and preventative action reviews, out of specification investigation review), as well as customer experience reviews (complaints, corrections, customer feedback, removals, recalls, etc.). Risk management process should be outlined from in-house Risk Management Processes and Design Control documents in concordance with regulatory standards in this area (e.g., 21 CFR Part 820.30 and ISO 14971) and should include Risk Management Plan, Preliminary Hazard Analysis Matrix, and System Level Hazard Analysis, Design and Process Failure Mode and Effects Analysis.

Marketing Statement. Describes the exact terminology of how the test will be marketed.

References. Any pertinent references on the system employed, any supporting studies, and published articles on the epitope related to the particular disease.

The Device Label. The label for a CDx should contain information about the diagnostic test which is necessary to "ensure the Safety and/or Efficacy of a specific targeted therapeutic treatment (drug) used in Personalized Medicine" according to the description in CFR21 CFR809.10. The goal of the label is to create clear instructions for use and describe the expected analytical and clinical performance of device. Specific guidelines include "the therapeutic drug is specifically identified in the diagnostic label intended use statement" and should ensure that notice of "where the need for an approved/cleared IVD is clearly and consistently indicated in the drug label." The label should also include the necessary information which is "to be used to assist physicians in making treatment decisions for their specific patient or a targeted patient sub-group." Finally, a prospective description of the performance in an intended use population should be included to allow laboratory proficiency monitoring.

10 Suggestions for Success

There are general administrative reasons for failure of a PMA, including:

Accuracy and organization:

- Table of contents missing, pages not numbered
- Check tables/figures/text for clarity, consistency, and accuracy
- "Put together in a hurry"-multiple cut-and-paste errors
- Lack of copies of IRB approval letters, informed consent, financial disclosure forms, list of investigators, etc.
- Lack of monitoring/auditing of clinical sites or poor BIMO inspection findings

Scientific reasons for PMA failure include:

- Poor statistical analysis of data
- No literature to support device
- Data not analyzed appropriately, not summarized, organized
- Lack of knowledge about the clinical disease

Strategic reasons for PMA failure include:

- The device does not demonstrate the Intended Use. The "Intended Use" is the driving force of the review, and studies which do not adequately support the claims of the Intended Use will guide the PMA submission review towards failure
- Issues with the Quality System
- Poor inspection of manufacturing facility or poorly written manufacturing sections

While the FDA is currently accepting applications without the use of image analysis and based on manual pathologist scores in the steps below, this is likely to change in the future. The statistical analysis and study results for the steps below are far superior with image analysis, generating larger data sets with less human subjectivity. For this reason, in our companion diagnostics programs we engage pathologists and image analysis scientists early in the development, so that we have a customized image analysis program for utilization in the steps below. The development of regulated image analysis companion diagnostics was reviewed previously (4).

Involve pathologists early in the process and throughout the process. It is easy to lose sight of the fact that the ultimate decision makers and end users of a tissue-based companion diagnostic are pathologists. It has been our experience that the pathologists' experience and opinions are highly sought after by the FDA, both informally and formally, during their review process. Pathologists will be the customers using any newly cleared test, and they continue to be highly effective gatekeepers of any new diagnostic technology adoption in the anatomic pathology laboratory.

Utilize image analysis as frequently as possible throughout the FDA clearance process. There has been a tendency to attempt to clear the antibody first with manual scoring, and only later utilize whole slide scanning and image analysis for a follow-on 510 k clearance. Partly this is due to unfamiliarity with regulated image analysis processes by autostainer companies and partly due to the poor flexibility of many existing commercial clinical image analysis software programs for new biomarkers. We utilize image analysis throughout the process to generate objective data that can be used at each step.

Apply the same de-risking strategy to diagnostics as to pharmaceutical drug development. Pharmaceutical executives are experts at de-risking the drug development process, starting with a large funnel of leads, and filtering out and narrowing down investments into fewer and fewer compounds. The further along the compound the higher the subsequent investment. Yet in companion diagnostics, most large pharmaceutical companies are too enamored by the large early-stage deal, signing multimillion, multiyear dollar vendor lock-in programs with diagnostics autostainer companies, that are not in the best interest of the pharmaceutical partner. While these programs fund the R&D of autostainer companies, they generate strong inflexibility for the companion diagnostics programs. Questions like "What will it cost to add a second tissue organ for clearance, e.g., add gastrointestinal to breast?" or "What if we want to bring forward a diagnostic antibody from a competitive vendor on this autostainer platform?" or "How many countries and patent locations do not have access to this particular autostainer platform?" are often not asked or well considered. These poor decisions result in years of delay and millions of dollars of additional costs on the part of the pharmaceutical sponsor. Much of this waste could be avoided by de-risking the process, and not tying up options until absolutely necessary.

The antibody, not the autostainer, is the essential core component of the IHC segment companion diagnostic. Developing a drug through IND to NDA requires absolutely focus and the ruthless ability to cut out distractions from this goal. The same is true of the development of a companion diagnostic. Success is intimately tied to the quality of the diagnostic antibody. An outstanding antibody will generally perform well on any major autostainer platform. A poor antibody will have problems on all of them. It is to the exclusive benefit of the autostainer partner to add protocols, licensing rules, and other barriers to ensure that only one automated platform is supported. Limiting the market opportunity to only one platform and its worldwide install base can be highly problematic in scaling a large therapeutic. Focusing on the antibody and preferably using more than one autostainer platform can avoid this common pitfall. However, the assays need to be validated and approved for each platform.

Both the interpretation/score and the scanner are essential components of the image analysis segment of the companion diagnostic. In the case of a whole slide scanner, an RUO device is not sufficient for PMA approval. The slide scanner must already be FDA cleared, demonstrating substantial equivalence to a predicate manual pathology read of a slide. This will demonstrate the device is capable of meeting the manufacturing and performance demands required to meet PMA requirements. This existing information can be incorporated into the PMA, provided that there are no differences between clinical and launch platform. If the use of prototype hardware or software is needed for the clinical trial, it must be justified and agreed upon by the FDA in advance of the registration trial. If off the shelf software is used as part of the device, it must be sufficiently documented.

The outcome of the test results need to link back to the functional requirements described for the device software. A summary of the software validation/verification testing is not sufficient;

full results of the complete analysis must be included. A hazard analysis must be performed, and any mitigations must be justified and traceable back to the original functional requirements.

Employ a complete diagnostics development team. At a minimum, the team needs to include pathologists, domain-specific biologists with diagnostic industry experience, image analysis scientists, a statistician, project managers, and regulatory experts. More than one pathologist is frequently required, as both pharmaceutical industry experience and the given disease domain (e.g., lung tumors, breast tumors, etc) are rarely available in the same individual. A mix of veterinary and human pathologists can be helpful.

References

1. Nassar A, Cohen C, Agersborg S (2011) Trainable immunohistochemical HER2/neu image analysis: a multisite performance study using 260 breast tissue specimens. Arch Pathol Lab Med 135(7):896–902
2. Nassar A, Cohen C, Albitar M (2011) Reading immunohistochemical slides on a computer monitor—a multisite performance study using 180 HER2-stained breast carcinomas. Appl Immun Mol Morph 19(3):212–7
3. Potts S, Johnson T, Voelker F (2011) Multiplexed measurement of proteins in tissue in a clinical environment. Appl Immun Mol Morph 19(6):494–8
4. Lange H (2011) Digital pathology: a regulatory overview. Lab Medicine 42:587–591

Methods in Pharmacology and Toxicology (2015): 345–358
DOI 10.1007/7653_2014_14

Published online: 25 October 2014

Implementing a Multi-analyte Immunohistochemistry Panel into a Drug Development Program

Carla Heise, Pierre Brousset, Tommy Fu, David A. Eberhard, Graham W. Slack, Camille Laurent and Randy D. Gascoyne

Abstract

Diffuse large B-cell lymphoma (DLBCL) consists of several molecular subtypes. Based on presumptive cell of origin, there are two primary molecularly distinct subtypes: germinal center B-cell (GCB) and activated B-cell (ABC)/non-GCB lymphoma. These subtypes were initially classified by gene expression profiling (GEP) by microarray, but immunohistochemistry (IHC) assays employing antibodies against antigens CD10, BCL6, MUM1/IRF4, FOXP1, and GCET1 have been utilized as surrogates, and may provide a more simple, cost-effective, and reliable means to differentiate between GCB and non-GCB subtypes. Herein we describe an approach to the adoption and optimization of multi-analyte IHC panel assays for DLBCL tumor subtyping, and their implementation for potential use as a companion diagnostic in a prospective, biomarker-stratified drug development program.

Key words Immunohistochemistry, Companion diagnostic, Multiple analyte, Algorithm, Assay validation, Diffuse large B-cell lymphoma, Tumor subtyping

1 Introduction

The molecular characterization of diffuse large B-cell lymphoma (DLBCL) indicates that it is a genetically heterogeneous disease composed of subtypes that can be defined in reference to normal B-cell types, possibly reflecting their pathogenesis associated with the transformation of the B cell of origin. Thus, DLBCL can be classified into primarily two subtypes based on the presumptive cell of origin, germinal center B-cell (GCB) and activated B-cell (ABC)/non-GCB lymphoma. These subtypes are distinctly different in terms of oncogenic pathways, tumor biology, and clinical prognosis [1–4]. The GCB subtype correlates with a significantly better prognosis than the ABC subtype, independent of the International Prognostic Index [5, 6]. Furthermore, the efficacy of some therapies under clinical investigation for treating DLBCL may be dependent on subtype [7–9].

Original classification of these subtypes in DLBCL was based on gene expression profiling (GEP) of mRNA transcripts using microarrays [5]. The approach using GEP stratifies DLBCL into a 3-class predictor, with approximately 14–22 % of cases remaining unclassified using this approach [3, 6]. An immunohistochemistry (IHC) assay panel to assess protein expression was developed to distinguish DLBCL subtypes by staining with antibodies specific for CD10, BCL6, and MUM1/IRF4 [10], providing a less complex and more accessible alternative to GEP [11–13]. The IHC approach consolidates subtyping into a 2-class predictor including GCB and non-GCB subtypes [10]. A subsequent study determined that including two additional markers, FOXP1 and GCET1, further improved concordance between IHC and GEP results [11]. An important aspect of these multi-analyte IHC panels is that the type of algorithm applied to assimilate the results from several individual assays is conceptually different than the approach to multi-analyte GEP of microarray data. For microarray data analysis, the expression level of each gene is treated as a data point independent of the other genes, and the spectrum of expression levels across the entire gene set is represented as a "signature" or "profile." With this approach, there can be some degree of redundancy where a failed or inaccurate result for one gene is compensated for by the collective strength of multiple other genes. By contrast, the IHC panels depend on a step-by-step decision-tree model where the expression result of each protein (gene product) then determines which protein marker to consider at the next step in the decision-making algorithm. In this decision-tree algorithm, the precision and accuracy of each protein expression assay in the panel is extremely important in arriving at the correct final result.

IHC is a widely used technique in surgical and clinical pathology because of its ease of use and broad applicability. However, the IHC technique is inherently semiquantitative, and therefore consistency, reproducibility, and a lack of universally accepted standardization guidelines present major challenges in developing IHC as a quantitative assay [14]. Variation in IHC results can be associated with differences in laboratory techniques, scoring criteria, and inter-observer microscopic interpretations [12]. Therefore, standard protocols that are validated for reproducibility are necessary for optimal clinical utility. In this chapter, we describe the development of a multi-analyte IHC panel into a robust subtyping assay for use in the clinical development program for a new therapeutic agent in DLBCL. The assay is intended to be used by central reference laboratories to select or stratify patients for treatment in pivotal phase III registrational trials of the drug to be developed, and it could be considered for co-development and regulatory approval as a companion diagnostic for the therapeutic indication in clinical practice.

2 Strategy for Development of a Robust, Qualified, Multi-panel IHC Assay to Be Used for Patient Stratification in Registrational Clinical Trials

This chapter describes our approach for adopting and validating previously published protocols for the performance and interpretation of two multi-analyte IHC assays, to establish reproducible methods and compare assay performance across independent testing laboratories. These steps included: (1) development of standard assay technical protocols shared across laboratories and initial intra-laboratory optimization; (2) interlaboratory concordance analyses using tissue microarrays (further interlaboratory assay optimization and technical validation); and (3) prospective performance of the assay by each laboratory in an independent, blinded fashion on "real-world" tissue samples obtained from patients enrolled in a phase II study to evaluate the cross-laboratory reproducibility of the assay in the clinical setting in which it is intended to be used. Figure 1 depicts the steps involved in assay development and the specific steps employed to optimize and implement the IHC assays for DLBCL subtyping in a clinical program, with the end goal of producing a companion diagnostic. Optimization and technical validation of the assay was accomplished using tissue microarrays (TMAs) for the initial concordance studies and GEP results as a subtyping reference. The assay was then applied by the same laboratories to tissue samples of patients with DLBCL enrolled in a phase II/III study. All subtyping, scoring, and re-scoring were performed by each laboratory under blinded conditions.

2.1 Investigational Sites

Investigational sites included two academic medical center laboratories and one central reference laboratory:

- British Columbia Cancer Agency, Vancouver, Canada.
- Purpan University Hospital, Toulouse, France.
- Laboratory Corporation of America, Center for Molecular Biology and Pathology, Research Triangle Park, North Carolina, USA.

2.2 Reagents and Equipment

- Participating laboratories used antibodies from the same commercial sources. Antibodies used for staining were specific for CD10 (Novocastra; Leica Biosystems, Buffalo Grove, IL), BCL6 (Ventana Medical Systems, Oro Valley, AZ), MUM1 (Dako, Carpinteria, CA), FOXP1 (Abcam, Cambridge, UK), and GCET1 (Abcam).
- Ventana Discovery XT or Benchmark XT platforms were employed for automated IHC staining.

2.3 Assay Optimization and Initial Interlaboratory Concordance

- Initial assay implementation and optimization was performed in each laboratory using their own in-house formalin-fixed, paraffin-embedded (FFPE) samples representing positive and negative normal tissue controls and DLBCL cases. A small series

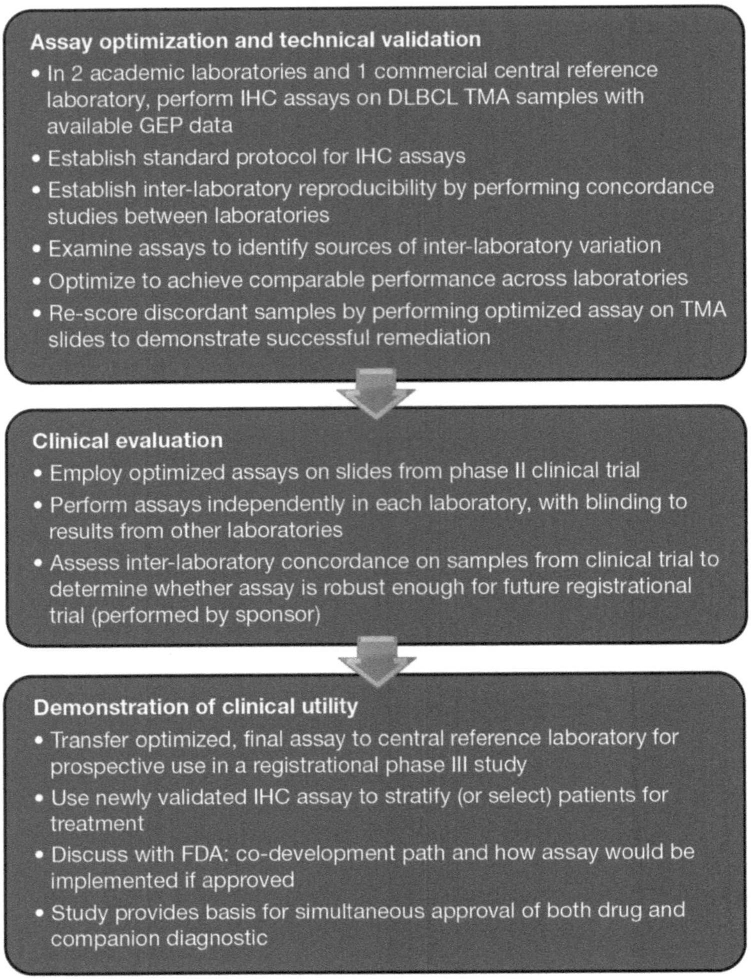

Fig. 1 Strategy for companion diagnostic assay development. *DLBCL* diffuse large B-cell lymphoma, *FDA* United States Food and Drug Administration, *GEP* gene expression profiling, *IHC* immunohistochemistry, *TMA* tissue microarray

of 5 DLBCL cases, cut as full sections on slides, was also shared between the laboratories to facilitate preliminary optimization.

- For the initial interlaboratory concordance study, TMAs were prepared from tumors collected from 51 patients with DLBCL who were associated with the Leukemia and Lymphoma Molecular Profiling Project (LLMPP). Frozen samples of these same tissues were previously profiled using GEP analyses [3].

2.4 Staining Evaluation

- Positive control tissue (tonsil) and internal cellular controls were used to validate the staining technique and assay performance for each antibody. Specificity of staining was determined by evaluating the staining patterns associated with particular subcellular components (e.g., membrane, cytoplasm, or nucleus) in the various cell types that have been previously reported to

Table 1
Antibody stains used in the IHC assay

Antibody	Protein function	Location of expression	Positive staining	Internal cellular control staining
CD10	Antigenic glycoprotein	Cell surface of GCBs	Distinct membrane staining	Stromal fibroblasts or granulocytes
BCL6	Transcriptional repressor	Nucleus of benign GCBs and subset of CD4+ T cells	Tumor cells with any degree of nuclear staining	Scattered CD4+ T cells
MUM1	Lymphocyte-specific transcription factor, late stages of B-cell differentiation	GCB nucleus	Tumor cells with any degree of nuclear staining	Activated T cells or plasma cells
FOXP1	Transcriptional repressor	Nuclei of B cells inside and outside of germinal centers	Distinct nuclear staining	Scattered small lymphocytes
GCET1	Protease inhibitor	Cytoplasm of GCBs	Perinuclear Golgi pattern	Residual benign GCBs

GCB germinal center B-cell, *IHC* immunohistochemistry

express or not express the antigen of interest (e.g., various B-cell populations or other cell types).

- Assay conditions for each antigen were optimized by each laboratory to produce minimal nonspecific background staining and maximal specific staining patterns.
- Details on the five proteins detected by the stains (CD10, BCL6, MUM1, FOXP1, GCET1) are included in Table 1.

2.5 IHC Scoring

- For each antibody stain, an intensity score of specific staining (0 = no staining; 1 = weak; 2 = moderate; 3 = strong) was assigned based on the staining examples in Fig. 2. Images of these examples were provided to each laboratory as references to promote consistency in stain evaluation across laboratories. The percentage of positively stained tumor cells was subjectively estimated by the evaluating pathologist. Necrotic or degenerating tumor cells were not included in the scoring.

2.6 Subtype Assignment

- Tumor subtyping was performed by the evaluating pathologist using both the Hans (CD10, BCL6, MUM1) [10] and Choi (CD10, BCL6, MUM1, FOXP1, GCET1) [11] criteria, as outlined in Fig. 3.

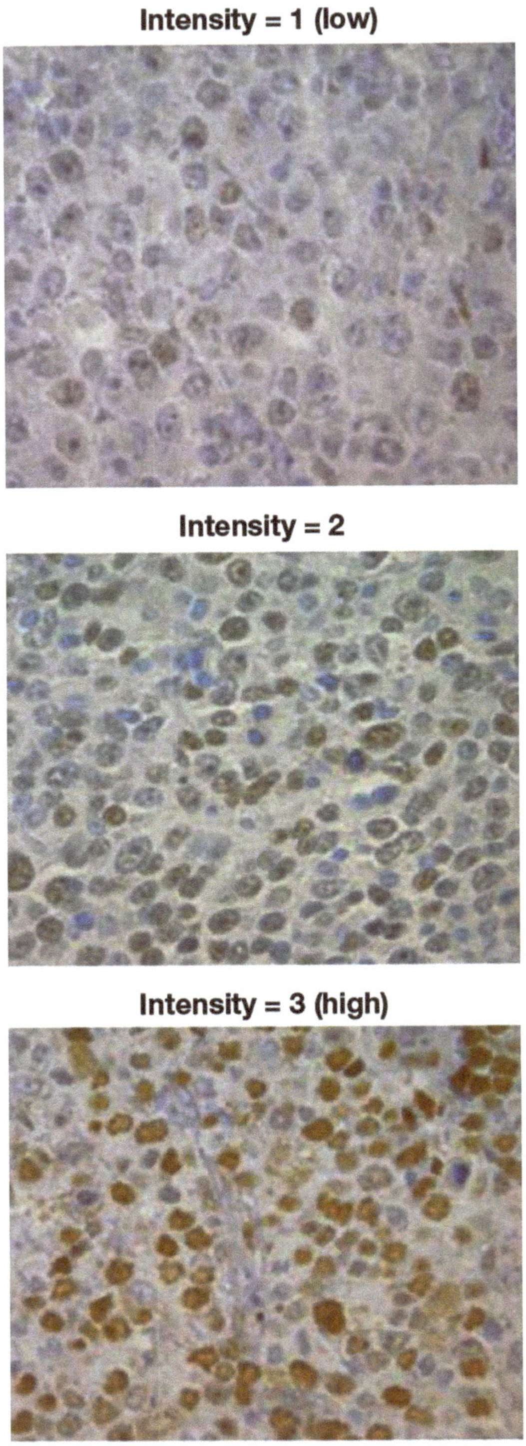

Fig. 2 Examples for immunohistochemical stain intensity score

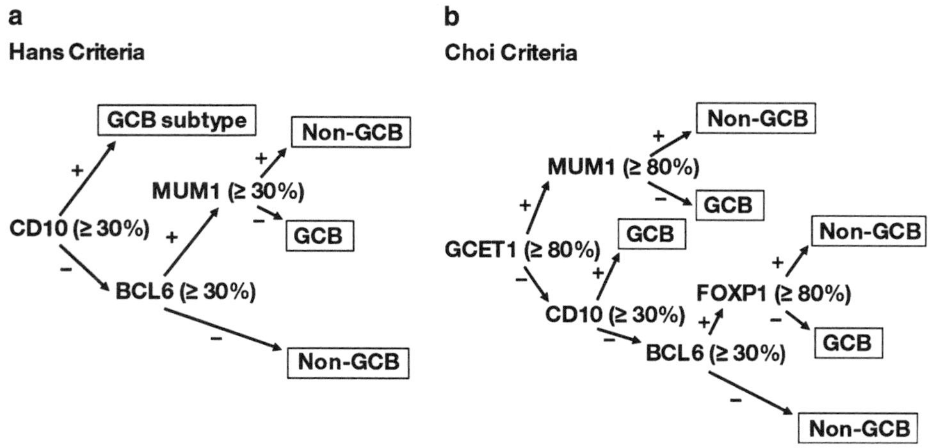

Fig. 3 Scoring criteria for determining lymphoma subtype based on antibody staining. (**a**) Hans criteria: Reprinted from Blood ©2004, vol. 103 issue 1, pages 275–82, Hans CP et al., Confirmation of the molecular classification of diffuse large B-cell lymphoma by immunohistochemistry using a tissue microarray [10], with permission of the American Society of Hematology. (**b**) Choi criteria: Reprinted from Clin Cancer Res, ©2009, vol. 15 issue 17, pages 5494–502, Choi WW et al., A new immunostain algorithm classifies diffuse large B-cell lymphoma into molecular subtypes with high accuracy [11], with permission from AACR. *GCB* germinal center B-cell subtype, + positive, – negative

2.7 Evaluation of Accuracy (Sensitivity and Specificity) of the Multi-analyte IHC Panels vs. GEP

- Tissue microarray data were used to evaluate the accuracy of results of tumor subtyping using the multi-analyte IHC assay algorithms.
- Data were compared with results from "gold standard" GEP classification to determine the sensitivity (true positive) and specificity (true negative) for IHC results.
- An overall accuracy (concordance) score was determined by combining specificity and sensitivity into a single measure of percentage of cases for which the IHC assay results agree with the GEP subtype.
- Samples that were scored as non-GCB by both IHC and GEP techniques were considered "true positive"; those scored as non-GCB by IHC and GCB by GEP were considered "false positive"; those scored as GCB by IHC and non-GCB by GEP as "false negative"; and those scored as GCB by both IHC and GEP as "true negative." Thus, sensitivity and specificity were calculated as follows:

$$\begin{aligned}\text{Sensitivity} &= \text{True Positive}/(\text{True Positive} + \text{False Negative})\\ &= \text{Probability (IHC-tested non-GCB}|\text{ among all ABC}\\ &\quad \text{cases by GEP)}\end{aligned}$$

$$\begin{aligned}\text{Specificity} &= \text{True Negative}/(\text{True Negative} + \text{False Positive})\\ &= \text{Probability (IHC-tested GCB}|\text{ among all GCB}\\ &\quad \text{cases by GEP)}\end{aligned}$$

2.8 Evaluation of Interlaboratory Reproducibility

- Concordance (percentage agreement) was calculated as the number of pairs with agreement divided by total number of pairs between any two laboratories, and was calculated for subtypes determined by either the Hans [10] or Choi [11] algorithm.
- Cohen's Kappa was calculated to measure agreement between any two pair-wise laboratories. To measure agreement among all three laboratories, Fleiss' Kappa, which generalizes Cohen's Kappa from two laboratories to multiple laboratories, was also calculated.

2.9 Interpretation of Results and Discordant Data Handling

- Following initial interlaboratory concordance analysis on the TMA samples, pathologists from the three laboratories reviewed and discussed discordant cases over a multi-head microscope to permit causal analysis of discrepancies, such as percentage of positively stained tumor cells near the cutoff value (i.e., borderline cases); cases showing weak tumor staining or high background; cases with poor histologic preservation or small numbers of tumor cells; or other technical or interpretive issues.
- Remediation plans were then made by the study pathologists to overcome these variations, to standardize the approach used for classification.
- Discordant samples were then re-scored where there were differences, to demonstrate successful remediation.

2.10 Case Reporting and Data Review for the Prospective Blinded Analysis of the Phase II Study

- Diagnostic pathology tumor tissue blocks were obtained from a subset of 50 patients enrolled in the phase II clinical trial. Unstained sections were provided to each of the three laboratories for IHC staining according to the optimized laboratory protocols.
- Results were recorded on report forms depicted in Fig. 4 and submitted to the study sponsor for database entry.
- After IHC results were obtained from the first 10 patients, an interim analysis was performed to assess interlaboratory concordance. Discrepant results were discussed by the pathologists to determine if further refinement of the assay or criteria was necessary for the remaining 40 samples. Following discrepancy remediation, the final interlaboratory reproducibility analyses of the 50 samples were performed.

3 Results

3.1 Assay Optimization and Technical Validation

Published protocols for DLBCL subtyping by IHC [10, 11] were used to develop standard protocols to be utilized by each testing laboratory. However, because instrumentation and personnel differ

Subtype by 3 Marker Assay (Hans): GCB ☐ Non-GCB ☐ Unclassifiable ☐ Not Done ☐
Subtype by 5 Marker Assay (Choi): GCB ☐ Non-GCB ☐ Unclassifiable ☐ Not Done ☐
Outside Report, Cell of Origin: GCB ☐ Non-GCB ☐ Unclassifiable ☐ Not Done ☐
Subtype for consensus: Yes ☐ No ☐

List of Stains

				Pos	Neg	Not Eval.
H&E						
CD20	%Pos____					
BCL6	%____	Intensity (0-3) ____	Histoscore ____	☐	☐	☐
CD10	%____	Intensity (0-3) ____	Histoscore ____	☐	☐	☐
MUM1	%____	Intensity (0-3) ____	Histoscore ____	☐	☐	☐
GCET1	%____	Intensity (0-3) ____	Histoscore ____	☐	☐	☐
FOXP1	%____	Intensity (0-3) ____	Histoscore ____	☐	☐	☐

Fig. 4 Case report form for reporting IHC staining results to the study sponsor for database entry. *GCB* germinal center B-cell subtype, *Eval* evaluated, *H&E* hematoxylin and eosin, *IHC* immunohistochemistry, *Neg* negative, *Pos* positive

among laboratories, each laboratory first optimized the technical staining performance and pathologist's interpretation of each of the antibody assays according to their own standards, using appropriately selected positive and negative tissue and cell type controls, including several DLBCL cases. Then, an initial evaluation of interlaboratory concordance of IHC staining and pathologist's interpretation was performed using DLBCL TMAs, with replicate TMA sections independently stained and interpreted by each laboratory. DLBCL subtype assignment was performed using both Hans [10] and Choi [11] decision-tree algorithms (Table 2). After the first round of independent, blinded scoring, IHC pair-wise concordance ranged from 74 to 92 % (Hans criteria) and 72 to 86 % (Choi criteria) (Table 3). The number of discordant samples from the 51 TMA cases ranged from 4 to 14. At the specific antibody level, discordance ranged from 2 to 25 %, although concordance was generally not higher or lower for any particular stain. Accuracy of IHC subtyping was assessed by comparison with GEP results from the same cases. Sensitivity ranged from 67 to 100 % and specificity ranged from 86 to 96 %, with Laboratory B recording the highest values.

Following this interlaboratory concordance analysis, discordant cases were discussed among pathologists from all three laboratories, noting observer differences and/or staining differences, with reconciliation discussions for re-scoring. Depending on the antibody, the primary source of discordance was noted as observer variability, stain intensity, or transcription error by the pathologist.

Table 2
IHC classifications of TMA samples compared with those of GEP, by round

	IHC Classifications	GEP	
		ABC	GCB
Round 1	Lab A Hans		
	Non-GCB	14	1
	GCB	7	21
	Lab A Choi		
	Non-CB	14	3
	GCB	7	19
	Lab B Hans		
	Non-GCB	20	3
	GCB	1	19
	Lab B Choi		
	Non-GCB	20	2
	GCB	0	20
	Lab C Hans		
	Non-GCB	15	1
	GCB	6	21
	Lab C Choi		
	Non-GCB	15	1
	GCB	6	21
Round 2	Lab A Hans		
	Non-GCB	17	2
	GCB	4	20
	Lab A Choi		
	Non-GCB	16	2
	GCB	5	20
	Lab B Hans		
	Non-GCB	17	2
	GCB	4	19
	Lab B Choi		
	Non-GCB	15	2
	GCB	5	19
	Lab C Hans		
	Non-GCB	20	2
	GCB	1	20
	Lab C Choi		
	Non-GCB	20	2
	GCB	1	20

[a]Subtype as determined by GEP
ABC activated B-cell, *GCB* germinal center B-cell, *GEP* gene expression profiling, *IHC* immunohistochemistry, *TMA* tissue microarray

All laboratories re-scored discordant samples on the TMA where there were differences, after which the number of discrepancies was reduced to 4–6 cases, and concordance for re-scored cases was improved (Tables 2 and 3). Percent sensitivity and specificity increased for both Laboratory A and C after the second round of

Table 3
Summary of interlaboratory IHC concordance between pair-wise laboratories for TMA (Total Pairs of Samples = 51)

	Lab A vs. Lab B	Lab A vs. Lab C	Lab B vs. Lab C
IHC concordance Round 1			
% Concordance Hans	74	92	78
Cohen's Kappa (95 % CI)	0.50 (0.28, 0.72)	0.83 (0.67, 0.99)	0.58 (0.37, 0.78)
% Concordance Choi	72	86	78
Cohen's Kappa (95 % CI)	0.45 (0.22, 0.69)	0.71 (0.51, 0.91)	0.57 (0.36, 0.79)
IHC concordance Round 2			
% Concordance Hans	90	94	92
% Concordance Choi	90	92	88

CI confidence interval, *IHC* immunohistochemistry, *TMA* tissue microarray

Table 4
Summary of interlaboratory IHC concordance between pair-wise laboratories for patient samples from a phase II study

	Lab A vs. Lab B	Lab A vs. Lab C	Lab B vs. Lab C
IHC concordance			
Total pairs of samples	48	48	50
% Concordance Hans	88	98	88
Cohen's Kappa (95 % CI)	0.75 (0.57, 0.93)	0.96 (0.88, 1.0)	0.76 (0.58, 0.94)
Total pairs of samples	48	36	34
% Concordance Choi	88	92	82
Cohen's Kappa (95 % CI)	0.75 (0.56, 0.94)	0.82 (0.64, 1.0)	0.63 (0.36, 0.89)

CI confidence interval, *IHC* immunohistochemistry

scoring, while sensitivity for Laboratory B decreased slightly. A consensus was reached that these IHC assay procedures and scoring criteria were adequately standardized for testing samples from patients in a phase II trial to determine their DLBCL cell of origin subtype.

3.2 Prospective Interlaboratory Concordance Studies Using Clinical Trial Tissue Samples

The assay was repeated for individual full-section tissue samples obtained from a subset of patients enrolled in the phase II study (Table 4). Samples were collected prospectively as patients were enrolled in the trial and were stained as received using appropriate positive and negative assay controls. Thus, staining in this trial was conducted in a similar manner to that of standard clinical diagnostic testing, in which each case is assayed independently. In contrast, cases analyzed by TMA as described above are stained batchwise in a single assay run; this approach does not represent a standard clinical testing format. In most cases, percent

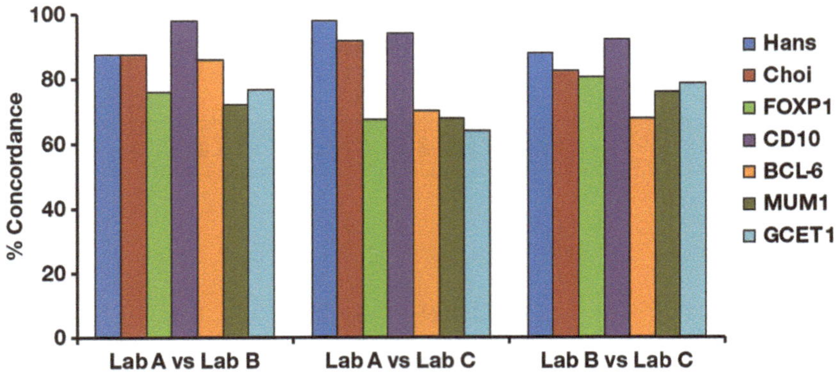

Fig. 5 Interlaboratory concordance of immunohistochemical assays on tissue samples from a phase II trial

concordance was higher than that for the TMA samples in round 1, and more similar to the concordance rates observed after round 2, demonstrating that the steps taken to optimize the assay and interpretation after the initial TMA concordance analysis were successful and could be applied to the single-case diagnostic testing situation. While no major differences between the two criteria (Hans and Choi) were noted, concordance rates for CD10 were consistently higher for all interlaboratory pairs when compared with the concordance rates of the other antibodies (Fig. 5).

4 Conclusions

- The cell of origin (GCB or non-GCB) subtype of DLBCL was accurately determined by IHC with high sensitivity compared to GEP. The results of the interlaboratory qualified IHC assays were very consistent among the three laboratories.
  - Previous studies have found a poor concordance among IHC results for markers of DLBCL subtypes [12]. However, data from that report were generated by eight individual laboratories with different methods and reagents. The results reported in the present study demonstrate that by harmonization of technical details and uniform scoring methodologies, a high rate of IHC concordance can be reached.
  - Although there has been some uncertainty regarding the correlation between GEP and IHC subtyping for GCB and non-GCB DLBCL cell of origin [15], previous publications using standardized methodologies have shown that it is possible to reach high concordance of IHC and GEP [16].

   - The results reported here demonstrate that it is possible to construct a reproducible assay with high intra-laboratory reproducibility, thus making it suitable for implementation in a clinical program.
- While it was previously shown that addition of FOXP1 and GCET1 to the IHC assay improved concordance with GEP subtyping results [11], no advantage was noted in the studies reported herein. No particular antibody stain was interpreted in a discordant manner.
- As described in Fig. 1, a next step in the development of this companion diagnostic would be transfer of the assay to a central reference laboratory and its prospective use in a phase III trial to select patients for therapy with an investigational treatment.
   - This study would determine whether the investigational treatment may be more effective in a particular subtype of DLBCL and potentially improve outcomes for these patients.

Acknowledgments

Financial support for medical editorial assistance was provided by Celgene. We thank Kyoko Kelly, Ph.D., ProEd Communications, Inc., for her medical editorial assistance with this chapter.

References

1. Bea S, Zettl A, Wright G et al (2005) Diffuse large B-cell lymphoma subgroups have distinct genetic profiles that influence tumor biology and improve gene-expression-based survival prediction. Blood 106(9):3183–3190
2. De Paepe P, De Wolf-Peeters C (2007) Diffuse large B-cell lymphoma: a heterogeneous group of non-Hodgkin lymphomas comprising several distinct clinicopathological entities. Leukemia 21(1):37–43
3. Lenz G, Wright G, Dave SS et al (2008) Stromal gene signatures in large-B-cell lymphomas. N Engl J Med 359(22):2313–2323
4. Lenz G, Wright GW, Emre NC et al (2008) Molecular subtypes of diffuse large B-cell lymphoma arise by distinct genetic pathways. Proc Natl Acad Sci U S A 105(36):13520–13525
5. Alizadeh AA, Eisen MB, Davis RE et al (2000) Distinct types of diffuse large B-cell lymphoma identified by gene expression profiling. Nature 403(6769):503–511
6. Rosenwald A, Wright G, Chan WC et al (2002) The use of molecular profiling to predict survival after chemotherapy for diffuse large-B-cell lymphoma. N Engl J Med 346 (25):1937–1947
7. Dunleavy K, Pittaluga S, Czuczman MS et al (2009) Differential efficacy of bortezomib plus chemotherapy within molecular subtypes of diffuse large B-cell lymphoma. Blood 113 (24):6069–6076
8. Hernandez-Ilizaliturri FJ, Deeb G, Zinzani PL et al (2011) Higher response to lenalidomide in relapsed/refractory diffuse large B-cell lymphoma in nongerminal center B-cell-like than in germinal center B-cell-like phenotype. Cancer 117(22):5058–5066
9. Wilson WH, Gerecitano JF, Goy A et al (2012) The Bruton's tyrosine kinase (BTK) inhibitor, ibrutinib (PCI-32765), has preferential activity in the ABC subtype of relapsed/refractory de novo diffuse large B-cell lymphoma (DLBCL): interim results of a multicenter, open-label, phase 2 study. Blood 120(21):abstr 686
10. Hans CP, Weisenburger DD, Greiner TC et al (2004) Confirmation of the molecular classification of diffuse large B-cell lymphoma by immunohistochemistry using a tissue microarray. Blood 103(1):275–282
11. Choi WW, Weisenburger DD, Greiner TC et al (2009) A new immunostain algorithm classifies diffuse large B-cell lymphoma into molecular

subtypes with high accuracy. Clin Cancer Res 15(17):5494–5502

12. de Jong D, Rosenwald A, Chhanabhai M et al (2007) Immunohistochemical prognostic markers in diffuse large B-cell lymphoma: validation of tissue microarray as a prerequisite for broad clinical applications–a study from the Lunenburg Lymphoma Biomarker Consortium. J Clin Oncol 25(7):805–812
13. Haarer CF, Roberts RA, Frutiger YM, Grogan TM, Rimsza LM (2006) Immunohistochemical classification of de novo, transformed, and relapsed diffuse large B-cell lymphoma into germinal center B-cell and nongerminal center B-cell subtypes correlates with gene expression profile and patient survival. Arch Pathol Lab Med 130(12):1819–1824
14. Goldstein NS, Hewitt SM, Taylor CR, Yaziji H, Hicks DG (2007) Recommendations for improved standardization of immunohistochemistry. Appl Immunohistochem Mol Morphol 15 (2):124–133
15. Vose JM (2011) Relapsed diffuse large B-cell lymphoma: clinical utility of cell of origin. J Clin Oncol 29(31):4065–4066
16. Meyer PN, Fu K, Greiner TC et al (2011) Immunohistochemical methods for predicting cell of origin and survival in patients with diffuse large B-cell lymphoma treated with rituximab. J Clin Oncol 29(2):200–207

Methods in Pharmacology and Toxicology (2015): 359–371
DOI 10.1007/7653_2014_34

Published online: 30 January 2015

Cutpoint Methods in Digital Pathology and Companion Diagnostics

Joshua C. Black, Mahipal V. Suraneni, and Steven J. Potts

Abstract

Digital image analysis is quickly transforming pathology into a quantitative discipline, where numerous cellular features from individual cells in a tissue samples can be measured and analyzed. Furthermore, digital image analysis is bringing immunohistochemical methods to the forefront of biomarker research making them an integral part of drug development and personalized medicine. As hundreds of quantitative measures generated by digital image analysis of tissues can be correlated to clinical outcomes, the need grows for statistical methods to evaluate these quantitative approaches. Cutpoint analysis is one such statistical method often requested for analyzing data from tissue biopsy evaluation in drug development. This frequently is driven by the need to analyze biomarker expression data in relation to patient survival to stratify patients into subsets of responders and nonresponders to a particular drug treatment. In this chapter we discuss data and outcome-driven approaches in cutpoint analysis, and the caution that should be observed in using cutpoint analysis and appropriate methods for avoiding type I errors. Finally, we provide two examples of how cutpoint analysis can be applied in pharmaceutical datasets.

Key words Immunohistochemistry, Biomarkers, Cutpoint analysis, Patient selection, Companion diagnostics

1 Introduction

Biomarkers should be evaluated using a fit-for-purpose approach. For example, in a typical study where the desire is to use overall survival (OS) as a primary endpoint, the cutpoint should be determined in a preliminary study using OS as an endpoint and the methodology should not be modified in subsequent studies. Cutpoint analysis should follow these fit-for-purpose guidelines. The cutpoint approach has had a significant impact on epidemiological studies in clinical and preclinical settings. Discretizing a continuous variable into two or more categories can be advantageous for several applications including assay development, eligibility criteria for drug study enrollment, discerning prognostic vs. predictive effects of biomarkers in drug development, stratifying patients based on expected treatment effectiveness, and imposing biologically assumed thresholds [1–7]. At times, this discretizing can simplify the presentation of results to recognizable descriptive statistics like risk ratios and reduce the complexity of analysis. If an underlying effect is suspected to be organized into categories, this method can statistically model the divisions between categories and give estimates for the location of those divisions [8]. This chapter will explore the general strategies for obtaining cutpoints and consider two examples where cutpoint analysis proved beneficial to the analysis of immune cells and two different cancer types. Each focuses on a different application of cutpoint analysis: survival analysis and logistic regression. Methods in cutpoint analysis fall into two overall strategies: data-oriented approaches and outcome-oriented approaches.

The decision to apply cutpoint analysis can be a complex one. In its simplest form, cutpoint analysis dichotomizes a discrete variable into two categories, such as "high" and "low." In an ideal situation, the underlying biology would support and predict such categorizations [9]. The statistical model resembles a step function where multiple values of the continuous predictor variable result in equivalent risk of outcome until the selected cutpoint, where the risk of outcome discontinuously changes for the new category. In nearly all biological situations, this is unrealistic, but the simplification could avoid the need for overcomplex nonlinear fitting of continuous covariates [2, 5, 10, 11]. Without clear biological justification, appropriateness of a cutpoint analysis can be assessed graphically [1, 11–13]. Another situation where cutpoint analysis is a clear option would be when a continuous variable is very highly skewed with many samples having an extreme value [9]. Some examples would be hours smoked per week or per cell expression of a rare biomarker. A large number of samples would have a value of zero, implying a data-oriented cutpoint of zero and nonzero. Beyond these situations, use of cutpoint analysis is often justified on the basis of the simplicity of the descriptive statistics. Odds ratios and relative risks are quite familiar to individuals in the biomedical industries, thus making cutpoint analysis potentially useful as a summary method. This simplicity must be weighed against the information loss inherent to cutpoint analysis.

The risks of discretizing continuous variables have been well documented [1, 3, 5–7, 9, 14–18]. The overriding concern and criticism of using cutpoint analysis is it may result in overfitting data and thus inflate the type I error rate. This is particularly problematic in large multiplexed studies common in biological assays that generate high dimensional datasets, where the number of variables can approach or surpass the number of patients observed. There have been notable overfitting examples with cutpoint analysis in mass spec data, in next gen sequencing, and in whole genome association studies [19–21]. Immunohistochemistry (IHC) tends to be at the opposite extreme of these multiplexed approaches. Only a small number of proteins, and frequently only one, are measured in many immunohistochemistry assay development projects. Immunohistochemistry, since its inception in 1940s, has remained as one of the important tools in clinical diagnosis [22], disease prognosis [23], and recently emerging as a tool in predicting treatment response and clinical outcomes [24]. The costs and time associated with working up an appropriately optimized IHC assay have been well discussed by other authors in this book. The time, cost, and lack of tumor samples prohibit the use of a large number of IHC markers. This can be an advantage in methodology from a statistical standpoint, when only one or only a few data measures from IHC assays in biomarker studies are evaluated for cutpoint analysis. This does not relieve the statistician from using caution in the application of cutpoint methods as well as appropriate reporting of significance. Furthermore, digital image analysis is rapidly changing the amount of data that can be extracted from IHC biomarker studies. With digital image analysis, it is possible to generate hundreds of measurements from millions of cells analyzed in a tissue section, making it imperative to choose and apply statistical methods such as cutpoint analysis carefully.

Evaluation of biomarker expression using IHC and subsequent correlation of IHC biomarker data to the patient outcome data using cutpoint analysis has been demonstrated in several oncology-related research studies [25–28]. In a breast cancer study, minimum *p*-value cutpoints derived based on Estrogen Receptor (ER) immunostaining have shown to predict patient survival [25]. Receiver operating characteristic (ROC) curve analysis was used in a different study to determine clinically important cut-offs based on immunostaining of receptor for hyaluronic acid-mediated motility (RHAMM) in 1,197 colorectal cancer samples [26]. This study shows that IHC assay in combination with ROC analysis can be used to identify cutpoints to stratify patients in relation to various clinicopathological endpoints. In addition, automated quantitative analysis (AQUA) and IHC measurements of epidermal growth factor receptor (EGFR) expression were used to derive cutpoints that could predict response to Gefitinib in non-small cell lung cancer (NSCLC) patients [27]. In most of these studies, one or a few data measures derived from the IHC assays were analyzed further using cutpoint analysis.

There are several prominent cautions to be aware of when using cutpoint analysis. By definition, these methods trade away useful information in exchange for any potential advantage [5]. Information pertaining to individuals is combined together into a single parameter describing the overall group. This loss of information will obscure completely any within-group differences. As mentioned, inflation of type I error is one of the most prominent dangers in outcome-oriented approaches, though methods for correcting this have been suggested [1, 5, 24–31]. Outcome-oriented methods also tend to overestimate the effect size in question. The ability to replicate the same cutpoint in subsequent studies can be limited. Divisions based on data-oriented approaches may be arbitrary and fail to capture the prognostic or predictive value of the variable [10]. Finally, simplicity of the results may not be desirable [5, 9]. Regression analysis is an equally simple technique that can convey the predictive power of a variable while retaining individual information. While audiences of statistical results may be more familiar with dichotomous descriptive statistics like risk ratios or odds ratios, some have suggested that, when appropriate, education of descriptive statistics in regression is preferable to simplifying the analytical results [9].

2 Cutpoint Analysis Methods

Discretizing a continuous variable can occur under two distinct paradigms: data-oriented or outcome-oriented. Under both paradigms, it is assumed that discrete categories do exist. Therefore, the step of assessing the validity of cutpoint analysis should not be overlooked. In addition to biological justification, graphical plots such as grouped data plots and Martingale residuals can be helpful in visualizing continuous variable clustering [1, 13]. Smoothing splines have also been suggested as a method to assess cutpoint validity, and a description of their use has been given previously [10, 32].

2.1 Data-Oriented Methods

A continuous variable can be discretized based on predefined criteria giving the analysis a data focused cutpoint. Splits along a percentile are common, particularly at the median. Splits based on standard deviations or an a priori designation are also used [1]. These cutpoints can be attractive due to their unbiased nature with respect to the outcome and the simplicity of their implementation. Oftentimes, a biologically driven *a priori* cutpoint, such as body mass index >25 indicating overweight individuals, will directly stratify patients into categories with strong clinical relevance and permit interpretations with direct clinical consequence. Evaluation of the quality of these cutpoints should be assessed via grouped data plots or residual comparisons. Survival analysis cutpoints can be assessed using Martingale residuals [13]. These residuals can visualize the influence of each sample on the parameter estimates, and clear groupings of sample residuals further support the presence of a cutpoint.

However, data-oriented approaches can be arbitrary. Splits chosen at percentiles, *a priori* designations, or any arbitrary point may not accurately represent the underlying biological categories, assuming such categories exist. As such, the loss of information and power inherent to cutpoint analysis could obscure the stratification if the cuts chosen are far from the "true" cutpoints. In these cases, outcome-oriented methods, complex regression analyses, or other suitable models would be desired.

2.2 Outcome-Oriented Methods

Techniques that search for an optimized outcome statistic or parameter are the focus of outcome-oriented methods. In most scenarios, a biologically derived cutpoint is not known and exploration of different stratifications is desired. The continuous covariate is partitioned into a selection interval, typically an inner interval where the outer 10–40 % of values are excluded. These outer values are excluded to prevent rapid power drops due to sample size, inflation of the type I error rate, or "cherry-picking" of the data near the extremes [3, 31]. The selection interval is divided into a number of potential cutpoints and an appropriate analysis is run for each of these cutpoints. The maximum number of potential cutpoints should not exceed $(k - 1)$, where k is the number of samples in the data set. Final selection of a cutpoint is based on optimizing a parameter of interest. Minimum p-value approaches, or alternatively a maximal statistic approach, have received the most attention, but maximizing other parameters such as effect size or estimation precision have also been used [1, 33]. Constructing a cutpoint curve, where p-value is plotted against the different cutpoints chosen, would be a method of visualizing the trends in cutpoint. Figures 1 and 2 show an example output for a cutpoint analysis. This data set was constructed from a uniform distribution of generic biomarker levels between 1 and 50. Outcome, here modeled as overall survival in days, is dependent on treatment in the presence of the biomarker. Figure 1a is a scatterplot showing survival plotted against biomarker level. At low biomarker levels (<25), survival is a Gaussian distribution centered on 70 days (SD = 35), and there is no difference between treatment and placebo. At high biomarker levels (>25), placebo survival is unchanged, but mean survival time in the treatment group increases to 100 days (SD = 12). Treatment in the presence of the biomarker produces a real treatment benefit of increasing survival. Outcome was compared with a Cox proportional hazards model for survival data. Figure 1b shows two cutpoint curves for this data, one starting at the 20th percentile and cutting up to the 80th (25 cuts), and another starting at the 80th percentile and cutting down to the 20th (25 cuts). The former, labeled "Greater Than" in red, models samples above the cut according to Cox proportional hazards. The latter, labeled "Less Than" in blue, models samples below the cut. The red curve identified a minimum p-value (vertical line) of 0.0013 at biomarker cut 25.2, indicating a hazard ratio of 0.32. This is reinforced by the scatterplot, where to the right of the cutpoint, the drug samples tend to have higher outcome values than placebo samples. The blue curve did not find a statistically significant cutpoint at the usual 0.05 level. A truly dichotomous variable would remain significant through the remainder of the selection interval, assuming sufficient power. In practice, as the sample number drops near the end of the selection interval, power also drops and upturns in the cutpoint curves are common. If a cutpoint is found for only a central portion of the selection

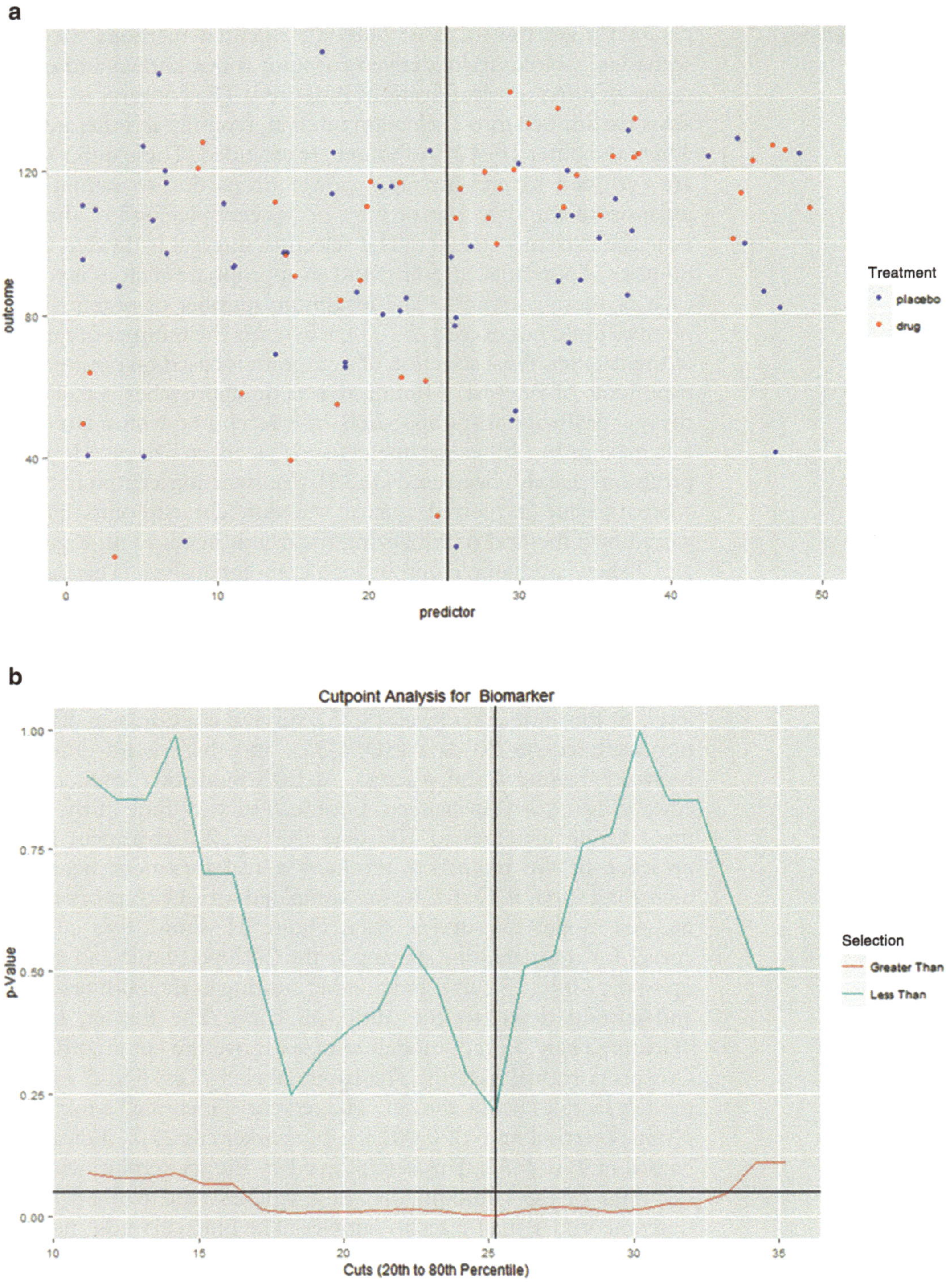

Fig. 1 Sample plots from cutpoint analysis. (**a**) Scatterplot from sample distribution. Above the cutpoint in biomarker levels (*vertical line*) patients with drug treatment have higher outcome values than placebo treatment. Below the cutpoint, the two groups are similarly distributed. (**b**) Cutpoint graph constructed for the sample distribution. Minimum *p*-value (*vertical line*) indicates most statistically significant separation between groups above the cutpoint. This cutpoint is in a flat region of the cutpoint curve, indicating a robust cutpoint selection. *Horizontal line* is $\alpha = 0.05$

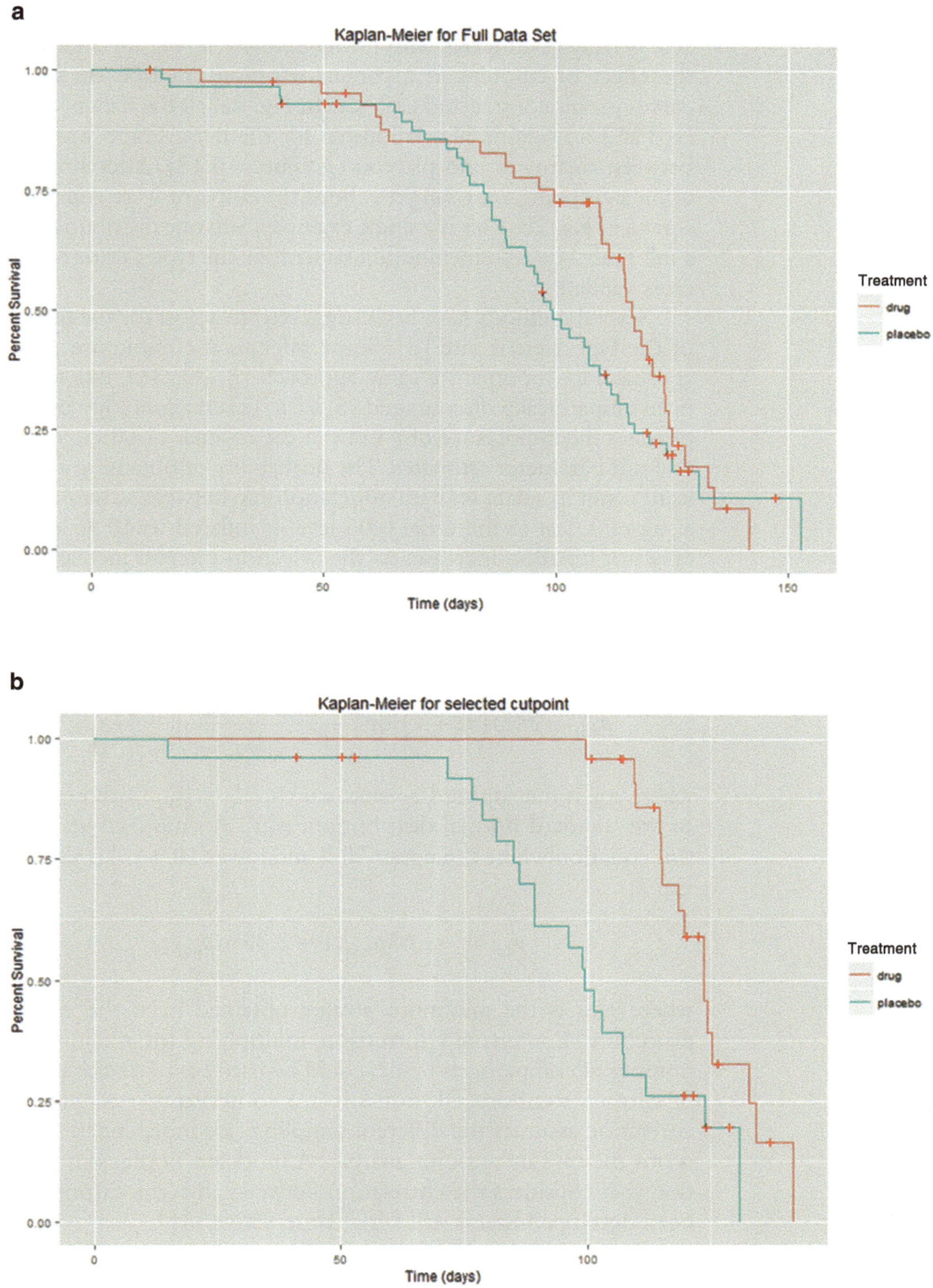

Fig. 2 Kaplan-Meier curves before and after cutpoint: (**a**) Without a cutpoint, no significant hazard ratio was present between treatment and placebo (p-value = 0.13). (**b**) After stratifying using minimum p-value cutpoint analysis, a significant ratio was present (HR = 0.32, p-value = 0.0013)

interval, this could indicate multiple categories or indicate the variable is truly continuous with no discrete levels. Figure 2 illustrates the effect of cutpoint as a patient selection criterion on survival curves in this constructed data set. In Fig. 2a, all the data is considered and no significant relationship for the hazard ratio was found between treatment and placebo (p-value = 0.13). After stratifying using a cutpoint, only samples above the cutpoint were considered, shown in Fig. 2b. This is a single example from one distribution, and a full analysis would include adjustment for the type I error rate and cross-validation.

Several methods have been suggested to adjust for the inflation of the type I error rate [3]. Some refer to the minimum p-value approach as "optimal" p-value approach [3, 30, 33], but this has been empathically discouraged [5, 15]. This terminology overemphasizes the importance of p-value at the expense of other clinically relevant parameter estimates. The probability of obtaining a significant result in a data set with no relationship between variables from a log-rank test at the usual 0.05 level is inflated to 40 %, and the effect of inflation increases as the selection interval increases [5]. For binary outcome variables, a straightforward correction based on the unadjusted p-value and the selection interval choices has been suggested [29], shown in Eq. 1.

$$p_{adj} = \phi(z)\left(z - \frac{1}{z}\right)\log\left(\frac{\varepsilon_{\text{high}}(1 - \varepsilon_{\text{low}})}{\left(1 - \varepsilon_{\text{high}}\right)\varepsilon_{\text{low}}}\right) + 4\frac{\phi(z)}{z}, \quad (1)$$

where p_{adj} is the adjusted p-value, z is the $(1 - p_{\text{adj}}/2)$th percentile of the standard normal distribution, $\phi(z)$. A simplified version of this equation where the upper 10 % and lower 10 % [5] is shown in Eq. 2.

$$p_{\text{adj}} = -1.63 p_{\text{min}}(1 + 2.35 \ln p_{\text{min}}), \quad (2)$$

where p_{min} is the minimum p-value obtained from the cutpoint procedure. Bonferroni's correction, multiplying the p-value by the number of comparisons made, could be used based on the number of cutpoints considered, but this is a conservative estimate. This correction assumes the different samplings are independent, which is not the case in cutpoint analysis. A modified Bonferroni correction that considers the correlation between adjacent cutpoints has been developed that is still fairly conservative [31].

Twofold cross-validation can be used to further verify the cutpoint determined from maximizing parameters. The data set is split into two equally sized subsets, and an optimization analysis is run where a cutpoint is individually determined for each set. The cutpoints are crossed, and values are assigned a category based on the opposite set's cutpoint. The analysis is run a third time where parameter estimates and p-values are determined using this assignment from the crossed cutpoints. The procedure can be scaled up to any number of n folds where the data is split into n subsets of approximately equal size. Each subset is omitted and a cutpoint assigned based on the rest of the data set. Categorization within the subset is based on this external cutpoint. After repeating n times, each data point has been assigned a category based on a cutpoint that was determined without that data point contributing to the optimization procedure. This has been found to approximate well the type I error rate and give an unbiased estimate of effect size [3]. This motif of training/test sets can avoid many of the optimization problems mentioned but has the drawback of requiring larger data sets with which to subdivide and train upon.

2.3 Combined Approach

For further confidence in a selected cutpoint, combination of data-oriented and outcome-oriented methods is strongly recommended. This allows validation of optimized cutpoints and observation of any trends present. Alongside residuals and group data plots mentioned earlier, simple scatterplots of the continuous variable against outcome can reveal interesting data structure or clustering of outcome above or below a threshold. It can also reveal hazards such as if multiple samples have an equivalent value in the middle of the selection interval, which would strongly influence the results of outcome-oriented methods. Dividing the continuous variable into quartiles can complement a minimum p-value method that dichotomizes the continuous variable. It is an unbiased way to approximate the outcome-oriented approach and confirm trends discovered. If a variable is truly dichotomous, parameter estimates from quartiles would trend in the same direction as the minimum p-value method. As always, it is useful to consider the drawbacks of each method. If sample size is low, quartile division could result in significant loss of power compared to dichotomization and obscure trends.

3 Application Example: Survival Analysis

The use of cutpoint in survival analysis is demonstrated in the context of a published study discussing predictors of patient outcome in colorectal cancer [34]. Alongside genomic analysis, the researchers investigated the influence of immune cells present at tumor sites on patient overall survival and disease free survival. Using digital imaging tools and immunohistostaining, immune cells were quantified according to type, density, and location. Four different biomarkers (CD3, CD8, GZMB, and CD45RO) were examined in two areas within the tumor environment (center of tumor and invasive margin). The presence of each biomarker in either tumor compartment was correlated significantly to increased disease free survival. In addition, the densities of these biomarkers within the tumor compartments were considered with cutpoint analysis.

Density, as a continuous variable, was dichotomized in several ways to demonstrate higher density was correlated with longer survival. Initially, the data was analyzed using a minimum *p*-value approach where each biomarker was individually stratified into high and low expression. The median survival times between the two expression levels were compared across a large range of cutpoints, and the difference in median survival time remained significant throughout most of the selection interval, even after correcting the *p*-values for type I inflation using Altman's method [5]. This large range of significant cutpoints supports the reproducibility of the work. The stratification was reanalyzed with two additional methods. Twofold cross-validation obtained a median *p*-value from 100 subdivisions, and the biomarkers were dichotomized at the median to obtain a third *p*-value. All three methods were found to be in concordance with each other. The median cutpoint was used to examine one biomarker, CD3, in two separate cohorts of patients, one from a different series of patients and another from a separate hospital. These data confirmed the results from the primary analysis.

Overall, the results suggest that the immune system is involved in preventing tumor resurgence and prolonging survival in colorectal cancer. Immunohistochemistry techniques were used to determine cell type, density, and location, of which all were found to correlate with improved survival times. The use of these combined cutpoint methods lends confidence to the conclusion that the expression of these biomarkers is prognostic of increased survival in patients. Cross-validation was conducted with a large number of subdivisions to improve the reliability of the estimated *p*-value. These prognostic biomarkers can be used to target further therapies, identify high-risk patients, and indicate clinical outcome.

4 Application Example: Logistic Regression

A second example study also investigated tumor-infiltrating lymphocytes but with respect to gastric cancer [35]. In addition to studying the influence on survival, this work also looked at lymphocyte influence on metastasis in the lymph nodes. Metastasis was defined categorically as either present or not, which lends itself to modeling via logistic regression. A similar group of biomarkers (CD3, CD8, CD20, and CD45RO) were stained using immunohistochemistry techniques, and cell density was quantified digitally. Cell density was found to be significant predictors of survival and tumor metastasis when dichotomized via cutpoint analysis techniques.

This work also used a combined approach to dichotomizing the continuous variables but instead begins with a data-oriented approach. As opposed to the traditional median, cell density was first dichotomized at the mean value. Assuming a Gaussian distribution, this choice is equivalent to the median, though the median is traditionally suggested as an unbiased dichotomization point [14]. Cell density was also cut at the 75th percentile. Both methods yielded odds ratios less than 1, indicating higher lymphocyte cell density was correlated with lower rates of tumor metastasis. This conclusion was confirmed using twofold cross-validation using the minimum *p*-value approach. Using a single subdivision, the minimum *p*-value was found for the inner 70 % selection interval and no further correction to the *p*-value was conducted. The researchers also investigated the prognostic ability of these biomarkers in a multivariate setting where tumor invasion (categorized by T stage) and lymphatic invasion were covariates. For most biomarkers, the prognostic ability of the dichotomization remained significant.

This study demonstrated that type and density of tumor-invading lymphocytes are prognostic to patient survival and tumor metastasis in gastric cancer. Notably, this work used digital quantification variables dichotomized via cutpoint analysis to identify the prognostic influence of lymphocytes where previous work did not. Previously [36], lymphocyte infiltration was assessed manually and categorized into "marked infiltration" and "slight infiltration." There was no significant difference in survival times between these manually assessed groups. The analysis followed a combined approach that built up to more complicated analyses. Initial dichotomization was shown to have significant stratification, which was confirmed using several approaches to stratification. Biologically predicted covariates were also included to support the prognostic value of the biomarkers independently. However, a more complete analysis could have included a cutpoint curve in the cross-validation analysis indicating the *p*-value at various cuts. This would have communicated information regarding the trends for various cuts.

5 Conclusion

This chapter discussed the usefulness, methods, and examples of cutpoint analysis in digital pathology. The method of discretizing continuous variables can be useful in demonstrating prognostic ability of biomarkers, determining eligibility of patients for treatment enrollment, or stratifying predictive capacity of biomarkers in conjunction with drug treatment, among other uses. A cutpoint analysis that combines several methods is a strongly recommended approach that brings the most confidence to final conclusions. Combined approaches address concerns such as bias, inflation of the type I error rate, and replication of results. Concerns regarding simplicity should be carefully considered before using cutpoint analysis. Simplicity is often not desired, and individual information could be retained using regression methods. Finally from reported literature, it is possible that digital quantization methods with cutpoint analysis may discover clinically relevant results where other manual methods failed to capture the effect.

References

1. Mazumdar M, Glassman J (2000) Categorizing a prognostic variable: review of methods, code for easy implementation and applications to decision-making about cancer treatments. Stat Med 113–132
2. Contal C, O'Quigley J (1999) An application of changepoint methods in studying the effect of age on survival in breast cancer. Comput Stat Data Anal 30:253–270
3. Faraggi D, Simon R (1996) A simulation study of cross-validation for selecting an optimal cutpoint in univariate survival analysis. Stat Med 15:2203–2213
4. Mazumdar M, Smith A, Bacik J (2003) Methods for categorizing a prognostic variable in a multivariable setting. Stat Med 22:559–571
5. Altman D, Lausen B, Sauerbrei W, Schumacher M (1994) Dangers of using "optimal" cutpoints in the evaluation of prognostic factors. J Natl Cancer Inst 86:829–835
6. Cumsille F, Bangdiwala S, Sen PK, Kupper L (2000) Effect of dichotomizinlg a continuous variable on the model structure in multiple linear regression models. Commun Stat 29:643–654
7. Liquet B, Commenges D (2001) Correction of the p-value after multiple coding of an explanatory variable in logistic regression. Stat Med 20:2815–2826
8. Boucher K, Slattery M, Berry T, Quesenberry C, Anderson K (1998) A comparison of statistical methods to analyze dose–response and trend analysis in epidemiologic studies. Stat Meth Epidemiol 51:1223–1233
9. MacCallum R, Zhang S, Preacher K, Rucker D (2002) On the practice of dichotomization of quantitative variables. Psychol Methods 7:19–40
10. Williams B, Mandrekar J, Cha S, Furth A (2006) Finding optimal cutpoints for continuous covariates with binary and time-to-event outcomes. Technical report. May Clinic, Rochester, MN
11. Schulgen G, Lausen B, Olsen J, Schumacher M (1994) Outcome-oriented cutpoints in analysis of quantitative exposures. Am J Epidemiol 140:172–184
12. Greenland S (1995) Avoiding power loss associated with categorization and ordinal scores in dose-response and trend analysis. Epidemiology 6:450–454
13. Therneau T, Grambsch P, Fleming T (1990) Martingale-based residuals for survival models. Biometrika 77:147–160
14. Royston P, Altman D, Sauerbrei W (2006) Dichotomizing continuous predictors in multiple regression: a bad idea. Stat Med 25:127–141
15. Altman D, Royston P (2006) The cost of dichotomising continuous variables. BMJ 332:1080
16. Hilsenbeck S, Clark G, McGuire W (1992) Why do so many prognostic factors fail to pan out? Breast Cancer Res Treat 22:197–206
17. Ragland D (1992) Dichotomizing continuous outcome variables: dependence of the magnitude of association and statistical power on the cutpoint. Epidemiology 3:434–440
18. Hollander N, Sauerbrei W, Schmacher M (2004) Confidence intervals for the effect of a prognostic factor after selection of an 'optimal' cutpoint. Stat Med 23:1701–1713
19. Coombes K, Morris J, Hu J, Edmonson S, Baggerly K (2005) Serum proteomics profiling – a young technology begins to mature. Nat Biotechnol 23:291–292
20. Braggerly K, Morris J, Coombes K (2004) Reproducibility of SELDI-TOF protein patterns in serum: comparing datasets from different experiments. Bioinformatics 20:777–785
21. Potti A, Dressman H, Bild A, Reidel R, Chan G, Sayer R, Cragun J, Cottrill H, Kelley M, Peterson R, Harpole D, Marks J, Berchuck A, Ginsburg G, Febbo P, Lancaster J, Nevins J (2011) Retraction: genomic signatures to guide the use of chemotherapeutics. Nat Med 17:135
22. Matos LL, Trufelli DC, de Matos MG, da Silva Pinhal MA (2010) Immunohistochemistry as an important tool in biomarkers detection and clinical practice. Biomark Insights 5:9–20
23. Bodey B (2002) The significance of immunohistochemistry in the diagnosis and therapy of neoplasms. Expert Opin Biol Ther 2(4):371–393
24. Alymani NA, Smith MD, Williams DJ, Petty RD (2010) Predictive biomarkers for personalised anti-cancer drug use: discovery to clinical implementation. Eur J Cancer 46(5):869–879

25. Jalava P, Kuopio T, Huovinen R, Laine J, Collan Y (2005) Immunohistochemical staining of estrogen and progesterone receptors: aspects for evaluating positivity and defining the cutpoints. Anticancer Res 25:2535–2542
26. Zlobec I, Steele R, Terracciano L, Jass JR, Lugli A (2007) Selecting immunohistochemical cutoff scores for novel biomarkers of progression and survival in colorectal cancer. J Clin Pathol 60(10):1112–1116
27. Mascaux C, Wynes MW, Kato Y, Tran C, Asuncion BR, Zhao JM, Gustavson M, Ranger-Moore J, Gaire F, Matsubayashi J, Nagao T, Yoshida K, Ohira T, Ikeda N, Hirsch FR (2011) EGFR protein expression in non-small cell lung cancer predicts response to an EGFR tyrosine kinase inhibitor – a novel antibody for immunohistochemistry or AQUA technology. Clin Cancer Res 17(24):7796–7807
28. Salles G, de Jong D, Xie W, Rosenwald A, Chhanabhai M, Gaulard P, Klapper W, Calaminici M, Sander B, Thorns C, Campo E, Molina T, Lee A, Pfreundschuh M, Horning S, Lister A, Sehn LH, Raemaekers J, Hagenbeek A, Gascoyne RD, Weller E (2011) Prognostic significance of immunohistochemical biomarkers in diffuse large B-cell lymphoma: a study from the Lunenburge Lymphoma Biomarker Consortium. Blood 117(26):7070–7078
29. Miller R, Siegmund D (1982) Maximally selected chi square statistics. Biometrics 38:1011–1016
30. Hilsenbeck S, Clark G (1996) Practical p-value adjustment for optimally selected cutpoints. Stat Med 15:103–112
31. Lausen B, Schumacher M (1996) Evaluating the effect of optimized cutoff values in the assessment of prognostic factors. Comput Stat Data Anal 21:307–326
32. Harrell F (2001) Regression modelling strategies with applications to linear models, logistic regression, and survival analysis. Springer, New York
33. Magder L, Fix A (2003) Optimal choice of a cut point for a quantitative diagnostic test performed for research purposes. J Clin Epidemiol 56:956–962
34. Galon J, Costes A, Sanchez-Cabo F, Kirilovsky A, Mlecnik B, Lagorce-Pagé C, Tosolini M, Camus M, Berger A, Wind P, Zinzindohoué F, Bruneval P, Cugnenc P, Trajanoski Z, Fridman W, Pagés F (2006) Type, density, and location of immune cells within human colorectal tumors predict clinical outcome. Science 313:1960–1964
35. Lee HE, Chae S, Lee Y, Kim M, Lee HS, Lee B, Kim W (2008) Prognostic implications of type and density of tumour-infiltrating lymphocytes in gastric cancer. Br J Cancer 99:1704–1711
36. Fukuda K, Tsujitani S, Maeta Y, Yamaguchi K, Ikeguchi M, Kaibara N (2002) The expression of RCAS1 and tumor infiltrating lymphocytes in patients with T3 gastric carcinoma. Gastric Cancer 5:220–227

Methods in Pharmacology and Toxicology (2015): 373–375
DOI 10.1007/7653_2015

Index

A

B

C

D

E

F

R

S

T

V

W

Kolibri 360 GmbH
Ferdinand-Jühlke-Straße 7
99095 Erfurt, Deutschland
produktsicherheit@kolibri360.de